Lecture Notes on Functional Analysis

With Applications to Linear Partial Differential Equations

Lecture Notes on Functional Analysis

With Applications to Linear Partial Differential Equations

Alberto Bressan

Graduate Studies
in Mathematics

Volume 143

American Mathematical Society
Providence, Rhode Island

2010 *Mathematics Subject Classification.* Primary 46–01; Secondary 35–01.

For additional information and updates on this book, visit
www.ams.org/bookpages/gsm-143

Library of Congress Cataloging-in-Publication Data

Bressan, Alberto, 1956–
[Lectures. Selections]
Lecture notes on functional analysis with applications to linear partial differential equations / Alberto Bressan.
pages cm. — (Graduate studies in mathematics ; volume 143)
Includes bibliographical references and index.
ISBN 978-0-8218-8771-4 (alk. paper)
1. Functional analysis. 2. Differential equations, Linear. I. Title.

QA321.B74 2012
515′.7—dc23

2012030200

∞ The paper used in this book is acid-free and falls within the guidelines
established to ensure permanence and durability.
Visit the AMS home page at http://www.ams.org/

10 9 8 7 6 5 4 3 2 1 18 17 16 15 14 13

To Wen, Luisa Mei, and Maria Lan

Contents

Preface xi

Chapter 1. Introduction 1

§1.1. Linear equations 1

§1.2. Evolution equations 4

§1.3. Function spaces 7

§1.4. Compactness 7

Chapter 2. Banach Spaces 11

§2.1. Basic definitions 11

§2.2. Linear operators 16

§2.3. Finite-dimensional spaces 20

§2.4. Seminorms and Fréchet spaces 23

§2.5. Extension theorems 26

§2.6. Separation of convex sets 30

§2.7. Dual spaces and weak convergence 32

§2.8. Problems 35

Chapter 3. Spaces of Continuous Functions 45

§3.1. Bounded continuous functions 45

§3.2. The Stone-Weierstrass approximation theorem 47

§3.3. Ascoli's compactness theorem 53

§3.4. Spaces of Hölder continuous functions 56

§3.5. Problems 57

Chapter 4. Bounded Linear Operators 61
§4.1. The uniform boundedness principle 61
§4.2. The open mapping theorem 63
§4.3. The closed graph theorem 64
§4.4. Adjoint operators 66
§4.5. Compact operators 68
§4.6. Problems 71

Chapter 5. Hilbert Spaces 77
§5.1. Spaces with an inner product 78
§5.2. Orthogonal projections 79
§5.3. Linear functionals on a Hilbert space 82
§5.4. Gram-Schmidt orthogonalization 84
§5.5. Orthonormal sets 85
§5.6. Positive definite operators 89
§5.7. Weak convergence 92
§5.8. Problems 95

Chapter 6. Compact Operators on a Hilbert Space 101
§6.1. Fredholm theory 101
§6.2. Spectrum of a compact operator 106
§6.3. Selfadjoint operators 107
§6.4. Problems 111

Chapter 7. Semigroups of Linear Operators 115
§7.1. Ordinary differential equations in a Banach space 115
§7.2. Semigroups of linear operators 120
§7.3. Resolvents 124
§7.4. Generation of a semigroup 128
§7.5. Problems 134

Chapter 8. Sobolev Spaces 139
§8.1. Distributions and weak derivatives 139
§8.2. Mollifications 146
§8.3. Sobolev spaces 151
§8.4. Approximations of Sobolev functions 157
§8.5. Extension operators 161
§8.6. Embedding theorems 163

§8.7. Compact embeddings 175
§8.8. Differentiability properties 179
§8.9. Problems 180

Chapter 9. Linear Partial Differential Equations 185
§9.1. Elliptic equations 185
§9.2. Parabolic equations 200
§9.3. Hyperbolic equations 207
§9.4. Problems 212

Appendix. Background Material 217
§A.1. Partially ordered sets 217
§A.2. Metric and topological spaces 217
§A.3. Review of Lebesgue measure theory 222
§A.4. Integrals of functions taking values in a Banach space 226
§A.5. Mollifications 228
§A.6. Inequalities 233
§A.7. Problems 237

Summary of Notation 241

Bibliography 245

Index 247

Preface

The first version of these lecture notes was drafted in 2010 for a course at the Pennsylvania State University. The book is addressed to graduate students in mathematics or other disciplines, who wish to understand the essential concepts of functional analysis and their application to partial differential equations. Most of its content can be covered in a one-semester course at the first-year graduate level.

In writing this textbook, I followed a number of guidelines:

- Keep it short, presenting all the fundamental concepts and results, but not more than that.
- Explain clearly the connections between theorems in functional analysis and familiar results of finite-dimensional linear algebra.
- Cover enough of the theory of Sobolev spaces and semigroups of linear operators as needed to develop significant applications to elliptic, parabolic, and hyperbolic PDEs.
- Include a large number of homework problems and illustrate the main ideas with figures, whenever possible.

In functional analysis one finds a wealth of beautiful results that could be included in a monograph. However, for a textbook of this nature one should resist such a temptation.

After the Introduction, Chapters 2 to 6 cover classical topics in linear functional analysis: Banach spaces, Hilbert spaces, and linear operators. Chapter 4 is devoted to spaces of continuous functions, including the Stone-Weierstrass approximation theorem and Ascoli's compactness theorem. In view of applications to linear PDEs, in Chapter 6 we prove some basic results on Fredholm operators and the Hilbert-Schmidt theorem on compact symmetric operators in a Hilbert space.

Chapter 7 provides an introduction to the theory of semigroups, extending the definition of the exponential function e^{tA} to a suitable class of (possibly unbounded) linear operators. We stress the connection with finite-dimensional ODEs and the close relation between the resolvent operators and backward Euler approximations.

After an introduction explaining the concepts of distribution and weak derivative, Chapter 8 develops the theory of Sobolev spaces. These spaces provide the most convenient abstract framework where techniques of functional analysis can be applied toward the solution of ordinary and partial differential equations.

The first three sections in Chapter 9 describe applications of the previous theory to elliptic, parabolic, and hyperbolic PDEs. Since differential operators are unbounded, it is often convenient to recast a linear PDE in a "weak form", involving only bounded operators on a Hilbert-Sobolev space. This new equation can then be studied using techniques of abstract functional analysis, such as the Lax-Milgram theorem, Fredholm's theory, or the representation of the solution in terms of a series of eigenfunctions.

The last chapter consists of an Appendix, collecting background material. This includes: definition and properties of metric spaces, the contraction mapping theorem, the Baire category theorem, a review of Lebesgue measure theory, mollification techniques and partitions of unity, integrals of functions taking values in a Banach space, a collection of inequalities, and a version of Gronwall's lemma.

These notes are illustrated by 41 figures. Nearly 180 homework problems are collected at the end of the various chapters. A complete set of solutions to the exercises is available to instructors. To obtain a PDF file of the solutions, please contact the author, including a link to your department's web page listing you as an instructor or professor.

It is a pleasure to acknowledge the help I received from colleagues, students, and friends, while preparing these notes. To L. Berlyand, G. Crasta, D. Wei, and others, who spotted a large number of misprints and provided many useful suggestions, I wish to express my gratitude.

Alberto Bressan
State College, July 2012

Chapter 1

Introduction

This book provides an introduction to linear functional analysis, extending techniques and results of classical linear algebra to infinite-dimensional spaces. With the development of a theory of function spaces, functional analysis yields a powerful tool for the study of linear ordinary and partial differential equations. It provides fundamental insights on the existence and uniqueness of solutions, their continuous dependence on initial or boundary data, the convergence of approximations, and on various other properties.

The following remarks highlight some key results of linear algebra and their infinite-dimensional counterparts.

1.1. Linear equations

Let A be an $n \times n$ matrix. Given a vector $\mathbf{b} \in \mathbb{R}^n$, a basic problem in linear algebra is to find a vector $\mathbf{x} \in \mathbb{R}^n$ such that

$$A\mathbf{x} = \mathbf{b}. \tag{1.1}$$

In the theory of linear PDEs, an analogous problem is the following. Consider a bounded open set $\Omega \subset \mathbb{R}^n$ and a linear partial differential operator of the form

$$Lu \doteq -\sum_{i,j=1}^{n} (a^{ij}(x)u_{x_i})_{x_j} + \sum_{i=1}^{n} b^i(x)u_{x_i} + c(x)u. \tag{1.2}$$

Given a function $f : \Omega \mapsto \mathbb{R}$, find a function u, vanishing on the boundary of Ω, such that

$$Lu = f. \tag{1.3}$$

There are fundamental differences between the problems (1.1) and (1.3). The matrix A yields a continuous linear transformation on the finite-dimensional space $\mathbb{R}^n$. On the other hand, the differential operator L can be regarded as an unbounded (hence discontinuous) linear operator on the infinite-dimensional space $\mathbf{L}^2(\Omega)$. In particular, the domain of L is not the entire space $\mathbf{L}^2(\Omega)$ but only a suitable subspace.

In spite of these differences, since both problems (1.1) and (1.3) are linear, there are a number of techniques from linear algebra that can be applied to (1.3) as well.

(I): Positivity

Assume that the matrix A is strictly positive definite, i.e., there exists a constant $\beta > 0$ such that

$$\langle Ax, x\rangle \ \geq\ \beta\,|x|^2 \qquad \text{for all } x \in \mathbb{R}^n.$$

Then A is invertible and the equation (1.1) has a unique solution for every $\mathbf{b} \in \mathbb{R}^n$.

This result has a direct counterpart for elliptic PDEs. Namely, assume that the operator L is strictly positive definite, in the sense that (after a formal integration by parts)

(1.4)

$$\begin{aligned}\langle Lu, u\rangle_{\mathbf{L}^2} \ &=\ \int_\Omega \left(\sum_{i,j=1}^n a^{ij}(x) u_{x_i} u_{x_j} + \sum_{i=1}^n b^i(x) u_{x_i} u + c(x) u^2\right) dx \\ &\geq\ \beta\,\|u\|^2_{H^1_0(\Omega)}\end{aligned}$$

for some constant $\beta > 0$ and all $u \in H^1_0(\Omega)$. Here $H^1_0(\Omega)$ is a space of functions which vanish on the boundary of Ω and such that

$$\|u\|_{H^1_0(\Omega)} \ \doteq\ \left(\int_\Omega |u|^2\,dx + \int_\Omega |\nabla u|^2\,dx\right)^{1/2} \ <\ \infty\,;$$

see Chapter 8 for precise definitions. If (1.4) holds, one can then prove that the problem (1.3) has a unique solution $u \in H^1_0(\Omega)$, for every given $f \in \mathbf{L}^2(\Omega)$.

A key assumption, in order for the inequality (1.4) to hold, is that the operator L should be elliptic. Namely, at each point $x \in \Omega$ the $n \times n$ matrix $(a^{ij}(x))$ should be strictly positive definite.

(II): Fredholm alternative

A well-known criterion in linear algebra states that the equation (1.1) has a unique solution for every given $\mathbf{b} \in \mathbb{R}^n$ if and only if the homogeneous equation

$$A\mathbf{x} = 0$$

has only the solution $\mathbf{x} = 0$. Of course, this holds if and only if the matrix A is invertible.

In general, continuous linear operators on an infinite-dimensional space X do not share this property. Indeed, one can construct a bounded linear operator $\Lambda : X \mapsto X$ which is one-to-one but not onto, or conversely.

Yet, the finite-dimensional theory carries over to an important class of operators, namely, those of the form $\Lambda = I - K$, where I is the identity and K is a compact operator. If Λ is in this class, then one can still prove the equivalence

$$\Lambda \text{ is one-to-one} \qquad \Longleftrightarrow \qquad \Lambda \text{ is onto.}$$

By an application of this theory it follows that, for a linear elliptic operator, the equation (1.3) has a unique solution $u \in H_0^1(\Omega)$ for every $f \in \mathbf{L}^2(\Omega)$ if and only if the homogeneous equation

$$Lu = 0$$

has only the zero solution.

(III): Diagonalization

If one can find a basis $\{\mathbf{v}_1, \ldots, \mathbf{v}_n\}$ of $\mathbb{R}^n$ consisting of eigenvectors of A, then with respect to this basis the system (1.1) takes a diagonal form and is thus easy to solve.

For a general matrix A with multiple eigenvalues, it is well known that such a basis of eigenvectors need not exists. A positive result in this direction is the following. If the $n \times n$ matrix A is *symmetric*, then one can find an orthonormal basis $\{\mathbf{v}_1, \ldots, \mathbf{v}_n\}$ of the Euclidean space $\mathbb{R}^n$ consisting of eigenvectors of A. Namely,

$$\langle \mathbf{v}_i, \mathbf{v}_j \rangle = \begin{cases} 1 & \text{if } i = j, \\ 0 & \text{if } i \neq j, \end{cases} \qquad\qquad A\mathbf{v}_k = \lambda_k \mathbf{v}_k .$$

Here $\lambda_1, \ldots, \lambda_n \in \mathbb{R}$ are the corresponding eigenvalues. The solution $\mathbf{x}$ of (1.1) can now be found by computing its coefficients $c_1, \ldots, c_n$ with respect to the orthonormal basis:

$$\mathbf{x} = \sum_{k=1}^{n} c_k \mathbf{v}_k, \qquad A\mathbf{x} = \sum_{k=1}^{n} \lambda_k c_k \mathbf{v}_k = \mathbf{b} = \sum_{k=1}^{n} \langle \mathbf{b}, \mathbf{v}_k \rangle \mathbf{v}_k .$$

Notice that, thanks to the basis of eigenvectors, the problem becomes decoupled. Instead of a large system of n equations in n variables, we only need to solve n scalar equations, one for each coefficient c_k. If all eigenvalues λ_k are nonzero, we thus have the explicit formula

$$\mathbf{x} = \sum_{k=1}^{n} \frac{1}{\lambda_k} \langle \mathbf{b}, \mathbf{v}_k \rangle \mathbf{v}_k . \tag{1.5}$$

One can adopt the same approach in the analysis of the elliptic operator L in (1.2), provided that $a^{ij} = a^{ji}$ and $b^i(x) = 0$. Indeed, these conditions make the operator "symmetric". One can then find a countable orthonormal basis $\{\phi_1, \phi_2, \ldots\}$ of the space $\mathbf{L}^2(\Omega)$ consisting of functions $\phi_k \in H_0^1(\Omega)$ such that

$$\langle \phi_i, \phi_j \rangle_{\mathbf{L}^2} = \begin{cases} 1 & \text{if } i = j, \\ 0 & \text{if } i \neq j, \end{cases} \qquad L\phi_k = \lambda_k \phi_k , \tag{1.6}$$

for a suitable sequence of real eigenvalues $\lambda_k \to +\infty$. Assuming that $\lambda_k \neq 0$ for all k, the unique solution of (1.3) can now be written explicitly as

$$u = \sum_{k=1}^{\infty} \frac{1}{\lambda_k} \langle f, \phi_k \rangle_{\mathbf{L}^2} \phi_k . \tag{1.7}$$

Notice the close resemblance between the formulas (1.5) and (1.7). In essence, one only needs to replace the Euclidean inner product on $\mathbb{R}^n$ by the inner product on $\mathbf{L}^2(\Omega)$.

1.2. Evolution equations

Let A be an $n \times n$ matrix. For a given initial state $\mathbf{b} \in \mathbb{R}^n$, consider the Cauchy problem

$$\frac{d}{dt}\mathbf{x}(t) = A\mathbf{x}(t), \qquad \mathbf{x}(0) = \mathbf{b}. \tag{1.8}$$

According to linear ODE theory, this problem has a unique solution:

$$\mathbf{x}(t) = e^{tA}\mathbf{b}, \tag{1.9}$$

where

$$e^{tA} = \sum_{k=0}^{\infty} \frac{t^k A^k}{k!} . \tag{1.10}$$

Notice that the right-hand side of (1.10) is defined as a convergent series of $n \times n$ matrices. Here $A^0 = I$ is the identity matrix. The family of matrices $\{e^{tA};\ t \in \mathbb{R}\}$ has the "group property", namely

$$e^{0\,A} = I\,, \qquad e^{tA}e^{sA} = e^{(t+s)A} \qquad \text{for all } t, s \in \mathbb{R}\,.$$

If A is symmetric, then it admits an orthonormal basis of eigenvectors $\{\mathbf{v}_1, \ldots, \mathbf{v}_n\}$, with corresponding eigenvalues $\lambda_1, \ldots, \lambda_n$. In this case, the solution (1.9) can be written more explicitly as

$$e^{tA}\mathbf{b} \;=\; \sum_{k=1}^{n} e^{t\lambda_k} \langle \mathbf{b},\, \mathbf{v}_k \rangle\, \mathbf{v}_k\,.$$

The theory of linear semigroups provides an extension of these results to unbounded linear operators in infinite-dimensional spaces. In particular, it applies to parabolic evolution equations of the form

$$\text{(1.11)} \qquad \frac{d}{dt}u(t) \;=\; -Lu(t), \qquad u(0) \;=\; g \;\in\; \mathbf{L}^2(\Omega)\,, \qquad u = 0 \text{ on } \partial\Omega\,,$$

where L is the partial differential operator in (1.2) and $\partial\Omega$ denotes the boundary of Ω. When $a^{ij} = a^{ji}$ and $b^i(x) = 0$, the elliptic operator L is symmetric and the solution can be decomposed along the orthonormal basis $\{\phi_1, \phi_2, \ldots\}$ of the space $\mathbf{L}^2(\Omega)$ considered in (1.6). This yields the representation

$$\text{(1.12)} \qquad u(t) \;=\; S_t g \;\doteq\; \sum_{k=1}^{\infty} e^{-t\lambda_k} \langle g,\, \phi_k \rangle_{\mathbf{L}^2}\, \phi_k, \qquad t \geq 0\,.$$

Notice that the operator L is unbounded (its eigenvalues satisfy $\lambda_k \to +\infty$ as $k \to \infty$). However, the operators S_t in (1.12) are bounded for every $t \geq 0$ (but not for $t < 0$). The family of linear operators $\{S_t\,;\ t \geq 0\}$ is called a *linear semigroup*, since it has the semigroup properties

$$S_0 \;=\; I\,, \qquad S_t \circ S_s \;=\; S_{t+s} \quad \text{for all } s, t \geq 0\,.$$

Intuitively, we could think of S_t as an exponential operator: $S_t \doteq e^{-Lt}$. However, since L is unbounded, one should be aware that an exponential formula such as (1.10) is no longer valid. When the explicit formula (1.12) is not available, the operators S_t must be constructed using some different approximation method. In the finite-dimensional case, the exponential of a matrix A can be recovered by

$$\text{(1.13)} \qquad e^{tA} \;=\; \lim_{n\to\infty} \left(I - \frac{t}{n}A\right)^{-n}$$

and also by

$$\text{(1.14)} \qquad e^{tA} \;=\; \lim_{\lambda\to\infty} e^{tA_\lambda}, \qquad A_\lambda \;\doteq\; A(I - \lambda^{-1}A)^{-1}\,.$$

Remarkably, the two formulas (1.13)–(1.14) retain their validity also for a wide class of unbounded operators on infinite-dimensional spaces.

The hyperbolic initial value problem

$$(1.15)\qquad u_{tt} + Lu = 0, \qquad \begin{cases} u(0) = g\,, \\ u_t(0) = h\,, \end{cases} \qquad u = 0 \text{ on } \partial\Omega\,,$$

can also be treated by similar methods.

The finite-dimensional counterpart of (1.15) is the system of second-order linear equations

$$(1.16)\qquad \frac{d^2}{dt^2}\mathbf{x}(t) + A\mathbf{x}(t) = 0, \qquad \mathbf{x}(0) = \mathbf{a}\,, \quad \frac{d}{dt}\mathbf{x}(0) = \mathbf{b}\,.$$

Here $\mathbf{x}\,,\mathbf{a},\,\mathbf{b} \in \mathbb{R}^n$ and A is an $n \times n$ matrix. Denoting time derivatives by an upper dot and setting $\mathbf{y} \doteq \dot{\mathbf{x}}$, (1.16) can be written as a first-order system:

$$(1.17)\qquad \begin{pmatrix}\dot{\mathbf{x}}\\ \dot{\mathbf{y}}\end{pmatrix} = \begin{pmatrix}0 & I\\ -A & 0\end{pmatrix}\begin{pmatrix}\mathbf{x}\\ \mathbf{y}\end{pmatrix}, \qquad \begin{pmatrix}\mathbf{x}(0)\\ \mathbf{y}(0)\end{pmatrix} = \begin{pmatrix}\mathbf{a}\\ \mathbf{b}\end{pmatrix}.$$

The same results valid for first-order linear ODEs can thus be applied here. If A is symmetric, then it has an orthonormal basis of eigenvectors $\{\mathbf{v}_1, \ldots, \mathbf{v}_n\}$ with corresponding eigenvalues $\lambda_1, \ldots, \lambda_n$. In this case, the solution of (1.16) can be written as

$$(1.18)\qquad \mathbf{x}(t) = \sum_{k=1}^{n} c_k(t)\mathbf{v}_k\,.$$

Each coefficient $c_k(\cdot)$ can be independently computed, by solving the second-order scalar ODE

$$\frac{d^2}{dt^2}c_k(t) + \lambda_k c_k(t) = 0\,, \qquad c_k(0) = \langle \mathbf{a},\, v_k\rangle\,, \qquad \frac{d}{dt}c_k(0) = \langle \mathbf{b},\, v_k\rangle\,.$$

Returning to the problem (1.15), if the elliptic operator L is symmetric, then the solution can again be decomposed along the orthonormal basis $\{\phi_1, \phi_2, \ldots\}$ of the space $\mathbf{L}^2(\Omega)$ considered in (1.6). This yields the entirely similar representation

$$(1.19)\qquad u(t) = \sum_{k=1}^{\infty} c_k(t)\,\phi_k \qquad t \geq 0\,,$$

where each function c_k is determined by the equations

$$\frac{d^2}{dt^2}c_k(t) + \lambda_k c_k(t) = 0\,, \qquad c_k(0) = \langle g,\, \phi_k\rangle_{\mathbf{L}^2}\,, \qquad \frac{d}{dt}c_k(0) = \langle h,\, \phi_k\rangle_{\mathbf{L}^2}\,.$$

1.3. Function spaces

In functional analysis, a key idea is to regard functions $f : \mathbb{R}^n \mapsto \mathbb{R}$ as points in an abstract vector space. All the information about a function is condensed in one single number $\|f\|$, which we call the *norm* of f. Typically, the norm measures the "size" of f and of its partial derivatives up to some order k. It is remarkable that so many results can be achieved in such an economical way, relying only on this single concept, coupled with the structure of vector space. This accounts for the wide success of functional analytic methods.

Toward all applications of functional analysis to integral or differential equations, one needs to develop a theory of function spaces. In this direction, it is natural to consider the spaces $\mathcal{C}^k$ of functions with bounded continuous partial derivatives up to order k. The "size" of a function $f \in \mathcal{C}^k(\mathbb{R}^n)$ is here measured by the norm

$$\|f\|_{\mathcal{C}^k} \doteq \max_{\alpha_1+\cdots+\alpha_n \leq k} \sup_{x\in\mathbb{R}^n} \left|\partial_{x_1}^{\alpha_1}\cdots\partial_{x_n}^{\alpha_n} f(x)\right|.$$

The spaces $\mathcal{C}^k$, however, are not always appropriate for the study of PDEs. Indeed, from physical or geometrical considerations one can often provide estimates not on the maximum value of a solution and its derivatives, but on their $\mathbf{L}^p$ norm, for some $p \geq 1$. This motivates the introduction of the Sobolev spaces $W^{k,p}$, containing all functions whose derivatives up to order k lie in $\mathbf{L}^p$. The "size" of a function $f \in W^{k,p}(\mathbb{R}^n)$ is now measured by the norm

$$\|f\|_{W^{k,p}} \doteq \left(\sum_{\alpha_1+\cdots+\alpha_n\leq k} \int_{\mathbb{R}^n} \left|\partial_{x_1}^{\alpha_1}\cdots\partial_{x_n}^{\alpha_n} f(x)\right|^p dx\right)^{1/p}.$$

Because of their fundamental role in PDE theory, all of Chapter 8 will be devoted to the study of Sobolev spaces.

1.4. Compactness

When solving an equation, if an explicit formula for the solution is not available, a common procedure relies on three steps:

(i) Construct a sequence of approximate solutions $(u_n)_{n\geq 1}$.

(ii) Extract a convergent subsequence $u_{n_j} \to \bar{u}$.

(iii) Prove that the limit $\bar{u}$ is a solution.

When we reach step (ii), a major difference between the Euclidean space $\mathbb{R}^N$ and abstract function spaces is encountered. Namely, in $\mathbb{R}^N$ all closed bounded sets are compact. Otherwise stated, in $\mathbb{R}^N$ the Bolzano-Weierstrass theorem holds:

- *From every bounded sequence $(u_n)_{n\geq 1}$ one can extract a convergent subsequence.*

As proved in Chapter 2 (see Theorem 2.22), this crucial property is valid in every finite-dimensional normed space but fails in every infinite-dimensional one. In a space of functions, showing that a sequence of approximate solutions is bounded, i.e., $\|u_n\| \leq C$ for some constant C and all $n \geq 1$, does not guarantee the existence of a convergent subsequence. To overcome this fundamental difficulty, two main approaches can be adopted.

(i) Introduce a weaker notion of convergence. Prove that every bounded sequence (also in an infinite-dimensional space) has a subsequence which converges in this weaker sense.
A key result in this direction, the Banach-Alaoglu theorem, will be proved at the end of Chapter 2. Weak convergence in Hilbert spaces is discussed in Chapter 5.

(ii) Consider two distinct norms, say $\|u\|_{weak} \leq \|u\|_{\text{strong}}$, with the following property. If a sequence $(u_n)_{n\geq 1}$ is bounded in the strong norm, i.e., $\|u_n\|_{\text{strong}} \leq C$, then there exists a subsequence that converges in the weak norm: $\|u_{n_j} - \bar{u}\|_{\text{weak}} \to 0$, for some limit $\bar{u}$. Ascoli's theorem, proved in Chapter 3, and the Rellich-Kondrahov compact embedding theorem, proved in Chapter 8, yield different settings where this approach can be implemented.

A large portion of the analysis of partial differential equations ultimately relies on the derivation of a priori estimates. It is the nature of the problem at hand that dictates what kind of a priori bounds one can expect, and hence in which function spaces the solution can be found. This motivates the variety of function spaces which are currently encountered in literature.

While the techniques of functional analysis are very general and yield results of fundamental nature in an intuitive and economical way, one should be aware that only some aspects of PDE theory can be approached by functional analytic methods alone. Typically, the solutions constructed by these abstract methods lie in a Sobolev space of functions that possess just the minimum amount of regularity needed to make sense of the equations. For several elliptic and parabolic equations, it is known that solutions enjoy a much higher regularity. However, this regularity can only be established by a more detailed analysis. Further properties, such as the maximum principle for elliptic or parabolic equations and the finite propagation speed for hyperbolic equations, also require additional techniques, specifically designed for PDEs. For these issues, which are not within the scope of the present lecture notes, we refer to the monographs [**E, GT, McO, P, PW, RR, S, T**].

Chapter 2

Banach Spaces

Given a vector space X over the real or complex numbers, we wish to introduce a distance $d(\cdot,\cdot)$ between points of X. This will allow us to define limits, convergent sequences and series, and continuous mappings.

Since X has the algebraic structure of a vector space, the distance d should be consistent with this structure. It is thus natural to require the following properties:

(p1) The distance d is invariant under translations. Namely: $d(x,y) = d(x+z,\, y+z)$ for every $x,y,z \in X$. In particular, $d(x,y) = d(x-y,\, 0)$.

(p2) The distance d is positively homogeneous. Namely: $d(\lambda x, \lambda y) = |\lambda|\, d(x,y)$, for any scalar number λ and any $x,y \in X$.

(p3) Every open ball $B(x_0, r) \doteq \{x \in X\,;\ d(x,x_0) < r\}$ is a convex set.

The invariance under translations implies that the distance $d(\cdot,\cdot)$ is entirely determined as soon as we specify the function $x \mapsto \|x\| = d(x,0)$, i.e., the distance of a point x from the origin. This is what we call the **norm** of a vector $x \in X$. It can be taken as the starting point for the entire theory.

2.1. Basic definitions

Let X be a (possibly infinite-dimensional) vector space over the field $\mathbb{K}$ of numbers. We shall always assume that $\mathbb{K}$ is either the field of real numbers $\mathbb{R}$ or the field of complex numbers $\mathbb{C}$. A **norm** on X is a map $x \mapsto \|x\|$ from X into $\mathbb{R}$, with the following properties.

(N1) For every $x \in X$ one has $\|x\| \geq 0$, with equality holding if and only if $x = 0$.

(N2) For every $x \in X$ and $\lambda \in \mathbb{K}$ one has $\|\lambda x\| = |\lambda| \, \|x\|$.

(N3) For every $x, y \in X$ one has $\|x + y\| \leq \|x\| + \|y\|$.

A vector space X with a norm $\|\cdot\|$ satisfying (N1)–(N3) is called a **normed space**. In turn, a norm determines a distance between elements of X.

Lemma 2.1 (Distance defined by a norm). *Let $\|\cdot\|$ be a norm on the vector space X. Then*

$$d(x, y) \doteq \|x - y\| \tag{2.1}$$

defines a distance between points of X. Moreover, this distance has the additional properties of translation invariance, positive homogeneity, and convexity stated in **(p1)–(p3)** *above.*

Proof. 1. We check that, for all $x, y, z \in X$, the three basic properties of a distance are satisfied:

(D1) $d(x, y) \geq 0$ for all $x, y \in X$ with equality holding if and only if $x = y$;

(D2) $d(x, y) = d(y, x)$;

(D3) $d(x, z) \leq d(x, y) + d(y, z)$.

Indeed, (D1) is an immediate consequence of (N1). To prove (D2) we write

$$d(y, x) = \|y - x\| = \|(-1)(x - y)\| = |-1| \, \|x - y\| = \|x - y\| = d(x, y).$$

The triangle inequality follows from (N3), replacing x, y with $x - y$ and $y - z$, respectively:

$$d(x, z) = \|x - z\| = \|(x - y) + (y - z)\| \leq \|x - y\| + \|y - z\| = d(x, y) + d(y, z).$$

2. The property **(p1)** of translation invariance follows immediately from the definition. The homogeneity property **(p2)** follows from

$$d(\lambda x, \lambda y) = \|\lambda(x - y)\| = |\lambda| \, \|x - y\| = |\lambda| \, d(x, y).$$

Finally, to check that every open ball is convex, by translation invariance it suffices to prove that every ball centered at the origin is convex. If $x, y \in B(0, r)$ and $0 \leq \theta \leq 1$, then the convex combination satisfies

$$\|\theta x + (1 - \theta) y\| \leq |\theta| \, \|x\| + |1 - \theta| \, \|y\| < \theta r + (1 - \theta) r = r.$$

Hence $\theta x + (1-\theta)y \in B(0,r)$, which proves that the open ball $B(0,r)$ is convex. $\square$

The distance $d(x,y) = \|x-y\|$ determines a topology on the vector space X. We can thus talk about open sets, closed sets, convergent sequences, and continuous mappings.

Throughout the following, the open and the closed balls centered at a point x with radius $r>0$ are denoted respectively by

$$B(x,r) \doteq \{y \in X\,;\ \|y-x\| < r\}, \qquad \overline{B}(x,r) \doteq \{y \in X\,;\ \|y-x\| \leq r\}.$$

We recall that a set $V \subseteq X$ is **open** if, for every $x \in V$, there exists $r>0$ such that $B(x,r) \subseteq V$. A set $U \subseteq X$ is **closed** if its complement $X \setminus U$ is open.

A **sequence** $(x_n)_{n\geq 1}$ converges to a point $\bar{x} \in X$ if

$$\lim_{n\to\infty} \|x_n - \bar{x}\| = 0.$$

Given a **series** of elements of X, we say that the series converges to $\bar{x}$, and write

$$\sum_{k=1}^{\infty} y_k = \bar{x},$$

if the sequence of partial sums converges to $\bar{x}$, namely

$$\lim_{n\to\infty} \left\| \bar{x} - \sum_{k=1}^{n} y_k \right\| = 0.$$

Given two normed spaces X, Y, we say that a map $f : X \mapsto Y$ is **continuous** if, for every $x \in X$ and $\varepsilon > 0$, there exists $\delta > 0$ such that

$$\|f(x') - f(x)\| < \varepsilon \qquad \text{whenever } x' \in X,\ \|x'-x\| < \delta.$$

A sequence $(x_n)_{n\geq 1}$ is a **Cauchy sequence** if, for every $\varepsilon > 0$, one can find an integer N large enough so that

$$\|x_m - x_n\| \leq \varepsilon \qquad \text{whenever } m,n \geq N.$$

A normed space X is **complete** if every Cauchy sequence converges to some limit point $\bar{x} \in X$. A complete normed space is called a **Banach space**.

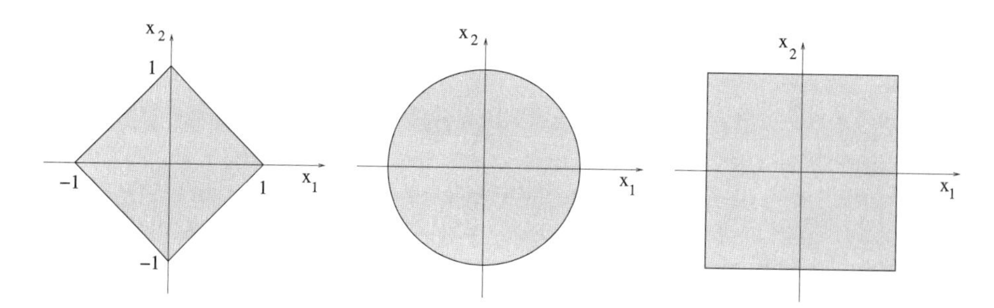

Figure 2.1.1. The unit balls in $\mathbb{R}^2$ with norms $\|\cdot\|_1$ (left), $\|\cdot\|_2$ (center), and $\|\cdot\|_\infty$ (right), described in Example 2.3.

2.1.1. Examples of normed and Banach spaces.

Example 2.2. The finite-dimensional space $\mathbb{R}^n = \{x = (x_1, \ldots, x_n),\ x_i \in \mathbb{R}\}$ with Euclidean norm

$$\|x\|_2 \doteq \sqrt{x_1^2 + \cdots + x_n^2} \tag{2.2}$$

is a Banach space over the real numbers. In particular, the field $\mathbb{R}$ of all real numbers can be regarded as a 1-dimensional Banach space. In this case, the norm of a number $x \in \mathbb{R}$ is provided by its absolute value.

Example 2.3. On the space $\mathbb{R}^n$ one can consider the alternative norms

$$\|x\|_p \doteq \Big(|x_1|^p + \cdots + |x_n|^p\Big)^{1/p}, \qquad \|x\|_\infty \doteq \max_{1\le i\le n} |x_i|.$$

Here $1 \le p < \infty$. Each of these norms also makes $\mathbb{R}^n$ into a finite-dimensional Banach space.

Example 2.4. For any closed bounded interval $[a,b]$, the space $\mathcal{C}^0([a,b])$ of all continuous functions $f : [a,b] \mapsto \mathbb{R}$, with norm

$$\|f\|_{\mathcal{C}^0} \doteq \max_{x\in[a,b]} |f(x)|, \tag{2.3}$$

is a Banach space.

Example 2.5. Let Ω be an open subset of $\mathbb{R}^n$. For every $1 \le p < \infty$, consider the space $\mathbf{L}^p(\Omega)$ of all Lebesgue measurable functions $f : \Omega \mapsto \mathbb{R}$ such that $\int_\Omega |f(x)|^p\,dx < \infty$. This is a Banach space, equipped with the norm

$$\|f\|_{\mathbf{L}^p} \doteq \left(\int_\Omega |f(x)|^p\,dx\right)^{1/p}. \tag{2.4}$$

Two functions $f, \tilde f$ are regarded here as the same element of $\mathbf{L}^p(\Omega)$ if they coincide almost everywhere, i.e., if $\text{meas}\Big(\{x \in \Omega;\ f(x) \neq \tilde f(x)\}\Big) = 0$.

Similarly, the space $\mathbf{L}^\infty(\Omega)$ of all essentially bounded, measurable functions $f : \Omega \mapsto \mathbb{R}$ is a Banach space with norm

$$\|f\|_{\mathbf{L}^\infty} \doteq \operatorname*{ess\,sup}_{x\in\Omega} |f(x)|. \tag{2.5}$$

Example 2.6. For a fixed $p \geq 1$, consider the space of all sequences of real numbers whose p-th powers are summable:

$$\ell^p \doteq \Big\{\mathbf{x} = (x_1, x_2, \ldots)\,;\ \sum_{k=1}^{\infty} |x_k|^p < \infty\Big\}. \tag{2.6}$$

This is a Banach space with norm

$$\|\mathbf{x}\|_p \doteq \left(\sum_{k=1}^{\infty} |x_k|^p\right)^{1/p}. \tag{2.7}$$

Example 2.7. The space ℓ^∞ of all bounded sequences of real numbers, with norm

$$\|\mathbf{x}\|_\infty \doteq \sup_{k\geq 1} |x_k|, \tag{2.8}$$

is a Banach space. Within this space, one can consider the subspace c_0 of all sequences $(x_k)_{k\geq 1}$ that converge to zero as $k \to \infty$. This is also a Banach space, for the same norm (2.8).

Remark 2.8. Within the space ℓ^p, $1 \leq p < \infty$, consider the family of unit vectors

$$\mathbf{e}_1 = (1, 0, 0, 0, \ldots), \quad \mathbf{e}_2 = (0, 1, 0, 0, \ldots), \quad \mathbf{e}_3 = (0, 0, 1, 0, \ldots), \quad \ldots. \tag{2.9}$$

These are linearly independent. The set of all linear combinations[1]

$$\operatorname{span}\{\mathbf{e}_k\,;\ k \geq 1\} = \Big\{\sum_{k=1}^{N} \theta_k \mathbf{e}_k\,;\ N \geq 1,\ \theta_k \in \mathbb{R}\Big\} \tag{2.10}$$

does not coincide with the entire space ℓ^p, but is a dense subset of ℓ^p. Indeed, it consists of all sequences of the form $x = (x_1, x_2, \ldots, x_N, 0, 0, 0, \ldots)$, having finitely many nonzero entries.

The set $\{\mathbf{e}_k\,;\ k \geq 1\}$ is not an algebraic basis for ℓ^p, but it provides a topological basis. Namely, every element $\mathbf{x} = (x_1, x_2, \ldots) \in \ell^p$ can be obtained as the sum of the convergent series

$$\mathbf{x} = \sum_{k=1}^{\infty} x_k\, \mathbf{e}_k\,.$$

[1]Notice that by a *linear combination* one always means a finite sum, not a series.

Remark 2.9. On the field of real or complex numbers, the absolute value $|\alpha|$ measures how big the number α is. Similarly, on a vector space X one can think of the norm $\|f\|$ as measuring the size of an element $f \in X$. We observe, however, that it is possible to adopt different norms (and hence different distances) on the same space of functions, obtaining quite different convergence results. This is illustrated by the next example.

Example 2.10. Let X be the space of all polynomial functions on the interval $[0,1]$. On X we can consider the two norms

$$\|f\|_{\mathcal{C}^0} \doteq \max_{x\in[0,1]} |f(x)|, \qquad \|f\|_{\mathbf{L}^1} \doteq \int_0^1 |f(x)|\,dx\,. \tag{2.11}$$

These norms yield different convergence results (see Figure 2.1.2). For example, consider the sequence of monomials $f_n(x) = x^n$. Letting $n \to \infty$ one has $\|f_n\|_{\mathbf{L}^1} \to 0$. Therefore the sequence $(f_n)_{n\geq 1}$ converges to zero (i.e., to the identically zero function) in the $\mathbf{L}^1$ norm. On the other hand, $\|f_n\|_{\mathcal{C}^0} = 1$ for all $n \geq 1$. In terms of the $\mathcal{C}^0$ norm, this same sequence is not a Cauchy sequence and does not have any limit.

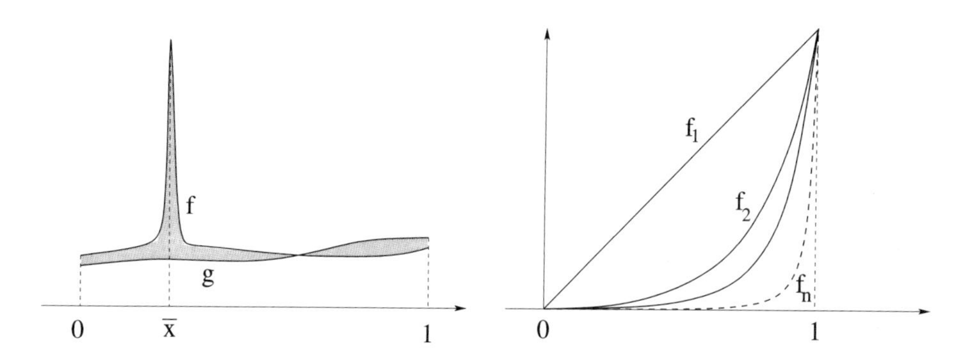

Figure 2.1.2. Left: the $\mathbf{L}^1$ distance between the two functions f, g, measured by the area of the shaded region, is small. However, their $\mathcal{C}^0$ distance, measured by $\|f-g\|_{\mathcal{C}^0} = |f(\bar{x}) - g(\bar{x})|$ is large. Right: the sequence of polynomials $f_n(x) = x^n$ converges to zero in the space $\mathbf{L}^1([0,1])$ but does not have any limit in the space $\mathcal{C}^0([0,1])$.

2.2. Linear operators

Let X, Y be normed spaces on the same field $\mathbb{K}$ of scalar numbers. A **linear operator** is a mapping Λ from a subspace $\text{Dom}(\Lambda) \subseteq X$ into Y such that

$$\Lambda(c_1x_1 + c_2x_2) = c_1\Lambda x_1 + c_2\Lambda x_2 \qquad \text{for all } x_1, x_2 \in \text{Dom}(\Lambda),\ c_1, c_2 \in \mathbb{K}.$$

Here $\text{Dom}(\Lambda)$ is the **domain** of Λ. The **range** of Λ is the subspace

$$\text{Range}(\Lambda) \doteq \{\Lambda x\,;\ x \in \text{Dom}(\Lambda)\} \subset Y.$$

The **null space** or **kernel** of Λ is the subspace

$$\mathrm{Ker}(\Lambda) \doteq \{x \in X\,;\ \Lambda x = 0\} \subseteq X.$$

Notice that Λ is one-to-one if and only if $\mathrm{Ker}(\Lambda) = \{0\}$.

Next, consider a linear operator $\Lambda : X \mapsto Y$ defined on the entire space X, i.e., with $\mathrm{Dom}(\Lambda) = X$. We say that Λ is **bounded** if

$$\|\Lambda\| \doteq \sup_{\|x\|\leq 1} \|\Lambda x\| < \infty. \tag{2.12}$$

Theorem 2.11 (Continuity of bounded operators). *A linear operator $\Lambda : X \mapsto Y$ is bounded if and only if it is continuous.*

Proof. 1. If Λ is continuous, then in particular it is continuous at the origin. Hence there exists $\delta > 0$ such that $\|x\| \leq \delta$ implies $\|\Lambda x\| \leq 1$. By linearity, this implies that

$$\|\Lambda x\| \leq \frac{1}{\delta} \qquad \text{whenever} \qquad \|x\| \leq 1\,.$$

Hence Λ is bounded.

2. Conversely, let Λ be bounded, so that (2.12) holds. By linearity, we obtain

$$\|\Lambda x_1 - \Lambda x_2\| = \|\Lambda(x_1 - x_2)\| = \|x_1 - x_2\| \left\|\Lambda\Big(\frac{x_1 - x_2}{\|x_1 - x_2\|}\Big)\right\| \leq \|x_1 - x_2\|\,\|\Lambda\|\,.$$

Hence Λ is uniformly Lipschitz continuous with Lipschitz constant $\|\Lambda\|$. □

If X, Y are normed spaces over the same field of scalar numbers, we denote by $\mathcal{B}(X;\, Y)$ the **space of bounded linear operators** from X into Y. Notice that here we require that the domain of these operators should be the entire space X.

Theorem 2.12 (The space of bounded linear operators). *The space $\mathcal{B}(X;\, Y)$ of all bounded linear operators from X into Y is a normed space, with norm defined at* (2.12)*. If Y is a Banach space, then $\mathcal{B}(X;\, Y)$ is a Banach space.*

Proof. If Λ_1, Λ_2 are linear operators, and $c_1, c_2 \in \mathbb{K}$, then their linear combination is defined as

$$(c_1\Lambda_1 + c_2\Lambda_2)(x) \doteq c_1\Lambda_1 x + c_2\Lambda_2 x\,.$$

We now check that the properties (N1)–(N3) of a norm are satisfied.

1. If $\Lambda = 0$ is the zero operator, then $\Lambda x = 0$ for all $x \in X$, and $\|\Lambda\| = 0$. On the other hand, if Λ is not the zero operator, then $\Lambda x \neq 0$ for some

$x \neq 0$. Hence $\Lambda(\frac{x}{\|x\|}) = \frac{1}{\|x\|}\Lambda x \neq 0$ and the supremum in (2.12) is strictly positive. This proves (N1).

2. If $\alpha \in \mathbb{K}$, then (N2) follows by the identities

$$\|\alpha\Lambda\| = \sup_{\|x\|\leq 1} \|\alpha\Lambda x\| = |\alpha| \sup_{\|x\|\leq 1} \|\Lambda x\| = |\alpha|\, \|\Lambda\|\,.$$

3. To check the triangle inequality (N3), for every $x \in X$ with $\|x\| \leq 1$ we write

$$\|(\Lambda_1 + \Lambda_2)x\| = \|\Lambda_1 x + \Lambda_2 x\| \leq \|\Lambda_1 x\| + \|\Lambda_2 x\| \leq \|\Lambda_1\| + \|\Lambda_2\|.$$

Taking the supremum over all $x \in X$ with $\|x\| \leq 1$ we obtain (N3).

4. Next, assume that Y is a Banach space. We need to show that $\mathcal{B}(X; Y)$ is complete. Let $(\Lambda_n)_{n\geq 1}$ be a Cauchy sequence of bounded linear operators. For each $x \in X$, this implies

$$\limsup_{m,n\to\infty} \|\Lambda_m x - \Lambda_n x\| \leq \limsup_{m,n\to\infty} \|\Lambda_m - \Lambda_n\|\, \|x\| = 0.$$

Therefore the sequence of points $(\Lambda_n x)_{n\geq 1}$ is Cauchy in Y. Since Y is complete, this sequence has a unique limit, which we call Λx.

Since every Λ_n is a linear operator, it is clear that the Λ is linear as well. We claim that Λ is also bounded (and hence continuous). By assumption, we can choose N large enough such that

$$\|\Lambda_k - \Lambda_N\| \leq 1 \qquad \text{for all } k \geq N.$$

Therefore, for any $x \in X$ with $\|x\| \leq 1$ one has

$$\|\Lambda x\| = \lim_{k\to\infty} \|\Lambda_k x\| \leq \|\Lambda_N x\| + \limsup_{k\to\infty} \|\Lambda_k - \Lambda_N\|\|x\| \leq \|\Lambda_N\| + 1\,.$$

Since x was an arbitrary point in the unit ball, this proves that the limit operator Λ is bounded, and hence $\Lambda \in \mathcal{B}(X; Y)$. □

2.2.1. Examples of linear operators.

Example 2.13 (Matrices as linear operators). Every $n \times m$ matrix $A = (a_{ij})$ determines a bounded linear operator $\Lambda : \mathbb{R}^m \mapsto \mathbb{R}^n$ defined by

$$\Lambda(x_1, \ldots, x_m) = (y_1, \ldots, y_n), \qquad \text{with} \qquad y_i \doteq \sum_{j=1}^{m} a_{ij} x_j\,.$$

Example 2.14 (Diagonal operators on a space of sequences). Let $1 \leq p \leq \infty$, and consider the space $X = \ell^p$ of all sequences $\mathbf{x} = (x_1, x_2, \ldots)$

of real numbers, with norm defined as (2.7) or (2.8). Let $(\lambda_1, \lambda_2, \ldots)$ be an arbitrary sequence of real numbers, and define the operator $\Lambda : X \mapsto X$ as

$$\Lambda(x_1, x_2, x_3, \ldots) \doteq (\lambda_1 x_1\,,\ \lambda_2 x_2\,,\ \lambda_3 x_3\,,\ \ldots\,). \tag{2.13}$$

With reference to the basis of unit vectors $\{\mathbf{e}_1, \mathbf{e}_2, \ldots\}$ in (2.9), we can think of Λ as an infinite matrix:

$$\begin{pmatrix} \lambda_1 & 0 & & \\ 0 & \lambda_2 & & \\ & & \lambda_3 & \\ & & & \ddots \end{pmatrix}$$

with $\lambda_1, \lambda_2, \ldots$ along the diagonal and 0 everywhere else. We now have two cases.

(i) If the sequence $(\lambda_k)_{k\geq 1}$ is bounded, then the operator Λ is bounded. Its norm is

$$\|\Lambda\| = \sup_k |\lambda_k|\,. \tag{2.14}$$

(ii) If the sequence $(\lambda_k)_{k\geq 1}$ is unbounded, then the operator Λ is not bounded. Its domain

$$\mathrm{Dom}(\Lambda) = \Big\{x \in \ell^p\,;\ \Lambda x \in \ell^p\Big\}$$

is a vector subspace, strictly contained in ℓ^p.

Example 2.15 (Differentiation operator). Consider the open interval $I =]0, \pi[$ and let $X = \mathcal{B}\mathcal{C}(I)$ be the space of bounded continuous real-valued functions on I, with norm

$$\|f\| = \sup_{0<x<\pi} |f(x)|\,.$$

The differential operator $\Lambda f = f'$ is clearly a linear operator on X. To see that this operator is not bounded, consider the sequence of functions $f_k(x) = \sin kx$. Then $f_k'(x) = k\,\cos kx$, hence

$$\|f_k\| = 1\,, \qquad \|\Lambda f_k\| = k \qquad \text{for all } k \geq 1.$$

The domain $\mathrm{Dom}(\Lambda)$ of this differential operator is the space of all bounded continuous functions $f : I \mapsto \mathbb{R}$ that are everywhere differentiable and have a bounded, continuous derivative. This is a proper subspace of $\mathcal{B}\mathcal{C}(I)$.

Example 2.16 (Shift operator on $\mathbf{L}^p(\mathbb{R})$). Let $1 \leq p \leq \infty$. Fix any $a \in \mathbb{R}$. Given a function $f \in \mathbf{L}^p(\mathbb{R})$, define $(\Lambda_a f)(x) \doteq f(x-a)$. Clearly $\|\Lambda f\|_{\mathbf{L}^p} = \|f\|_{\mathbf{L}^p}$. Therefore, $\Lambda_a : \mathbf{L}^p \mapsto \mathbf{L}^p$ is a bounded linear operator, with norm $\|\Lambda_a\| = 1$. Notice that the operator Λ_a is one-to-one and onto.

Example 2.17 (Shift operators on ℓ^p). Let $1 \le p \le \infty$. Define the operators $\Lambda_+ : \ell^p \mapsto \ell^p$ and $\Lambda_- : \ell^p \mapsto \ell^p$ as

$$\begin{aligned}\Lambda_+(x_1, x_2, x_3, \ldots) &\doteq (0, x_1, x_2, \ldots)\,,\\ \Lambda_-(x_1, x_2, x_3, \ldots) &\doteq (x_2, x_3, x_4, \ldots)\,.\end{aligned}$$

Observe that these are linear continuous operators, with $\|\Lambda_+\| = \|\Lambda_-\| = 1$. Moreover, Λ_+ is one-to-one but not onto, while Λ_- is onto, but not one-to-one.

Example 2.18 (Multiplication operator). Let $\Omega \subset \mathbb{R}^n$ and let $g : \Omega \mapsto \mathbb{R}$ be a bounded, measurable function. For any $1 \le p \le \infty$, on the space $\mathbf{L}^p(\Omega)$ consider the multiplication operator: $(M_g f)(x) \doteq g(x)\, f(x)$. This is a continuous operator, with norm

$$\|M_g\| \doteq \sup_{\|f\|_{\mathbf{L}^p} \le 1} \|gf\|_{\mathbf{L}^p} = \|g\|_{\mathbf{L}^\infty}\,. \tag{2.15}$$

Example 2.19 (Integral operator). Let $a < b$ and consider the space $X = \mathcal{C}^0([a,b])$ of real-valued continuous functions defined on the closed interval $[a,b]$. Consider the integral operator

$$(\Lambda f)(x) \doteq \int_a^x f(y)\, dy\,.$$

Then $\Lambda : X \mapsto X$ is a bounded linear operator. Indeed,

$$\Big|(\Lambda f)(x)\Big| = \left|\int_a^x f(y)\, dy\right| \le \int_a^x |f(y)|\, dy \le (b-a) \cdot \max_{x \in [a,b]} |f(x)|\,.$$

Hence $\|\Lambda f\| \le (b-a)\|f\|$ and $\|\Lambda\| \le (b-a)$.

2.3. Finite-dimensional spaces

We say that two norms $\|\cdot\|_\diamond$ and $\|\cdot\|_\spadesuit$ on the same vector space X are **equivalent** if there exists a constant $C \ge 1$ such that

$$\frac{1}{C}\,\|x\|_\diamond \le \|x\|_\spadesuit \le C\,\|x\|_\diamond \qquad \text{for all } x \in X\,. \tag{2.16}$$

Equivalent norms yield the same Cauchy sequences and the same topology on X. In general, an infinite-dimensional space X can have many nonequivalent norms. As shown in Example 2.10, on the space of all polynomials in one real variable, the $\mathbf{L}^1$ norm and the $\mathcal{C}^0$ norm defined in (2.11) are not equivalent.

The next results show that, for a finite-dimensional vector space X, all norms are equivalent. Indeed, every finite-dimensional normed space of dimension N is equivalent to the Euclidean space $\mathbb{K}^N$. We recall that the Euclidean norm of a vector $\alpha = (\alpha_1, \ldots, \alpha_N) \in \mathbb{K}^N$ is $\|\alpha\| \doteq \sqrt{\sum_k |\alpha_k|^2}$.

Theorem 2.20 (A finite-dimensional normed space is homeomorphic to $\mathbb{K}^N$). *Let X be a finite-dimensional normed space over the field $\mathbb{K}$ of real or complex numbers. Let $\mathcal{B} = \{u_1, u_2, \ldots, u_N\}$ be a basis of X. Then*

(i) *X is complete, and hence a Banach space.*

(ii) *For every $\alpha = (\alpha_1, \alpha_2, \ldots, \alpha_N) \in \mathbb{K}^N$, define*

$$\Lambda\alpha \doteq \alpha_1 u_1 + \alpha_2 u_2 + \cdots + \alpha_N u_N . \tag{2.17}$$

Then the linear operator $\Lambda : \mathbb{K}^N \mapsto X$ is bijective and bounded. Moreover, its inverse $\Lambda^{-1} : X \mapsto \mathbb{K}^N$ is also bounded.

Proof. 1. The fact that $\{u_1, \ldots, u_N\}$ is a basis implies that Λ is one-to-one and onto. Hence the inverse operator $\Lambda^{-1} : X \mapsto \mathbb{K}^N$ is well defined.

2. Writing

$$\|\Lambda\alpha\| \leq \sum_{k=1}^{N} \|\alpha_k u_k\| \leq \|\alpha\| \sum_{k=1}^{N} \|u_k\| ,$$

we see that $\Lambda : \mathbb{K}^N \mapsto X$ is a bounded linear operator, hence continuous.

3. We claim that Λ^{-1} is also bounded. Otherwise there exists a sequence $(x_n)_{n\geq 1}$ in X with $\|x_n\| \leq 1$ for every n and such that

$$\|\Lambda^{-1}x_n\| \to \infty \qquad \text{as } n \to \infty.$$

Consider the normalized vectors

$$\beta_n \doteq \frac{\Lambda^{-1}x_n}{\|\Lambda^{-1}x_n\|} \in \mathbb{K}^N.$$

Then $\|\beta_n\| = 1$ and $\Lambda\beta_n \to 0$ as $n \to \infty$. Since $(\beta_n)_{n\geq 1}$ is a bounded sequence in the Euclidean space $\mathbb{K}^N$, it admits a convergent subsequence, say $\beta_{n_k} \to \beta \in \mathbb{K}^N$. Clearly

$$\|\beta\| = \lim_{k\to\infty} \|\beta_{n_k}\| = 1, \qquad \Lambda\beta = \lim_{k\to\infty} \Lambda\beta_{n_k} = \lim_{k\to\infty} \frac{x_{k_n}}{\|\Lambda^{-1}x_{k_n}\|} = 0 ,$$

because Λ is continuous. This contradicts the fact that Λ is one-to-one. We thus conclude that Λ^{-1} is bounded, hence continuous. This proves (ii).

4. To prove that X with norm $\|\cdot\|$ is complete, let $(x_n)_{n\geq 1}$ be a Cauchy sequence in X. Then $\Lambda^{-1}x_n$ defines a Cauchy sequence in $\mathbb{K}^N$, which converges to some point $\beta \in \mathbb{K}^N$ because $\mathbb{K}^N$ is complete. Since Λ is continuous, the sequence x_n converges to $\Lambda\beta$. □

Corollary 2.21. *In a finite-dimensional space, all norms are equivalent.*

Proof. Let $\|\cdot\|_\diamondsuit$ and $\|\cdot\|_\spadesuit$ be any two norms on the finite-dimensional vector space X. Let $\mathcal{B} = \{u_1, \ldots, u_N\}$ be a basis of X and define the linear map $\Lambda : \mathbb{K}^N \mapsto X$ as in (2.17). By the previous theorem, both Λ and Λ^{-1} are bounded linear operators (in each of the two norms on X). Hence there exist constants C', C'' such that

$$\frac{1}{C'}\,\|\Lambda^{-1}x\| \;\le\; \|x\|_\diamondsuit \;\le\; C'\,\|\Lambda^{-1}x\|\,,$$

$$\frac{1}{C''}\,\|\Lambda^{-1}x\| \;\le\; \|x\|_\spadesuit \;\le\; C''\,\|\Lambda^{-1}x\|\,,$$

for all $x \in X$. This implies (2.16). □

The classical theorem of Bolzano and Weierstrass states that every bounded sequence in $\mathbb{R}^N$ admits a convergent subsequence. The next theorem shows that this compactness property is true for all finite-dimensional normed spaces, and fails for all infinite-dimensional ones.

Theorem 2.22 (Locally compact normed spaces are finite-dimensional). *Let X be a normed space. The following are equivalent:*

(i) *X is finite-dimensional.*

(ii) *The closed unit ball $B_1 \doteq \overline{B(0,1)} = \{x \in X\,;\ \|x\| \le 1\}$ is compact.*

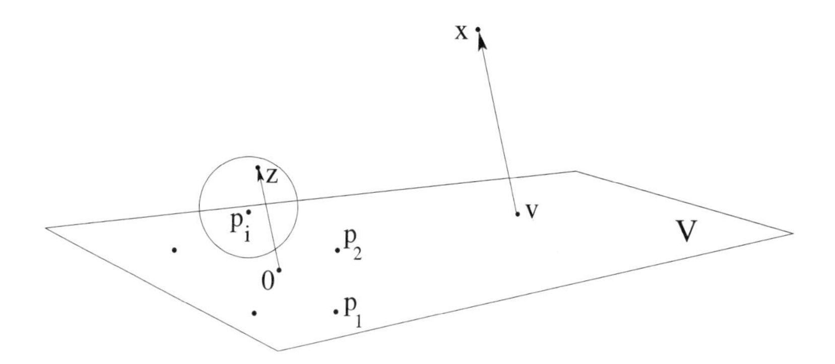

Figure 2.3.1. The construction used to prove Theorem 2.22.

Proof. (i)$\Longrightarrow$(ii). Let X have dimension N. Then by Theorem 2.20 there exists a linear homeomorphism $\Lambda : \mathbb{K}^N \mapsto X$ with bounded inverse. Since the unit ball $B_1 \subset X$ is closed and bounded, the same is true of $K \doteq \Lambda^{-1}(B_1) \subset \mathbb{K}^N$. By the Bolzano-Weierstrass theorem, the closed bounded set K is compact. Being the continuous image of a compact set, $B_1 = \Lambda(K)$ is compact as well.

(ii)$\Longrightarrow$(i). Assume that the closed unit ball $B_1 \subset X$ is compact. Then it is pre-compact and can be covered with finitely many open balls $B(p_i, 1/2)$ centered at points p_i, $i = 1, \ldots, n$, and with radius $1/2$. Consider the finite-dimensional subspace $V = \text{span}\{p_1, \ldots, p_n\}$ (see Figure 2.3.1). Observe that V is closed in X, because by Theorem 2.20 every finite-dimensional normed space is complete.

We claim that $V = X$, hence X itself is finite-dimensional. If not, we could find a point $x \in X \setminus V$. Let $\rho \doteq d(x, V) \doteq \inf_{y \in V} \|y - x\|$. Notice that $\rho > 0$ because V is closed. Hence there exists a point $v \in V$ such that

$$\rho \leq \|x - v\| \leq \frac{3}{2}\rho. \tag{2.18}$$

Consider the unit vector

$$z \doteq \frac{x - v}{\|x - v\|} \in B_1.$$

By construction, there exists a point $p_i \in B_1$ such that $\|z - p_i\| < \frac{1}{2}$. We thus have

$$x = v + \|x - v\|\, z = v + \|x - v\| p_i + \|x - v\|(z - p_i).$$

Since $v + \|x - v\| p_i \in V$, we must have

$$\|x - v\|\, \|z - p_i\| \geq d(x, V) \geq \rho.$$

Hence $\|x - v\| \geq 2\rho$, in contradiction to (2.18). This shows that $X = V$, completing the proof. □

2.4. Seminorms and Fréchet spaces

To motivate the introduction of *seminorms* we observe that, for certain spaces of functions, there is no natural way to define a norm.

Example 2.23. Consider the space $X \doteq \mathcal{C}(]0,1[)$ of all continuous (possibly unbounded) functions on the open interval $]0,1[$. Then, by setting

$$p(f) \doteq \sup_{0<x<1} |f(x)|,$$

we do not obtain a norm on the entire space X. Indeed, the right-hand side takes the value $+\infty$ whenever f is unbounded.

On the other hand, given a closed subinterval $[a,b] \subset\,]0,1[$, the "semi-norm"

$$p^{a,b}(f) \doteq \max_{x \in [a,b]} |f(x)| \tag{2.19}$$

is always a well-defined real number. Notice that the functional in (2.19) does not satisfy all requirements of a norm, because there exist functions

$f \in \mathcal{C}(]0,1[)$ that vanish on $[a,b]$ but are not identically zero. In this case, $p^{a,b}(f) = 0$ but $f \neq 0$.

Now let X be a vector space on the field $\mathbb{K}$. A real-valued map $x \mapsto p(x)$ is called a **seminorm** on X if it satisfies the following properties:

(SN1) For every $x \in X$ one has $p(x) \geq 0$.

(SN2) For every $x \in X$ and $\lambda \in \mathbb{K}$ one has $p(\lambda x) = |\lambda|\, p(x)$.

(SN3) For every $x, y \in X$, $p(x+y) \leq p(x) + p(y)$.

Notice that (SN2)–(SN3) are exactly the same as in the definition of a norm. The only difference is that in (SN1) we do not require the strict positivity condition. In other words, we allow for $p(x) = 0$ even if $x \neq 0$.

If $p(\cdot)$ is a seminorm, by setting $d(x,y) \doteq p(x-y)$ we are not guaranteed to obtain a distance on the space X. Indeed, we may have $d(x,y) = p(x-y) = 0$ even if $x \neq y$. There are, however, interesting cases where a distance can be obtained by using not just one but countably many seminorms.

We say that a sequence $(p_k)_{k\geq 1}$ of seminorms on X is **separating** if, for every $x \in X$ with $x \neq 0$, there exists at least one index k such that $p_k(x) > 0$.

Lemma 2.24 (Distance generated by seminorms). *Let $(p_k)_{k\geq 1}$ be a separating sequence of seminorms on the vector space X. Then*

$$d(x,y) \doteq \sum_{k=1}^{\infty} 2^{-k} \frac{p_k(x-y)}{1+p_k(x-y)} \tag{2.20}$$

defines a distance on X.

Proof. The identities

$$d(x,x) = 0, \qquad d(x,y) = d(y,x)$$

are an immediate consequence of (SN2). The assumption that the sequence (p_k) is separating guarantees that $d(x,y) > 0$ as soon as $x \neq y$.

To prove the triangle inequality, we observe that, if $a, b, c \geq 0$ and $c \leq a+b$, then

$$\frac{c}{1+c} \leq \frac{a+b}{1+a+b} \leq \frac{a}{1+a} + \frac{b}{1+b}$$

because the function $s \mapsto \frac{s}{1+s}$ is increasing and concave down. By (SN3) we can use the above inequalities with $a = p_k(x-y)$, $b = p_k(y-z)$, $c = p_k(x-z)$

and obtain

$$\begin{aligned} d(x,z) &= \sum_{k=1}^{\infty} 2^{-k} \frac{p_k(x-z)}{1+p_k(x-z)} \\ &\leq \sum_{k=1}^{\infty} 2^{-k} \left(\frac{p_k(x-y)}{1+p_k(x-y)} + \frac{p_k(y-z)}{1+p_k(y-z)} \right) \\ &= d(x,y) + d(y,z) . \end{aligned}$$

This proves that $d(\cdot,\cdot)$ is indeed a distance. □

From the definition, it is clear that the distance (2.20) is invariant under translations, namely

$$d(x,y) \;=\; d(x+z,\, y+z) \qquad \text{for all } x,y,z \in X .$$

If the vector space X with the distance (2.20) is a *complete* metric space, then we say that X is a **Fréchet space**.

Example 2.25. Let $\Omega \subseteq \mathbb{R}^n$ be any open set, with boundary $\partial\Omega$. Then the space $\mathcal{C}(\Omega)$ of all (possibly unbounded) continuous functions $f : \Omega \mapsto \mathbb{R}$ does not have a natural norm. However, it can be given the structure of a Fréchet space as follows. For every $k \geq 1$, consider the compact subset (Figure 2.4.1)

$$A_k \doteq \Big\{ x \in \Omega;\ |x| \leq k\,,\ d(x,\partial\Omega) \leq k^{-1} \Big\}. \tag{2.21}$$

Define the seminorms

$$p_k(f) \doteq \max_{x \in A_k} |f(x)| \tag{2.22}$$

and let $d(\cdot,\cdot)$ be the corresponding distance as in (2.20).

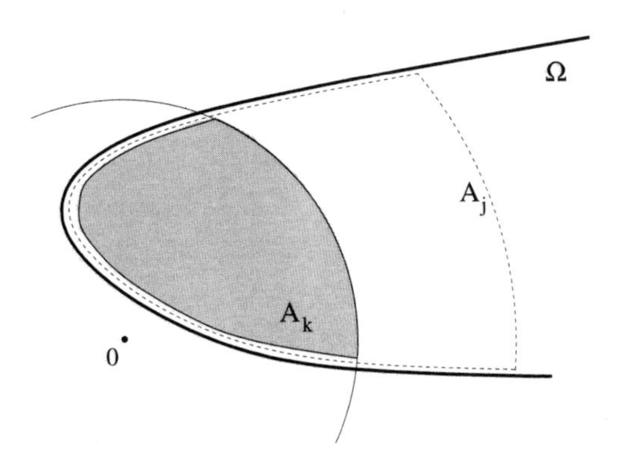

Figure 2.4.1. The compact subsets $A_k \subset \Omega$.

We show that $\mathcal{C}(\Omega)$ with the above distance is a complete metric space, hence a Fréchet space. Let $(f_j)_{j\geq 1}$ be a Cauchy sequence. For every k this implies

$$\limsup_{i,j\to\infty} p_k(f_i - f_j) = \limsup_{i,j\to\infty} \sup_{x\in A_k} |f_i(x) - f_j(x)| = 0. \tag{2.23}$$

Since every point $x \in \Omega$ is contained in one of the sets A_k, by (2.23) the sequence $f_j(x)$ is Cauchy and hence converges to some limit $f(x)$. More generally, every compact subset $K \subset \Omega$ is contained in one of the sets A_k. Again by (2.23), the convergence $f_j \to f$ must be uniform on compact subsets of Ω, hence the limit f is continuous.

It now remains to prove that $d(f_j, f) \to 0$ as $j \to \infty$. For any fixed $m \geq 1$, using the fact that $f_j \to f$ uniformly on the compact set A_m, we find

$$\begin{aligned}
&\limsup_{j\to\infty} d(f_j, f)\\
&\leq \limsup_{j\to\infty} \sum_{k=1}^{m} 2^{-k} \frac{p_k(f_j - f)}{1 + p_k(f_j - f)} + \limsup_{j\to\infty} \sum_{k=m+1}^{\infty} 2^{-k} \frac{p_k(f_j - f)}{1 + p_k(f_j - f)}\\
&\leq 0 + 2^{-m}.
\end{aligned}$$

Since m is arbitrary, the convergence is proved.

Example 2.26 (The spaces $\mathbf{L}^p_{\mathrm{loc}}$). As in the previous example, let $\Omega \subseteq \mathbb{R}^n$ be an open set. We say that an open set Ω' is **compactly contained in** Ω, and write $\Omega' \subset\subset \Omega$, if the closure $\overline{\Omega}'$ is a compact subset of Ω.

For $p \in [1, \infty[$ we define $\mathbf{L}^p_{\mathrm{loc}}(\Omega)$ to be the space of all measurable functions f such that $\int_{\Omega'} |f|^p\, dx < \infty$ for every open set Ω' compactly contained in Ω. This space is not endowed with a natural norm. However, for each $k \geq 1$ we can consider the seminorm

$$p_k(f) \doteq \left(\int_{A_k} |f|^p\, dx\right)^{1/p} = \|f\|_{\mathbf{L}^p(A_k)}, \tag{2.24}$$

where A_k is the set in (2.21). The corresponding distance (2.20) renders $\mathbf{L}^p_{\mathrm{loc}}(\Omega)$ a Fréchet space.

2.5. Extension theorems

Let X be a vector space over the field $\mathbb{K}$. A linear map $F : X \mapsto \mathbb{K}$ is called a **linear functional** on X. The eventual goal of this section is to prove the existence of a large number of continuous linear functionals. In this direction, we first prove an extension theorem: given a linear functional

$f : V \mapsto \mathbb{K}$ defined on a subspace $V \subset X$, one can extend it to a functional $F : X \mapsto \mathbb{K}$ defined on the entire space, preserving some additional properties.

In the following, we consider a vector space X over the real numbers and let $p : X \mapsto \mathbb{R}$ be a function such that

$$p(x+y) \leq p(x)+p(y), \quad p(tx) = t\,p(x) \quad \text{for all } x,y \in X,\ t \geq 0. \tag{2.25}$$

Remark 2.27. If X is a normed space, the function $p(x) = \kappa\,\|x\|$ satisfies the above properties for any $\kappa > 0$. More generally, any seminorm also satisfies (2.25).

Observe that (2.25) implies that the function p is convex. Indeed

$$p(\theta x + (1-\theta)y) \leq p(\theta x) + p((1-\theta)y) = \theta p(x) + (1-\theta)p(y)$$

for all $x, y \in X$, $\theta \in [0,1]$. However, compared with a seminorm, here we also allow for $p(x) < 0$. Moreover, we do not require p to be symmetric with respect to the origin. In other words, one may well have $p(x) \neq p(-x)$.

Example 2.28. Let X be a normed space and let $\Omega \subset X$ be a bounded, open, convex set containing the origin. Then the functional

$$p(x) \doteq \inf \{\lambda \geq 0;\ x \in \lambda\Omega\} \tag{2.26}$$

satisfies the assumptions (2.25).

Theorem 2.29 (Hahn-Banach extension theorem). *Let X be a vector space over the reals and let $p : X \mapsto \mathbb{R}$ be a map with the properties (2.25). Consider a subspace $V \subseteq X$ and let $f : V \mapsto \mathbb{R}$ be a linear functional such that*

$$f(x) \leq p(x) \qquad \textit{for all } x \in V. \tag{2.27}$$

Then there exists a linear functional $F : X \mapsto \mathbb{R}$ such that $F(x) = f(x)$ for all $x \in V$ and

$$-p(-x) \leq F(x) \leq p(x) \qquad \textit{for all } x \in X. \tag{2.28}$$

Proof. 1. If $V = X$, observing that $f(x) = -f(-x) \geq -p(-x)$, the conclusion is clear. If $V \neq X$, choose any vector $x_0 \notin V$ and consider the strictly larger subspace

$$V_0 \doteq \{x + tx_0;\ x \in V,\ t \in \mathbb{R}\}.$$

For every $x, y \in V$, the bound on f yields

$$f(x) + f(y) = f(x+y) \leq p(x+y) \leq p(x - x_0) + p(x_0 + y).$$

Therefore

$$f(x) - p(x - x_0) \leq p(y + x_0) - f(y) \qquad \text{for all } x, y \in V.$$

Choosing $\beta \doteq \sup_{x \in V} \Big\{ f(x) - p(x - x_0) \Big\}$, we have

$$(2.29) \quad f(x) - p(x - x_0) \leq \beta \leq p(y + x_0) - f(y) \qquad \text{for all } x, y \in V.$$

2. We now extend f to a linear functional defined on the larger space V_0, by setting

$$f(x + tx_0) \doteq f(x) + \beta t, \qquad x \in V, \ t \in \mathbb{R}.$$

We claim that this extension still satisfies

$$(2.30) \qquad f(x + tx_0) \leq p(x + tx_0) \qquad \text{for all } x \in V, \ t \in \mathbb{R}.$$

Indeed, if $t = 0$, the above inequality follows from our initial assumptions. If $t > 0$, replacing both x and y by x/t in (2.29) we obtain

$$t \left[f\left(\frac{x}{t}\right) - p\left(\frac{x}{t} - x_0\right) \right] \leq t\beta \leq t \left[p\left(\frac{x}{t} + x_0\right) - f\left(\frac{x}{t}\right) \right],$$

$$f(x) - p(x - tx_0) \leq t\beta \leq p(x + tx_0) - f(x).$$

Therefore, for $x \in V$ and $t \geq 0$ we have

$$\begin{aligned} f(x - tx_0) &= f(x) - \beta t \leq p(x - tx_0), \\ f(x + tx_0) &= f(x) + \beta t \leq p(x + tx_0), \end{aligned}$$

proving (2.30).

3. The previous two steps show that every bounded linear functional f defined on a proper subspace $V \subset X$ can be extended to a strictly larger subspace, still satisfying the inequality (2.27). To complete the proof, we shall use a maximality principle.

Let $\mathcal{F}$ be the family of all couples (V, ϕ), where V is a subspace of X and $\phi : V \mapsto \mathbb{R}$ is a linear functional such that

$$\phi(x) \leq p(x) \qquad \text{for all } x \in V.$$

This family can be partially ordered by setting

$(V, \phi) \preceq (\widetilde{V}, \widetilde{\phi})$ if and only if $V \subseteq \widetilde{V}$ and ϕ coincides with the restriction of $\widetilde{\phi}$ to V.

By the Hausdorff Maximality Principle (see Theorem A.1 in the Appendix), the partially ordered family $\mathcal{F}$ contains a maximal element, say $(V^{\max}, F)$. If $V^{\max} \neq X$, then by the previous step the linear functional F

could be extended to a strictly larger subspace, in contradiction to the maximality assumption. Hence $V^{\max} = X$ and $F : X \mapsto \mathbb{R}$ is a linear functional such that

$$F(x) \leq p(x) \qquad \text{for all } x \in X.$$

By linearity, this implies that

$$-p(-x) \leq -F(-x) = F(x),$$

completing the proof. □

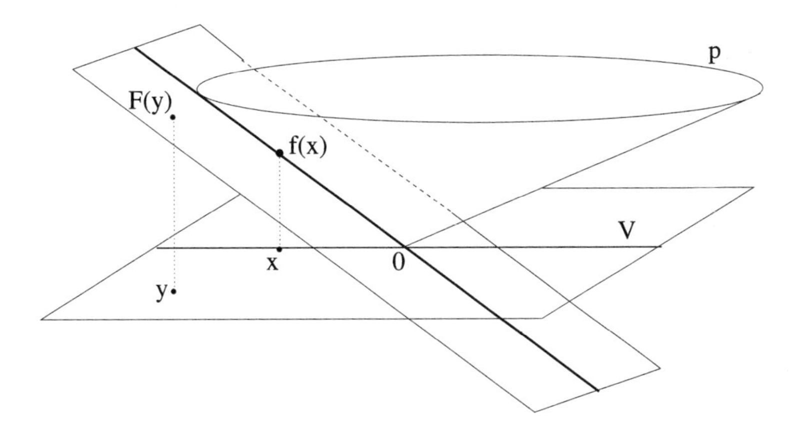

Figure 2.5.1. By the Hahn-Banach theorem, a linear functional $f : V \mapsto \mathbb{R}$ defined on a subspace $V \subset X$ and such that $f(x) \leq p(x)$ for all $x \in V$ can be extended to a linear functional $F : X \mapsto \mathbb{R}$ satisfying $F(y) \leq p(y)$ for all $y \in X$.

The previous theorem has a natural application to the case where $p(x) = \|x\|$ is a norm.

Theorem 2.30 (Extension theorem for bounded linear functionals). *Let X be a normed space over the field $\mathbb{K}$ of real or complex numbers. Let $f : V \mapsto \mathbb{K}$ be a bounded linear functional defined on a subspace $V \subseteq X$. Then f can be extended to a bounded linear functional $F : X \mapsto \mathbb{K}$ having the same norm:*

$$\|F\| \doteq \sup_{x \in X, \|x\| \leq 1} |F(x)| = \sup_{x \in V, \|x\| \leq 1} |f(x)| \doteq \|f\|.$$

Proof. 1. First assume that $\mathbb{K} = \mathbb{R}$, so f is real-valued. Set $\kappa \doteq \|f\|$, and define $p(x) \doteq \kappa\|x\|$. Then the result follows immediately from the previous theorem.

2. Next, consider the case where $\mathbb{K} = \mathbb{C}$ is the field of complex numbers. Notice that in this case, V and X can also be regarded as vector spaces over the real numbers. The functional $F : X \mapsto \mathbb{C}$ will be obtained by constructing separately its real and imaginary parts.

For $x \in V$, define $u(x) \doteq \mathrm{Re}\, f(x)$. This is a real-valued linear functional on V with norm $\|u\| \leq \kappa \doteq \|f\|$. Hence by the previous steps it admits an extension $U : X \mapsto \mathbb{R}$ with norm $\|U\| \leq \kappa$. We claim that the map

$$F(x) \doteq U(x) - iU(ix)$$

satisfies all requirements. Indeed, for $x \in V$,

$$F(x) = \mathrm{Re}\, f(x) - i\,\mathrm{Re}\, f(ix) = \mathrm{Re}\, f(x) + i\,\mathrm{Im}\, f(x) = f(x).$$

Moreover, let $\alpha \in \mathbb{C}$ be such that $|\alpha| = 1$, $\alpha F(x) = |F(x)|$. Then

$$|F(x)| = \alpha F(x) = U(\alpha x) \leq \kappa\|\alpha x\| = \kappa\|x\|.$$

Hence $\|F\| \leq \kappa \doteq \|f\|$. □

Corollary 2.31. *Let X be a Banach space. For any two vectors $x, y \in X$ with $x \neq y$, there exists a continuous linear functional $\phi : X \mapsto \mathbb{K}$ such that $\phi(x) \neq \phi(y)$.*

Corollary 2.32. *Let X be a Banach space. For every vector $x \in X$, there exists a continuous linear functional $\phi : X \mapsto \mathbb{K}$ such that $\phi(x) = \|x\|$ and $\|\phi\| = 1$.*

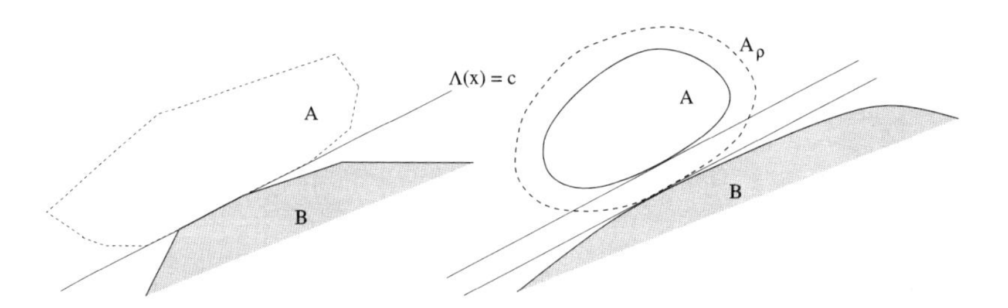

Figure 2.5.2. Left: if A is open, the disjoint convex sets A, B can be separated by a hyperplane. Right: If A is compact and B is closed, the disjoint convex sets A, B can be strictly separated.

2.6. Separation of convex sets

Consider the following problem. Given two disjoint convex sets A, B in a normed space X, can one find a bounded linear functional $\phi : X \mapsto \mathbb{R}$ such that the images $\phi(A)$ and $\phi(B)$ are disjoint? The following theorem provides a positive answer, relying on the Hahn-Banach extension theorem.

Theorem 2.33 (Separation of convex sets). *Let X be a normed space over the reals, and let A, B be nonempty, disjoint convex subsets of X.*

(i) *If A is open, then there exists a bounded linear functional $\phi : X \mapsto \mathbb{R}$ and a number $c \in \mathbb{R}$ such that*

$$\phi(a) \ < \ c \ \leq \ \phi(b) \qquad \text{for all } \ a \in A\,, \ b \in B\,. \tag{2.31}$$

(ii) *If A is compact and B is closed, then there exists a bounded linear functional $\phi : X \mapsto \mathbb{R}$ and numbers $c_1, c_2 \in \mathbb{R}$ such that*

$$\phi(a) \ \leq \ c_1 \ < \ c_2 \ \leq \ \phi(b) \qquad \text{for all } \ a \in A\,, \ b \in B\,. \tag{2.32}$$

Proof. 1. Choose points $a_0 \in A$ and $b_0 \in B$ and set $x_0 \doteq b_0 - a_0$. Consider the open set

$$\Omega \ \doteq \ A - B + x_0 \ \doteq \ \Big\{(a - a_0) + (b_0 - b)\,; \ a \in A, \ b \in B\Big\}.$$

Since A, B are convex and A is open, it is clear that Ω is an open, convex neighborhood of the origin. Moreover, $x_0 \notin \Omega$, because otherwise

$$x_0 = a - b + x_0\,, \qquad a - b = 0, \qquad \text{for some } a \in A, \ b \in B.$$

This is a contradiction because $A \cap B = \emptyset$.

2. Consider the functional

$$p(x) \ \doteq \ \inf\,\{\lambda \geq 0\,; \ x \in \lambda\Omega\}. \tag{2.33}$$

Since Ω is a neighborhood of the origin, we have $B(0, \rho) \subseteq \Omega$ for some $\rho > 0$. Hence

$$p(x) \ \leq \ \frac{\|x\|}{\rho} \qquad \text{for all } \ x \in X. \tag{2.34}$$

Moreover, the convexity of Ω implies that

$$p(x + y) \ \leq \ p(x) + p(y)\,, \qquad p(tx) \ = \ t\,p(x) \qquad \text{for all } \ x, y \in X\,, \ t \geq 0. \tag{2.35}$$

Notice that $p(x_0) \geq 1$ because $x_0 \notin \Omega$.

3. On the one-dimensional subspace $V \doteq \{tx_0\,; \ t \in \mathbb{R}\}$, define the linear functional f by setting $f(tx_0) \doteq t$. Observe that

$$f(x_0) \ = \ 1\,, \qquad f(tx_0) \ = \ t \ \leq \ tp(x_0) \ \leq \ p(tx_0)\,.$$

By the Hahn-Banach extension theorem, there exists a linear functional $\phi : X \mapsto \mathbb{R}$ such that

$$-p(-x) \ \leq \ \phi(x) \ \leq \ p(x) \qquad \text{for all } x \in X\,.$$

By (2.34) it is clear that ϕ is bounded. Indeed, $\|\phi\| \leq \rho^{-1}$.

4. If now $a \in A$ and $b \in B$, we have

$$\phi(a) - \phi(b) + 1 \;=\; \phi(a - b + x_0) \;\leq\; p(a - b + x_0) \;<\; 1$$

because $\phi(x_0) = f(x_0) = 1$ while $a - b + x_0 \in \Omega$ and Ω is open. Therefore

$$\phi(a) \;<\; \phi(b) \qquad \text{for all } \; a \in A\,, \; b \in B\,.$$

The sets $\phi(A)$ and $\phi(B)$ are nonempty, disjoint convex sets of $\mathbb{R}$, with $\phi(A)$ open. Taking $c \doteq \sup_{a\in A} \phi(a)$, the conclusion (2.31) is satisfied. This proves (i).

5. To prove (ii), observe that the assumptions on A, B imply

$$d(A,B) \;\doteq\; \inf \Big\{ \|a-b\|\,;\; a \in A,\; b \in B \Big\} \;>\; 0\,.$$

If we choose $\rho \doteq d(A,B)$, then the open neighborhood $A_\rho \;\doteq\; \{x \in X;\; d(x,A) < \rho\}$ of radius ρ around A does not intersect B. We can thus apply part (i) to the disjoint convex sets A_ρ and B. This yields a continuous linear functional $\phi : X \mapsto \mathbb{R}$ and a constant c_2 such that

$$\phi(x) \;<\; c_2 \;\leq\; \phi(y) \qquad \text{for all } \; x \in A_\rho\,, \; y \in B\,.$$

We now observe that the set $\phi(A)$ is compact, being the image of a compact set by a continuous map. Hence $c_1 \doteq \sup_{x\in A} \phi(x) < c_2$. This achieves the proof of (ii). □

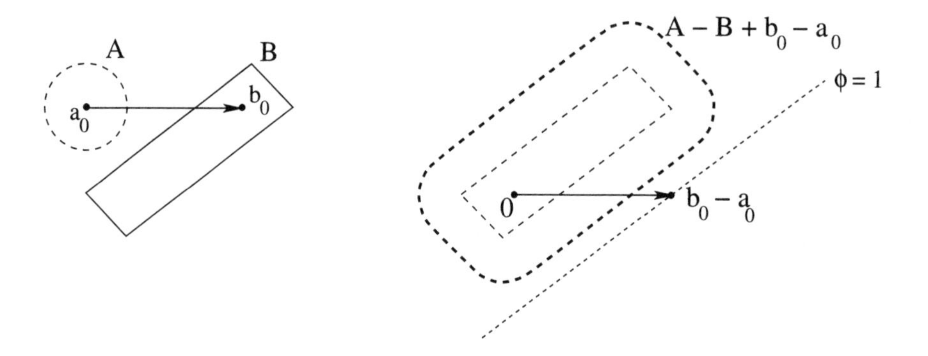

Figure 2.6.1. The construction used in the proof of the separation theorem.

2.7. Dual spaces and weak convergence

Let X be a Banach space over the field $\mathbb{K}$ of real or complex numbers. The set of all continuous linear functionals $\varphi : X \mapsto \mathbb{K}$ is called the **dual space** of X and denoted by X^*.

Observe that a linear functional $\varphi : X \mapsto \mathbb{K}$ can be regarded as a linear operator $\varphi : X \mapsto Y$, in the special case where $Y = \mathbb{K}$. We can thus use

Theorem 2.12 and conclude that $X^* = \mathcal{B}(X, \mathbb{K})$ is a Banach space, with norm

$$\|\varphi\|_* \doteq \sup_{\|x\|\leq 1} |\varphi(x)| . \tag{2.36}$$

2.7.1. Weak convergence. A sequence $x_1, x_2, \ldots$ in a Banach space X is **weakly convergent** if there exists $x \in X$ such that

$$\lim_{n\to\infty} \varphi(x_n) = \varphi(x) \qquad \text{for all } \varphi \in X^*.$$

In this case, we say that x is the **weak limit** of the sequence x_n and write $x_n \rightharpoonup x$. We recall that the sequence x_n converges strongly to x if $\|x_n - x\| \to 0$. Since by definition every $\varphi \in X^*$ is continuous, the strong convergence $x_n \to x$ clearly implies the weak convergence $x_n \rightharpoonup x$.

If a weak limit exists, then it is necessarily unique: $x_n \rightharpoonup x$ and $x_n \rightharpoonup y$ imply $x = y$. Indeed, assume $y \neq x$. Then by Corollary 2.31 there exists a continuous linear functional $\phi \in X^*$ such that $\phi(x) \neq \phi(y)$. This leads to a contradiction, because

$$\phi(x) = \lim_{n\to\infty} \phi(x_n) = \phi(y) .$$

2.7.2. Weak-star convergence. One can take a different perspective, and observe that each vector $x \in X$ determines a linear functional on X^*, namely

$$\varphi \mapsto \varphi(x), \qquad \varphi \in X^*. \tag{2.37}$$

By Corollary 2.32, the norm of this functional is

$$\|x\|_{**} \doteq \sup_{\|\varphi\|_*\leq 1} |\varphi(x)| = \|x\| . \tag{2.38}$$

We thus have a canonical embedding $\iota : X \mapsto (X^*)^*$. To each element $x \in X$ there corresponds a bounded linear functional $\iota(x)$ on the dual space X^*, namely, the map $\varphi \mapsto \varphi(x)$. By (2.38), this embedding is isometric, i.e., it preserves the norm.

If *every* bounded linear functional on X^* is of the form (2.37) for some $x \in X$, then $\iota(X) = (X^*)^*$ and we say that the space X is **reflexive**. Examples of reflexive spaces include all finite-dimensional spaces, and the spaces $\mathbf{L}^p(\Omega)$, ℓ^p in Examples 2.5 and 2.6, with $1 < p < \infty$.

One should be aware that, in general, the space $(X^*)^*$ can be strictly larger than $\iota(X)$. Examples of spaces which are not reflexive include the spaces $\mathbf{L}^1(\Omega)$, $\mathbf{L}^\infty(\Omega)$, ℓ^1, and ℓ^∞.

The embedding $\iota : X \mapsto (X^*)^*$ can be used to introduce a weak topology on the dual space X^*. We say that a sequence of bounded linear functionals $\varphi_n \in X^*$ **weak-star converges** to $\varphi \in X^*$, and write $\varphi_n \overset{*}{\rightharpoonup} \varphi$, if

$$\lim_{n\to\infty} \varphi_n(x) = \varphi(x) \qquad \text{for every } x \in X . \tag{2.39}$$

This is a much weaker property than the convergence $\varphi_n \to \varphi$ in norm. Indeed

$$\varphi_n \overset{*}{\rightharpoonup} \varphi \qquad \text{means that} \qquad |\varphi_n(x) - \varphi(x)| \to 0 \quad \text{for each } x \in X ,$$

$$\varphi_n \to \varphi \qquad \text{means that} \qquad \sup_{x\in X,\, \|x\|\le 1} |\varphi_n(x) - \varphi(x)| \to 0 .$$

By Theorem 2.22, if the space X^* is infinite-dimensional, then the closed unit ball $B_1 \subset X^*$ is not compact: one can find a sequence of linear functionals $\varphi_n \in X^*$, with $\|\varphi_n\|_* \le 1$ for every $n \ge 1$ but without any convergent subsequence (with respect to the norm topology of X^*). On the other hand, if instead of strong convergence we only ask for weak-star convergence, a positive result can be achieved.

Theorem 2.34 (Banach-Alaoglu). *Let X be a separable Banach space. Then every bounded sequence of linear functionals $\varphi_n \in X^*$ admits a weak-star convergent subsequence.*[2]

Proof. 1. Consider a bounded sequence of linear functionals $\varphi_n \in X^*$, say, with $\|\varphi_n\|_* \le C$ for all $n \ge 1$. Since X is separable, there exists a dense countable set $S = \{x_1, x_2, \ldots\} \subset X$.

2. We claim that there exists a subsequence $(\varphi_{n_j})_{j\ge1}$ which converges pointwise at each point x_k, so that

$$\lim_{j\to\infty} \varphi_{n_j}(x_k) = \varphi(x_k) \tag{2.40}$$

for some values $\varphi(x_k)$ and all $k \ge 1$. This subsequence will be constructed by a standard diagonalization procedure.

Since the sequence of numbers $(\varphi_n(x_1))_{n\ge1}$ is bounded, there exists an infinite set of indices $I_1 \subset \mathbb{N}$ such that the subsequence $(\varphi_n(x_1))_{n\in I_1}$ converges to some limit $\varphi(x_1)$.

Similarly, the sequence of numbers $(\varphi_n(x_2))_{n\ge1}$ is bounded. Hence there exists an infinite set of indices $I_2 \subset I_1$ such that the subsequence $(\varphi_n(x_2))_{n\in I_2}$ converges to some limit $\varphi(x_2)$.

[2] The Banach-Alaoglu theorem remains valid even without the assumption that X is separable. For a proof in this more general case we refer to [**C, R**].

By induction, for each k we can find an infinite set of indices $I_k \subset I_{k-1}$ such that the subsequence $(\varphi_n(x_k))_{n\in I_k}$ converges to some limit $\varphi(x_k)$.

We now choose a subsequence $n_1 < n_2 < n_3 < \cdots$, with $n_j \in I_j$ for every j. As $j \to \infty$, this yields the convergence $\varphi_{n_j}(x_k) \to \varphi(x_k)$ for every $k \geq 1$.

3. We claim that the limit function $\varphi : S \mapsto \mathbb{K}$ is Lipschitz continuous with constant $C = \sup_n \|\varphi_n\|_*$. Indeed, for every $x_h, x_k \in S$ we have

$$|\varphi(x_h) - \varphi(x_k)| = \lim_{j\to\infty} |\varphi_{n_j}(x_h) - \varphi_{n_j}(x_k)|$$

$$\leq \limsup_{j\to\infty} \|\varphi_{n_j}\|_* \|x_h - x_k\| \leq C \|x_h - x_k\| .$$

Therefore, the map φ can be uniquely extended by continuity to the closure of S, i.e., to the entire space X. This continuous extension will still be denoted by $\varphi : X \mapsto \mathbb{K}$.

4. It remains to show that the subsequence φ_{n_j} weak-star converges to φ. Let any $x \in X$ and $\varepsilon > 0$ be given. Since S is dense, we can choose a point $x_k \in S$ such that $\|x_k - x\| < \varepsilon$. Recalling that all functions φ_n and φ are Lipschitz continuous with constant C, we obtain

$$\limsup_{j\to\infty} |\varphi_{n_j}(x) - \varphi(x)|$$

$$\leq \limsup_{j\to\infty} |\varphi_{n_j}(x) - \varphi_{n_j}(x_k)|$$

$$+ \limsup_{j\to\infty} |\varphi_{n_j}(x_k) - \varphi(x_k)| + |\varphi(x_k) - \varphi(x)|$$

$$\leq C \|x - x_k\| + 0 + C \|x_k - x\| \leq 2C\varepsilon .$$

Since $\varepsilon > 0$ was arbitrary, this implies that $|\varphi_{n_j}(x) - \varphi(x)| \to 0$ as $j \to \infty$, proving the weak-star convergence $\varphi_{n_j} \overset{*}{\rightharpoonup} \varphi$.

Being the pointwise limit of a uniformly bounded sequence of linear functionals, it is clear that the map φ is bounded and linear as well. □

2.8. Problems

1. Check if the following are normed spaces. In the negative case, identify which of the properties (N1)–(N3) fails. In the positive case, decide if they are Banach spaces.

(i) Let $X = \mathbb{R}$, with

$$\|x\| = \begin{cases} x & \text{if } x \geq 0, \\ -2x & \text{if } x < 0. \end{cases}$$

(ii) Let X be the vector space whose elements are the sequences of real numbers $x = (x_1, x_2, x_3, \ldots)$ such that $x_k = 0$ for all except finitely many k. On X consider the norm (2.8).

(iii) Let X be the space of all polynomials (of any degree), with norm $\|p\| = \max_{x \in [0,1]} |p(x)|$.

(iv) Let X be the space of all polynomials of degree ≤ 2, with norm

$$\|p\| \doteq |p(0)| + |p'(0)| + |p''(0)|.$$

(v) Let X be the space of all continuous functions on the interval $[0,1]$, with

$$\|f\| \doteq \int_0^1 |f(x)|\,dx.$$

(vi) Fix $\kappa \in \mathbb{R}$ and let X be the space of all continuous functions $f : [0, \infty[\mapsto \mathbb{R}$ such that

$$\|f\| \doteq \sup_{t \geq 0} e^{\kappa t} |f(t)| < \infty.$$

(vii) Let $X = \mathbb{R}^2$. Given $x = (x_1, x_2)$, for a fixed $0 < p < 1$ define

$$\|x\| \doteq \Big(|x_1|^p + |x_2|^p\Big)^{1/p}.$$

Note: in connection with (vii), it is interesting to check whether

$$d(x,y) \doteq |x_1 - y_1|^{1/2} + |x_2 - y_2|^{1/2}$$

is a distance on $\mathbb{R}^2$. Are the open balls $B(x,r) = \{y\,;\ d(y,x) < r\}$ convex? Is $d(\cdot,\cdot)$ translation invariant?

2. Let X, Y be Banach spaces. Prove that the Cartesian product

$$X \times Y = \Big\{(x,y)\,;\ x \in X,\ y \in Y\Big\}$$

is also a Banach space, with norm

$$\|(x,y)\| \doteq \max\{\|x\|,\ \|y\|\}. \tag{2.41}$$

3. Let X be a Banach space over the field $\mathbb{K}$ of real or complex numbers. Notice that $X \times X$ and $\mathbb{K} \times X$ are Banach spaces, with product norms as in (2.41). Prove:

(i) The mapping $(x,y) \mapsto x + y$ from $X \times X$ into X is continuous.

(ii) The mapping $(\alpha, x) \mapsto \alpha x$ from $\mathbb{K} \times X$ into X is continuous.

4. Let X be a normed space with norm $\|\cdot\|$. Prove that every subspace $V \subset X$ is also a normed space, with the same norm. If X is a Banach space and V is closed, then V is also a Banach space.

5. Prove that a normed space X is complete if and only if every absolutely convergent series has a sum:

$$\sum_{k\geq 1} \|x_k\| < \infty \qquad \text{implies that} \qquad \sum_{k\geq 1} x_k \doteq \lim_{n\to\infty} \sum_{k=1}^{n} x_k \quad \text{exists.}$$

6. Let X be a vector space. The **convex hull** of a set $A \subset X$ is defined as the set of all convex combinations of elements of A, namely

$$\text{co}\, A \doteq \left\{ \sum_{k=1}^{N} \theta_k a_k \,;\; N \geq 1,\; a_k \in A\,,\; \theta_k \in [0,1],\; \sum_{k=1}^{N} \theta_k = 1 \right\}.$$

Prove that (i) $\text{co}\, A$ is convex, and (ii) $\text{co}\, A$ is the intersection of all convex sets that contain A.

7. Let X be a Banach space and let $A, B \subset X$. Prove the following statements.

(i) If A is open, then $\text{co}\, A$ is open as well.

(ii) If A is bounded, then $\text{co}\, A$ is bounded.

(iii) If A, B are bounded, then the set $A + B \doteq \{a + b;\; a \in A,\; b \in B\}$ is bounded as well.

(iv) If A is closed and B is compact, then $A + B$ is closed.

(v) The sum of two closed sets may not be closed.

(vi) If A is convex, then $A+A \doteq \{x+y;\; x \in A\,,\; y \in A\} = 2A \doteq \{2x\,;\; x \in A\}$.

(vii) If A is closed and $A + A = 2A$, then A must be convex.

8. Let X be a normed space. We say that a set $S \subseteq X$ is **symmetric** if $a \in S$ implies $-a \in S$. Prove that

(i) If S is convex, then its closure is convex as well.

(ii) If S is symmetric, then its closure is symmetric as well.

9. Let X, Y be Banach spaces over the real numbers, and let $\Lambda : X \mapsto Y$ be a bounded linear operator. Prove that

(i) If S is convex, then its image $\Lambda(S)$ is convex.

(ii) If S is symmetric, then its image $\Lambda(S)$ is symmetric.

10. Let $\Lambda : X \mapsto Y$ be a linear operator. Assume that, for every sequence $x_n \to 0$, the sequence $(\Lambda x_n)_{n\geq 1}$ is bounded. Prove that Λ is continuous.

11. Let X be an infinite-dimensional Banach space, and let S be a set of linearly independent vectors. By span(S) we denote the set of all (finite) linear combinations of elements of S. Prove that

(i) If $S = \{v_1, \ldots, v_N\}$ is a finite set, then span(S) is a closed subspace of X.

(ii) If $S = \{v_k\,;\ k \geq 1\}$ is an infinite sequence, then the vector space span(S) cannot be closed in X.

12 (Spaces over the real and over the complex numbers). Let X be a vector space over the complex numbers. Then X is also a vector space over the real numbers. If $\Phi : X \mapsto \mathbb{C}$ is a complex linear functional and $\phi(x) = \text{Re}\,\Phi(x)$ is its real part, prove that ϕ is a real linear functional, and $\Phi(x)$ can be reconstructed from ϕ as

$$\Phi(x) \;=\; \phi(x) - i\phi(ix).$$

13. Show that a normed space X is finite-dimensional if and only if its dual X^* is finite-dimensional.

14. Consider the spaces ℓ^p of sequences of real numbers, as in Examples 2.6 and 2.7. If $1 \leq p \leq q \leq \infty$, prove that $\ell^p \subseteq \ell^q$, with equality holding only if $p = q$. Moreover, prove that the identity operator $\Lambda : \ell^p \mapsto \ell^q$ defined as $\Lambda\mathbf{x} = \mathbf{x}$ is continuous, for every $p \leq q$.

15. For $1 \leq p < \infty$, prove that the subspace $V \doteq \text{span}\{\mathbf{e}_k\,;\ k \geq 1\}$ introduced in (2.10) is dense in the space ℓ^p. For $p = \infty$, show that the closure of V coincides with the subspace c_0 of all sequences that converge to zero.

16. (Properties of diagonal operators) For $1 \leq p \leq \infty$, consider the operator $\Lambda : \ell^p \mapsto \ell^p$ defined in (2.13). Prove the claims (i)–(ii) in Example 2.14.

17. Work out the details of Example 2.18. Namely, fix $1 \leq p \leq \infty$ and let $g : \Omega \mapsto \mathbb{R}$ be any measurable function.

(i) If $g \in \mathbf{L}^\infty(\Omega)$, prove that the multiplication operator $M_g : f \mapsto gf$ from $\mathbf{L}^p(\Omega)$ into itself is bounded and has norm given by (2.15).

(ii) If $g \notin \mathbf{L}^\infty(\Omega)$, prove that the linear operator $f \mapsto gf$ from $\mathbf{L}^p(\Omega)$ into itself is unbounded. Give a direct proof that $\text{Dom}(M_g) \neq \mathbf{L}^p(\Omega)$.

18. Fix $1 \leq p \leq \infty$. Let $(f_n)_{n\geq 1}$ be a sequence of functions in $\mathbf{L}^p(\mathbb{R})$, converging weakly to a function $f \in \mathbf{L}^p(\mathbb{R})$. Prove that

$$\lim_{n\to\infty} \int_a^b f_n(x)\,dx \;=\; \int_a^b f(x)\,dx \qquad \text{for all } a < b.$$

19. Consider the sequence of functions

$$f_n(x) = \begin{cases} 1/n & \text{if } x \in [0, n]\,, \\ 0 & \text{otherwise.} \end{cases}$$

(i) Prove that $f_n \to 0$ strongly (i.e., $\|f_n\| \to 0$) in every space $\mathbf{L}^p(\mathbb{R})$ with $1 < p \leq \infty$.

(ii) On the other hand, show that in the space $\mathbf{L}^1(\mathbb{R})$ this sequence is not strongly convergent. In fact, this sequence does not even admit any weakly convergent subsequence.

20. Let Y be a subspace of a Banach space X and let $\Lambda : Y \mapsto \mathbb{R}^n$ be a bounded linear operator from Y into the Euclidean space $\mathbb{R}^n$. Show that Λ can be extended to a linear operator $\widetilde{\Lambda} : X \mapsto \mathbb{R}^n$ with norm $\|\widetilde{\Lambda}\| \leq \sqrt{n}\,\|\Lambda\|$.

21. Let $\phi : \mathbb{R}^2 \mapsto \mathbb{R}$ be a linear functional, say $\phi(x_1, x_2) = ax_1 + bx_2$. Give a direct proof that

(i) If $\mathbb{R}^2$ is endowed with the norm $\|x\|_1 = |x_1| + |x_2|$, then the corresponding operator norm (2.36) is $\|\phi\|_\infty = \max\{|a|, |b|\}$.

(ii) If $\mathbb{R}^2$ is endowed with the norm $\|x\|_\infty = \max\{|x_1|, |x_2|\}$, then the corresponding norm (2.36) is $\|\phi\|_1 = |a| + |b|$.

(iii) If $\mathbb{R}^2$ is endowed with the norm $\|x\|_p = (|x_1|^p + |x_2|^p)^{1/p}$, with $1 < p < \infty$, then the corresponding norm (2.36) is $\|\phi\|_q = (|a|^q + |b|^q)^{1/q}$, with $\frac{1}{p} + \frac{1}{q} = 1$.

22. Let X be a vector space. Let $B \subset X$ be a convex subset such that, for every nonzero vector $x \in X$, there exists a positive number $\theta_x > 0$ such that

$$\alpha x \in B \quad \text{if and only if} \quad |\alpha| \leq \theta_x .$$

For $r \geq 0$, call $rB \doteq \{rb\,;\ b \in B\}$. Prove that

$$\|x\| \doteq \min\,\{r \geq 0\,;\ x \in rB\}$$

is a norm on X, and $B = \{x \in X\,;\ \|x\| \leq 1\}$ is the unit ball in this norm.

23. Let $\Omega \subseteq \mathbb{R}^n$ be an open set. On the space $\mathcal{C}(\Omega)$ of (possibly unbounded) continuous functions $f : \Omega \mapsto \mathbb{R}$, consider the seminorms $p_k(\cdot)$ and the distance $d(\cdot,\cdot)$ defined in (2.22), (2.20).

Next, consider any sequence of open sets A'_k compactly contained in Ω and such that $\bigcup_{k\geq 1} A'_k = \Omega$. Define the seminorms $p'_k(\cdot)$ and the distance $d'(\cdot,\cdot)$ as before, but replacing the sets A_k in (2.21) with the sets A'_k.

Prove that the two distances d, d' are equivalent. Namely, given any sequence of continuous functions $(f_j)_{j\geq 1}$, the following statements are equivalent:

(i) the sequence is Cauchy for the distance d,

(ii) the sequence is Cauchy for the distance d',

(iii) the functions f_j converge uniformly on each compact subset of Ω.

24. On the space $\mathcal{C}(\Omega)$, consider the seminorms $p_k(\cdot)$ defined in (2.22) and the distance $d(\cdot,\cdot)$ constructed in (2.20).

(i) Explain why $\|f\| \doteq \sum_{k=1}^\infty 2^{-k} p_k(f)$ does not yield a norm on $\mathcal{C}(\Omega)$.

(ii) Consider the open unit ball $B \doteq \{f \in \mathcal{C}(\Omega)\,;\ d(f,0) < 1\}$, where 0 stands for the identically zero function. Explain why

$$\|f\|_\diamond \ \doteq\ \inf\Big\{\lambda > 0\,;\ \lambda^{-1} f \in B\Big\}$$

does not yield a norm on $\mathcal{C}(\Omega)$.

25. Let X be a Banach space. Consider any set $S \subset X$ and assume $x \in \mathrm{span}(S)$. Prove that there exist points $x_j \in S$ and coefficients $c_j \in \mathbb{K}$, $j = 1, \ldots, N$, such that

$$x \ =\ \sum_{j=1}^N c_j x_j\,, \qquad \left\|\sum_{j=1}^k c_j x_j\right\| \ \leq\ 2\|x\| \qquad \text{for all } k = 1, \ldots, N\,. \tag{2.42}$$

26. Let S be a subset of a Banach space X. Prove that the following statements are equivalent:

(i) $x \in \overline{\mathrm{span}(S)}$.

(ii) $x = \sum_{j=1}^\infty c_j x_j \doteq \lim_{n\to\infty} \sum_{j=1}^n c_j x_j$, for some points $x_j \in S$ and numbers $c_j \in \mathbb{K}$.

27. Consider the Banach space ℓ^∞ consisting of all bounded sequences $x = (x_1, x_2, x_3, \ldots)$ of real numbers.

(i) Prove that there exists a bounded linear functional $F : \ell^\infty \mapsto \mathbb{R}$ such that

$$|F(x)| \ \leq\ \|x\|_{\ell^\infty} \ \doteq\ \sup_{n\geq 1} |x_n|\,, \tag{2.43}$$

$$F(x) \ =\ \lim_{n\to\infty} x_n \quad \text{if the limit exists.} \tag{2.44}$$

(ii) Show that, if $F : \ell^\infty \mapsto \mathbb{R}$ satisfies the above properties (2.43)–(2.44), then

$$\liminf_{n\to\infty} x_n \ \leq\ F(x) \ \leq\ \limsup_{n\to\infty} x_n \qquad \text{for all } x \in \ell^\infty\,.$$

(iii) Using (ii) prove that, if there exists an integer N such that $x_n \leq y_n$ for all $n \geq N$, then $F(x) \leq F(y)$.

(iv) Show that, for any sequence $a = (a_1, a_2, \ldots) \in \ell^1$, the continuous linear functional

$$F_a(x) \ \doteq\ \sum_{n\geq 1} a_n x_n$$

cannot satisfy (2.44). Hence the space of all continuous linear functionals on ℓ^∞ cannot be identified with ℓ^1.

(v) For every bounded sequence of real numbers $x = (x_1, x_2, \ldots)$, define

$$\widetilde{F}(x) \doteq \frac{1}{2}\left(\liminf_{n\to\infty} x_n\right) + \frac{1}{2}\left(\limsup_{n\to\infty} x_n\right).$$

Does $\widetilde{F}$ satisfy (2.43)–(2.44)? Is $\widetilde{F}$ a bounded linear functional on the Banach space ℓ^∞?

28. In the Banach space $X = \mathbf{L}^\infty(\mathbb{R})$, consider the subspace V consisting of all bounded continuous functions.

Prove that there exists a bounded linear functional $\Lambda : \mathbf{L}^\infty(\mathbb{R}) \mapsto \mathbb{R}$ with $\|\Lambda\| = 1$ such that $\Lambda f \doteq f(0)$ for every bounded continuous function f. However, show that there exists no function $g \in \mathbf{L}^1(\mathbb{R})$ such that $\Lambda f = \int fg\,dx$ for every $f \in \mathbf{L}^\infty(\mathbb{R})$.

Conclude that the dual space of $\mathbf{L}^\infty(\mathbb{R})$ cannot be identified with $\mathbf{L}^1(\mathbb{R})$.

29. Let X be a normed space and let $\Omega \subset X$ be an open, convex set containing the origin. Consider the functional

$$p(x) \doteq \inf\{\lambda \geq 0\,;\ x \in \lambda\Omega\}. \tag{2.45}$$

(i) Prove that $p(\cdot)$ satisfies the conditions

$$p(x+y) \leq p(x) + p(y)\,, \quad p(tx) = t\,p(x) \quad \text{for all } x, y \in X,\ t \geq 0\,.$$

(ii) Assuming that $B_r \doteq \{x \in X\,;\ \|x\| < r\} \subseteq \Omega$, prove that $p(x) \leq \|x\|/r$.

(iii) Assuming that $\Omega = \{x \in X\,;\ \|x\| < 1\}$ is the open unit ball, prove that $p(x) = \|x\|$.

30. Let $\mathcal{C}^0([0,1])$ be the Banach space of all real-valued continuous functions $f : [0,1] \mapsto \mathbb{R}$, with norm $\|f\| = \max_{x\in[0,1]}|f(x)|$.

(i) Show that $X = \{f \in \mathcal{C}^0([0,1])\,;\ f(0) = 0\}$ is a closed subspace of $\mathcal{C}^0$, hence a Banach space.

(ii) Prove that the map $f \mapsto \Lambda f \doteq \int_0^1 f(x)\,dx$ is a continuous linear functional on X. Compute its norm $\|\Lambda\| \doteq \sup_{\|f\|\leq 1} |\Lambda f|$. Is this supremum over the closed unit ball actually attained as a maximum?

31. Let X be a Banach space over the reals and let X^* be its dual. Let $\Omega \subset X$ be a convex set containing the origin. Define

$$\Omega^* \doteq \{\phi \in X^*\,;\ \phi(x) \leq 1 \text{ for all } x \in \Omega\}\,,$$

$$\Omega^{**} \doteq \{x \in X\,;\ \phi(x) \leq 1 \text{ for all } \phi \in \Omega^*\}\,.$$

Prove that Ω^{**} is the closure of Ω.

32. Let ξ be a nonzero vector in a Banach space X over the reals. Call $U = \text{span}\{\xi\} = \{t\xi\,;\ t \in \mathbb{R}\}$.

Prove that there exists a closed subspace $V \subset X$ such that $X = U \oplus V$. Namely, every element $x \in X$ can be written uniquely as a sum

$$x \;=\; u + v \quad \text{with} \quad u \in U\,,\ v \in V\,.$$

Moreover, the projections $x \mapsto u = \pi_U\, x$ and $x \mapsto v = \pi_V\, x$ are continuous linear operators.

33. On the Banach space $X = \mathbf{L}^1([-1,1])$, prove that there exists a continuous linear functional $\varphi : X \mapsto \mathbb{R}$ with norm $\|\varphi\| \;=\; 1$, having the following property:

If f is a polynomial of degree 1, then $\varphi(f) = f'(0)$.

34. Given an open set $\Omega \subset \mathbb{R}^n$, we denote by $\mathcal{C}^m(\Omega)$ the space of all continuous functions $f : \Omega \mapsto \mathbb{R}$ with continuous partial derivatives up to order m. Let A_k be the sets defined in (2.21). Show that the sequence of seminorms

$$p_k(f) \;\doteq\; \sum_{|\alpha|\leq m} \max_{x\in A_k} |D^\alpha f(x)| \tag{2.46}$$

makes $\mathcal{C}^m(\Omega)$ into a Fréchet space. Here $\alpha = (\alpha_1, \ldots, \alpha_n)$ is a multi-index of length $|\alpha| \doteq \alpha_1 + \cdots + \alpha_n$, and $D^\alpha f \doteq \left(\frac{\partial}{\partial x_1}\right)^{\alpha_1} \cdots \left(\frac{\partial}{\partial x_n}\right)^{\alpha_n} f$.

35. Let Y be a closed subspace of a Banach space X and assume $x_0 \in X \setminus Y$. Show that there exists a bounded linear functional $\varphi \in X^*$ such that $\|\varphi\| \leq 1$ and

$$\varphi(x_0) \;=\; d(x_0, Y) \;=\; \inf_{y\in Y} \|x_0 - y\|\,, \qquad \varphi(y) \;=\; 0 \quad \text{for all } y \in Y\,.$$

36. In the space ℓ^p of sequences of real numbers, consider the unit vectors $\mathbf{e}_k$ as in (2.9). Prove that

(i) If $1 < p < \infty$, the sequence $(\mathbf{e}_k)_{k\geq 1}$ converges weakly to zero in ℓ^p.

(ii) In the space ℓ^1, the sequence $(\mathbf{e}_k)_{k\geq 1}$ does not admit any weakly convergent subsequence.

37. Let $\phi : \mathbb{R} \mapsto \mathbb{R}$ be a smooth, increasing function, and consider the operator $(\Lambda f)(x) \;\doteq\; f(\phi(x))$. Derive a condition on ϕ which implies that Λ is a bounded linear operator from $\mathbf{L}^p(\mathbb{R})$ into itself. Consider the cases $1 \leq p < \infty$ and $p = \infty$ separately.

38. Show that the concept of Lebesgue measure cannot be extended to infinite-dimensional spaces. More precisely, let X be an infinite-dimensional Banach space.

Prove that there there cannot be a measure μ, defined on the sigma-algebra of Borel subsets of X, with the following properties:

(i) $\mu(\Omega) > 0$ for every nonempty open set $\Omega \subset X$;

(ii) μ is translation-invariant: $\mu(x + S) = \mu(S)$ for every $x \in X$ and $S \subset X$;

(iii) there exists a nonempty open set Ω_0 such that $\mu(\Omega_0) < \infty$.

39. Let X be a Banach space over the reals.

(i) Let $S \subset X$ be a closed convex set and assume $y \notin S$. Prove that there exists a bounded linear functional $\varphi \in X^*$ such that

$$\varphi(y) < \inf_{x \in S} \varphi(x).$$

(ii) Consider a weakly convergent sequence: $x_n \rightharpoonup y$. Let $S \doteq \overline{\text{co}\{x_n\,;\ n \geq 1\}}$ be the smallest closed convex set containing all points x_n. Prove that $y \in S$.

40. Given a function $f \in \mathbf{L}^\infty(\mathbb{R})$, we say that

$$\text{ess}\lim_{x \to 0} f(x) = \lambda$$

if there exists a function $\tilde{f}$ such that $\tilde{f}(x) = f(x)$ for a.e. $x \in \mathbb{R}$, and moreover $\lim_{x \to 0} \tilde{f}(x) = \lambda$.

(i) Prove that there exists a bounded linear functional $\Phi : \mathbf{L}^\infty(\mathbb{R}) \mapsto \mathbb{R}$ such that

$$\Phi(f) = \text{ess}\lim_{x \to 0} f(x)$$

whenever the limit exists.

(ii) Prove that the above conclusion fails if the space $\mathbf{L}^\infty(\mathbb{R})$ is replaced by $\mathbf{L}^1(\mathbb{R})$.

Chapter 3

Spaces of Continuous Functions

3.1. Bounded continuous functions

Let E be a metric space. By $\mathcal{C}(E)$ we denote the space of all continuous real-valued (possibly unbounded) functions $f : E \mapsto \mathbb{R}$. In general, this space does not have a natural norm. For this reason, we shall also consider the space $\mathcal{BC}(E)$ of all bounded continuous functions $f : E \mapsto \mathbb{R}$, with norm

$$\|f\| \doteq \sup_{x \in E} |f(x)| . \tag{3.1}$$

Most of this chapter will be concerned with the case where E is compact. In this case every continuous function $f : E \mapsto \mathbb{R}$ is necessarily bounded, hence $\mathcal{C}(E) = \mathcal{BC}(E)$.

Lemma 3.1. *$\mathcal{BC}(E)$ is a Banach space.*

Proof. 1. Let $(f_n)_{n \geq 1}$ be a Cauchy sequence in $\mathcal{BC}(E)$. Then for every fixed $x \in E$ the sequence of numbers $f_n(x)$ is Cauchy and hence converges to some limit, which we call $f(x)$.

2. By assumption, for every $\varepsilon > 0$ there exists N large enough so that

$$\sup_{x \in E} |f_n(x) - f_m(x)| \leq \varepsilon \qquad \text{for all } n, m \geq N .$$

Letting $m \to \infty$, since $f_m(x) \to f(x)$ we obtain

$$\sup_{n \geq N} \sup_{x \in E} |f_n(x) - f(x)| \leq \varepsilon .$$

In turn, this implies

$$\sup_{n\geq N} \|f_n - f\| \leq \varepsilon, \qquad \sup_{x\in E} |f(x)| \leq \varepsilon + \sup_{x\in E} |f_N(x)| < \infty .$$

Since $\varepsilon > 0$ was arbitrary, the first inequality shows the convergence $\|f_n - f\| \to 0$. The second inequality shows that f is bounded.

3. Finally, we prove that f is continuous. Let any $x \in E$ and $\varepsilon > 0$ be given. By uniform convergence, there exists an integer N such that $|f_N(x) - f(x)| < \varepsilon/3$ for every $x \in E$. Since f_N is continuous, there exists $\delta > 0$ such that $|f_N(y) - f_N(x)| < \varepsilon/3$ whenever $d(y,x) < \delta$. Putting together the above inequalities, when $d(y,x) < \delta$ we have

$$\begin{aligned}|f(y) - f(x)| &\leq |f(y) - f_N(y)| + |f_N(y) - f_N(x)| + |f_N(x) - f(x)| \\ &< \tfrac{\varepsilon}{3} + \tfrac{\varepsilon}{3} + \tfrac{\varepsilon}{3} = \varepsilon ,\end{aligned}$$

proving that f is continuous at the point x. □

Remark 3.2 (Pointwise vs. uniform convergence). The previous argument shows that, if a sequence of continuous functions f_n converges *uniformly* to a function f, then f is continuous as well. On the other hand, a sequence of continuous functions can converge *pointwise* to a discontinuous limit. For example, on the interval $E = [0,1]$, the sequence of functions $f_n(x) = x^n$ converges pointwise to the discontinuous function

$$f(x) = \begin{cases} 0 & \text{if } \ 0 \leq x < 1 , \\ 1 & \text{if } \ x = 1 . \end{cases}$$

Clearly, here the convergence is not uniform on the whole interval $[0,1]$.

The following theorem describes a case where pointwise convergence implies uniform convergence. We recall that a sequence of functions $f_n : E \mapsto \mathbb{R}$ is *increasing* if $m < n$ implies $f_m(x) \leq f_n(x)$ for all $x \in E$.

Theorem 3.3 (Dini). *Let E be a compact metric space. If $(f_n)_{n\geq 1}$ is an increasing sequence of functions in $\mathcal{C}(E)$, converging pointwise to a continuous limit function f, then $f_n \to f$ uniformly on E.*

Proof. Fix any $\varepsilon > 0$. By the assumption of pointwise convergence, for every $x \in E$ there exists an integer $N(x)$ such that $|f_{N(x)}(x) - f(x)| < \varepsilon$.

Since $f_{N(x)}$ and f are continuous, there exists an open neighborhood V_x of x such that $|f_{N(x)}(y) - f_{N(x)}(x)| < \varepsilon$ and $|f(y) - f(x)| < \varepsilon$ for every $y \in V_x$.

Since E is compact, we can cover E with finitely many of these neighborhoods, say $E \subseteq V_{x_1} \cup \cdots \cup V_{x_m}$.

Choose the integer $N = \max\{N(x_1), \ldots, N(x_m)\}$. For every $n \geq N$ and $y \in E$, assuming that $y \in V_{x_i}$ we have

$$f_{N(x_i)}(y) \leq f_N(y) \leq f_n(y) \leq f(y),$$

because the sequence is increasing. Therefore

$$\begin{aligned}|f_n(y) - f(y)| &\leq |f_{N(x_i)}(y) - f(y)| \\ &\leq |f_{N(x_i)}(y) - f_{N(x_i)}(x_i)| + |f_{N(x_i)}(x_i) - f(x_i)| + |f(x_i) - f(y)| \\ &< \varepsilon + \varepsilon + \varepsilon.\end{aligned}$$

Since $y \in E$ and $\varepsilon > 0$ were arbitrary, this establishes the uniform convergence $f_n \to f$. □

3.2. The Stone-Weierstrass approximation theorem

Given a domain $E \subset \mathbb{R}^n$, for computational purposes it can be useful to approximate a continuous, real-valued function $f \in \mathcal{BC}(E)$ with special functions: say, polynomials, exponential functions, or trigonometric polynomials. It is thus important to understand whether every function $f \in \mathcal{BC}(E)$ can be uniformly approximated by such functions. In this section we will prove a key result in this direction.

As a preliminary, observe that the space $\mathcal{BC}(E)$ is an **algebra**. Namely, it is closed under multiplication:

if $f, g \in \mathcal{BC}(E)$, then also $fg \in \mathcal{BC}(E)$.

Moreover, the norm of the product satisfies

$$\|fg\| \leq \|f\| \, \|g\| .$$

We say that a subspace $\mathcal{A} \subseteq \mathcal{BC}(E)$ is a **subalgebra** if $f, g \in \mathcal{A}$ implies $fg \in \mathcal{A}$.

Lemma 3.4 (Closure of a subalgebra). *If $\mathcal{A} \subseteq \mathcal{BC}(E)$ is a subalgebra, then its closure $\overline{\mathcal{A}}$ is a subalgebra as well.*

Proof. Indeed, assume $f, g \in \overline{\mathcal{A}}$. Then there exist uniformly convergent sequences $f_n, g_n \in \mathcal{A}$ with $f_n \to f$ and $g_n \to g$. One has

$$\|fg - f_n g_n\| \leq \|fg - f_n g\| + \|f_n g - f_n g_n\| \leq \|f - f_n\| \, \|g\| + \|f_n\| \, \|g - g_n\| .$$

Since the sequence $(f_n)_{n \geq 1}$ is uniformly bounded, the right-hand side approaches zero as $n \to \infty$. This shows the convergence $f_n g_n \to fg$. Since $\mathcal{A}$ is an algebra, $f_n g_n \in \mathcal{A}$ for every n. Hence $fg \in \overline{\mathcal{A}}$. □

We say that a subset $\mathcal{A} \subseteq \mathcal{BC}(E)$ **separates points** if, for every couple of distinct points $x, y \in E$, there exists a function $f \in \mathcal{A}$ such that $f(x) \neq f(y)$.

Theorem 3.5 (Stone-Weierstrass). *Let E be a compact metric space. If $\mathcal{A}$ is a subalgebra of $\mathcal{C}(E)$ that separates points and contains the constant functions, then $\overline{\mathcal{A}} = \mathcal{C}(E)$.*

Otherwise stated, let $\mathcal{A}$ be a family of continuous, real-valued functions $f : E \mapsto \mathbb{R}$ with the following properties:

(i) If $f, g \in \mathcal{A}$ and $a, b \in \mathbb{R}$, then the linear combination $af + bg$ lies in $\mathcal{A}$.

(ii) If $f, g \in \mathcal{A}$, then the product fg lies in $\mathcal{A}$ as well.

(iii) The constant function $f(x) \equiv 1$ lies in $\mathcal{A}$.

(iv) For every two distinct points $x, y \in E$, there exists $f \in \mathcal{A}$ such that $f(x) \neq f(y)$.

Then every continuous function $f : E \mapsto \mathbb{R}$ on the compact domain E can be uniformly approximated by functions in $\mathcal{A}$.

Proof. 1. There exists a sequence of polynomials $(p_n)_{n\geq 1}$ such that $p_n(t) \to \sqrt{t}$ uniformly for $t \in [0, 1]$.

To prove the above claim, the underlying idea is to construct approximate solutions to the equation $t - p^2(t) = 0$ by iteration. We thus set $p_0(t) \equiv 0$ and, by induction on $n = 0, 1, 2, \ldots$,

$$p_{n+1}(t) = p_n(t) + \frac{1}{2}(t - p_n^2(t)) \tag{3.2}$$

(see Figure 3.2.1). By induction, one checks that $p_n(t) \leq p_{n+1}(t) \leq \sqrt{t}$ for every $t \in [0, 1]$. Indeed,

$$\begin{aligned} \sqrt{t} - p_{n+1}(t) &= \sqrt{t} - p_n(t) - \frac{1}{2}(t - p_n^2(t)) \\ &= (\sqrt{t} - p_n(t))\left(1 - \frac{1}{2}(\sqrt{t} + p_n(t))\right) \geq 0. \end{aligned}$$

For every fixed $t \in [0, 1]$, the sequence $p_n(t)$ is increasing and bounded above. Hence it has a unique limit, say $g(t)$. By (3.2), this limit satisfies $t - g^2(t) = 0$. Since $g(t) \geq 0$ we conclude that $g(t) = \sqrt{t}$.

Finally, by Dini's theorem, the convergence is uniform for $t \in [0, 1]$.

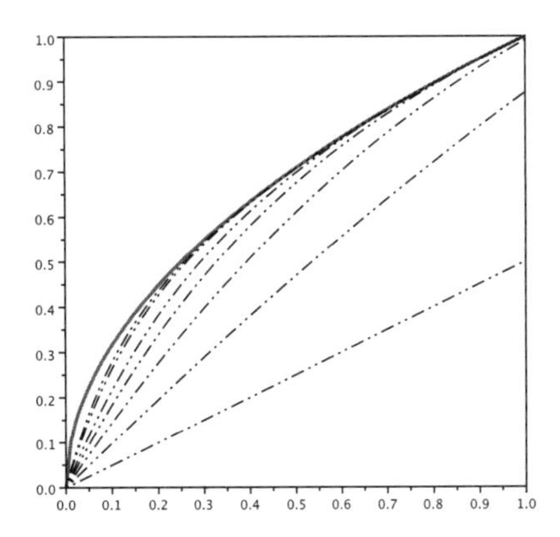

Figure 3.2.1. The first few polynomials in the sequence defined in (3.2).

2. For every function $f \in \mathcal{A}$, one has $|f| \in \overline{\mathcal{A}}$.

Indeed, let $\kappa \doteq \max_{x\in E}|f(x)|$. We can assume $\kappa \neq 0$. Then all functions $f_n(x) = p_n(f^2(x)/\kappa^2)$ lie in $\mathcal{A}$, because $\mathcal{A}$ is an algebra. Since $f^2(x)/\kappa^2 \in [0,1]$, the previous step yields the convergence

$$f_n(x) \;\to\; \sqrt{\frac{f^2(x)}{\kappa^2}} \;=\; \frac{|f(x)|}{\kappa},$$

uniformly for $x \in E$. Therefore $\frac{|f|}{\kappa} \in \overline{\mathcal{A}}$, and hence $|f| \in \overline{\mathcal{A}}$ as well.

3. We now apply the previous argument to the subalgebra $\overline{\mathcal{A}}$ and conclude that, if $f, g \in \overline{\mathcal{A}}$, then the functions

$$\max\{f,g\} \;=\; \frac{1}{2}(f+g+|f-g|)\,, \qquad \min\{f,g\} \;=\; \frac{1}{2}(f+g-|f-g|)$$

also lie in $\overline{\mathcal{A}}$.

4. For any two distinct points $y_1, y_2 \in E$ and any couple of real numbers a_1, a_2, there exists a function $f \in \mathcal{A}$ such that $f(y_1) = a_1$ and $f(y_2) = a_2$.

Indeed, by assumption there exists a continuous function $g \in \mathcal{A}$ such that $g(y_1) \neq g(y_2)$. Since $\mathcal{A}$ is an algebra and contains all constant functions, the function

$$f(x) \;\doteq\; a_1 + (a_2 - a_1)\frac{g(x) - g(y_1)}{g(y_2) - g(y_1)}$$

lies in $\mathcal{A}$ and satisfies our requirements.

5. Given any continuous function f, a point $y \in E$, and $\varepsilon > 0$, there exists a function $g_y \in \overline{\mathcal{A}}$ such that

$$g_y(y) \;=\; f(y), \qquad g_y(x) \;\leq\; f(x) + \varepsilon \quad \text{for every } x \in E\,. \tag{3.3}$$

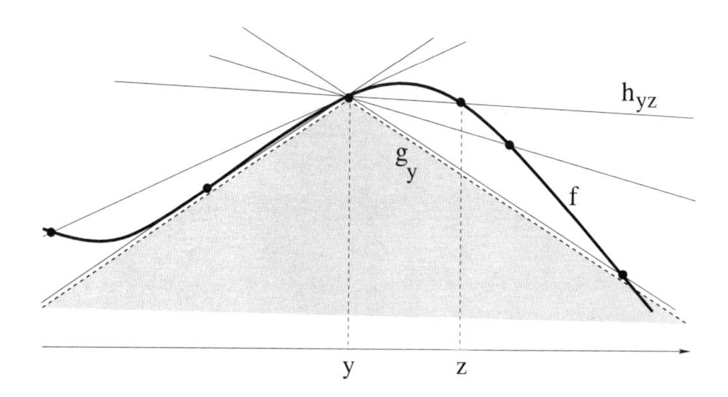

Figure 3.2.2. For a fixed y, taking the infimum of finitely many functions h_{yz} (here drawn as affine functions) we obtain a continuous function $g_y \leq f + \varepsilon$, with $g_y(y) = f(y)$.

Indeed (see Figure 3.2.2), by the previous step, for every point $z \in E$, there exists a function $h_{yz} \in \mathcal{A}$ such that $h_{yz}(y) = f(y)$ and $h_{yz}(z) = f(z)$.

Since f and h_{yz} are both continuous, there exists an open neighborhood V_z of z such that $h_{yz}(x) < f(x) + \varepsilon$ for every $x \in V_z$.

We can cover the compact set E with finitely many such neighborhoods: $E \subseteq V_{z_1} \cup \cdots \cup V_{z_m}$. Then the function

$$g_y(x) \doteq \min\left\{h_{yz_1}(x), \ldots, h_{yz_m}(x)\right\}$$

lies in $\overline{\mathcal{A}}$ and satisfies the conditions in (3.3).

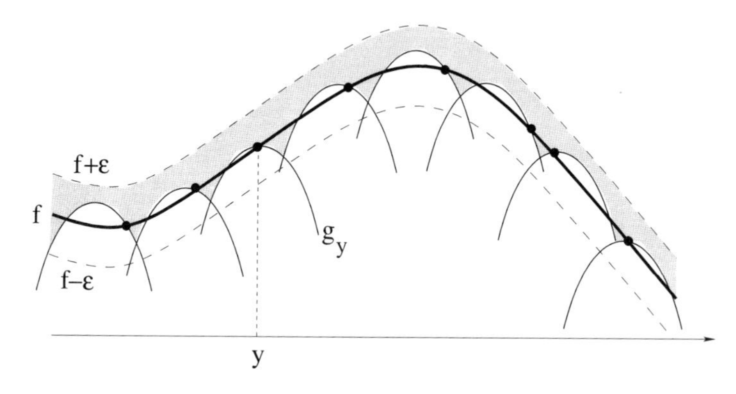

Figure 3.2.3. Taking the supremum of finitely many functions g_y we obtain a continuous function g such that $f - \varepsilon \leq g \leq f + \varepsilon$.

6. The closure of $\mathcal{A}$ is the entire space $\mathcal{C}(E)$.

Indeed, let $f \in \mathcal{C}(E)$ be any continuous function and let $\varepsilon > 0$ be given. For each $y \in E$, by the previous step there exists a function $g_y \in \overline{\mathcal{A}}$ (see

Figure 3.2.3) such that

$$g_y(y) = f(y), \qquad g_y(x) \leq f(x) + \varepsilon \quad \text{for every } x \in E.$$

By the continuity of f and g_y, there exists a neighborhood U_y of y such that

$$g_y(x) \geq f(x) - \varepsilon \qquad \text{for all } x \in U_y.$$

We now cover the compact set E with finitely many neighborhoods: $E \subseteq U_{y_1} \cup \cdots \cup U_{y_\nu}$. Then the function

$$g(x) \doteq \max\left\{g_{y_1}(x), \ldots, g_{y_\nu}(x)\right\}$$

lies in $\overline{\mathcal{A}}$ and satisfies

$$f(x) - \varepsilon \leq g(x) \leq f(x) + \varepsilon \quad \text{for all } x \in E. \qquad \square$$

A natural example of an algebra that satisfies all the assumptions in the Stone-Weiertrass theorem is provided by the polynomial functions.

Corollary 3.6 (Uniform approximation by polynomials). *Let E be a compact subset of $\mathbb{R}^n$. Let $\mathcal{A}$ be the family of all real-valued polynomials in the variables $(x_1, \ldots, x_n)$. Then $\mathcal{A}$ is dense in $\mathcal{C}(E)$.*

Indeed, the family of all real-valued polynomials in $(x_1, \ldots, x_n)$ is an algebra that contains the constant functions and separates points in $\mathbb{R}^n$. Hence, by Theorem 3.5 every continuous function $f : E \mapsto \mathbb{R}$ can be uniformly approximated by polynomials.

3.2.1. Complex-valued functions.

A key ingredient in the proof of Theorem 3.5 was the fact that, if two real-valued functions f, g lie in a subalgebra $\mathcal{A}$, then $\max\{f, g\}$ and $\min\{f, g\}$ lie in $\overline{\mathcal{A}}$. Clearly, such a statement would be meaningless for complex-valued functions. In fact, in its original form the Stone-Weierstrass theorem is NOT valid for complex-valued functions.

In order to obtain an approximation result valid for functions $f : E \mapsto \mathbb{C}$, the main idea is to regard $\mathbb{C} = \mathbb{R} \oplus i\mathbb{R}$ as a two-dimensional space over the reals. In the following, given a compact metric space E, we shall denote by $\mathcal{C}_{\mathbb{R}}(E; \mathbb{C})$ the space of all continuous complex-valued functions on E, regarded as a vector space over the real numbers.

Theorem 3.7. *Let E be a compact metric space. Let $\mathcal{A}$ be a subalgebra of $\mathcal{C}_{\mathbb{R}}(E; \mathbb{C})$ that separates points and contains the constant functions. Moreover, assume that whenever $f \in \mathcal{A}$, then also the complex conjugate function $\bar{f}$ lies in $\mathcal{A}$. Then $\mathcal{A}$ is dense in $\mathcal{C}_{\mathbb{R}}(E; \mathbb{C})$.*

Proof. 1. By the assumptions, if $f \in \mathcal{A}$, then its real and imaginary parts

$$\text{Re}(f) = \frac{f+\bar{f}}{2}, \qquad \text{Im}(f) = \frac{f-\bar{f}}{2}$$

also lie in $\mathcal{A}$. Let $\mathcal{A}_0$ be the subalgebra of $\mathcal{A}$ (over the real numbers), consisting of all functions $f \in \mathcal{A}$ with real values. Applying the Stone-Weierstrass theorem to $\mathcal{A}_0$, we conclude that $\mathcal{A}_0$ is dense in $\mathcal{C}(E) = \mathcal{C}_{\mathbb{R}}(E;\, \mathbb{R})$.

2. Given any $f \in \mathcal{C}_{\mathbb{R}}(E;\, \mathbb{C})$, we write f as a sum of its real and imaginary parts $f = \text{Re}(f) + i\, \text{Im}(f)$. By the previous step, there exist two sequences of real-valued functions $g_n, h_n \in \mathcal{A}_0$ such that

$$g_n \to \text{Re}(f), \qquad h_n \to \text{Im}(f) \tag{3.4}$$

as $n \to \infty$, uniformly on E.

Consider the sequence $f_n \doteq g_n + ih_n \in \mathcal{A}_0 + i\mathcal{A}_0 = \mathcal{A}$. By (3.4), we have the uniform convergence $f_n \to f$. Hence $\overline{\mathcal{A}} = \mathcal{C}_{\mathbb{R}}(E;\, \mathbb{C})$. □

Example 3.8 (Complex trigonometric polynomials). Let E be the unit circumference $\{x^2 + y^2 = 1\}$ in $\mathbb{R}^2$. Points on E will be parameterized by the angle $\theta \in [0, 2\pi]$. Let $\mathcal{A}$ be the algebra of all complex trigonometric polynomials:

$$p(\theta) = \sum_{n=-N}^{N} c_n e^{in\theta}, \tag{3.5}$$

where $N \geq 0$ is any integer and the coefficients c_n are complex numbers. It is clear that $\mathcal{A}$ is an algebra, contains the constant functions, and separates points. Moreover, $p \in \mathcal{A}$ implies $\bar{p} \in \mathcal{A}$ as well. By Theorem 3.7, the family of all these complex trigonometric polynomials is dense in $\mathcal{C}_{\mathbb{R}}(E;\, \mathbb{C})$.

Relying on the previous example, we now show that a real-valued, continuous periodic function can be uniformly approximated with trigonometric polynomials of the form

$$q(x) = \sum_{k=0}^{N} \alpha_k \cos kx + \sum_{k=1}^{N} \beta_k \sin kx\,. \tag{3.6}$$

Here $N \geq 1$ is any integer, while α_n, β_n are real numbers.

Corollary 3.9 (Approximation of periodic functions by trigonometric polynomials). *Let $f : \mathbb{R} \mapsto \mathbb{R}$ be a continuous function, periodic of period 2π. Then for any $\varepsilon > 0$ there exists a trigonometric polynomial q as in (3.6) such that*

$$|q(x) - f(x)| \leq \varepsilon \qquad \textit{for all } x \in \mathbb{R}\,. \tag{3.7}$$

Proof. By assumption, $f(x + 2\pi) = f(x)$ for every $x \in \mathbb{R}$. As shown in Example 3.8, there exists a complex trigonometric polynomial p of the form (3.5) such that

$$|p(x) - f(x)| \leq \varepsilon \qquad \text{for all } x \in \mathbb{R}\,. \tag{3.8}$$

Consider the complex coefficients $c_n = a_n + ib_n$, with $a_n, b_n \in \mathbb{R}$. Calling $q(x) \doteq \text{Re } p(x)$ the real part of p, we compute

$$q(x) = \sum_{n=-N}^{N} a_n \cos nx - \sum_{n=-N}^{N} b_n \sin nx = \sum_{n=0}^{N} \alpha_n \cos nx + \sum_{n=1}^{N} \beta_n \sin nx\,,$$

with

$$\alpha_0 = a_0\,, \qquad \alpha_n = a_{-n} + a_n\,, \qquad \beta_n = b_{-n} - b_n \qquad \text{for } n \geq 1.$$

Since f is real-valued, by (3.8) we have

$$\begin{aligned} |q(x) - f(x)| &= |\,\text{Re } p(x) - \text{Re } f(x)| \\ &\leq |p(x) - f(x)| \leq \varepsilon \quad \text{for all } x \in \mathbb{R}\,. \end{aligned}$$

□

Remark 3.10. If the periodic function f is even, i.e., $f(x) = f(-x)$, then it can be approximated with a trigonometric polynomial of the form (3.6) with $\beta_k = 0$ for every $k \geq 1$, that is, with a finite sum of cosine functions.

If f is odd, i.e., $f(x) = -f(-x)$, then in (3.6) one can take $\alpha_k = 0$ for every $k \geq 0$. In other words, f can be approximated by a finite sum of sine functions.

3.3. Ascoli's compactness theorem

In a finite-dimensional space, by the Bolzano-Weierstrass theorem every bounded sequence has a convergent subsequence. On the other hand, as shown in Theorem 2.22 of Chapter 2, this compactness property fails in every infinite-dimensional normed space. For example, in the space $\mathcal{C}([0,1])$ the sequences of continuous functions $f_n(x) = x^n$ or $f_n(x) = \sin nx$ are bounded but do not admit any uniformly convergent subsequence. It is thus natural to ask: what additional property of the functions f_n can guarantee the existence of a uniformly convergent subsequence? An answer is provided by Ascoli's theorem, relying on the concept of *equicontinuity*.

Let E be a metric space. A family of continuous functions $\mathcal{F} \subset \mathcal{C}(E)$ is called **equicontinuous** if, for every $x \in E$ and $\varepsilon > 0$, there exists $\delta > 0$ such that

$$d(y, x) < \delta \quad \text{implies} \quad |f(y) - f(x)| < \varepsilon \tag{3.9}$$

for all functions $f \in \mathcal{F}$. Notice that here $\delta > 0$ can depend on x and ε, but not on the particular function $f \in \mathcal{F}$.

Lemma 3.11. *Let E be a compact metric space and let $\mathcal{F} \subset \mathcal{C}(E)$ be equicontinuous. Then $\mathcal{F}$ is* **uniformly equicontinuous**. *Namely, for every $\varepsilon > 0$ there exists $\delta > 0$ such that*

$$d(x,y) < \delta \quad \textit{implies} \quad |f(x) - f(y)| < \varepsilon \quad \textit{for all} \quad x, y \in E, \ f \in \mathcal{F}. \tag{3.10}$$

Proof. Let $\varepsilon > 0$ be given. For each $x \in E$, choose $\delta = \delta(x) > 0$ such that (3.9) holds simultaneously for all functions $f \in \mathcal{F}$. By compactness, we can cover the space E with finitely many balls:

$$E \subseteq B(x_1, \delta_1) \cup \cdots \cup B(x_n, \delta_n),$$

where $\delta_i = \delta(x_i)$. Choose $\rho > 0$ so small that, for every $x \in E$, the ball $B(x, \rho)$ is entirely contained inside one of the balls $B(x_i, \delta_i)$.

Now assume $d(x,y) < \rho$. Then there exists an index $k \in \{1, \ldots, n\}$ such that $x, y \in B(x_k, \delta_k)$. This implies

$$|f(x) - f(y)| \le |f(x) - f(x_k)| + |f(y) - f(x_k)| \le \varepsilon + \varepsilon$$

for every function $f \in \mathcal{F}$. Since $\varepsilon > 0$ was arbitrary, this proves the lemma. □

If E is a compact metric space, then $\mathcal{C}(E) = \mathcal{BC}(E)$ is a Banach space. By completeness, it follows that for a subset $\mathcal{F} \subset \mathcal{C}(E)$ the following properties are equivalent:

(i) $\mathcal{F}$ is relatively compact, i.e., the closure $\overline{\mathcal{F}}$ is compact.

(ii) $\mathcal{F}$ is precompact, i.e., for every $\varepsilon > 0$ it can be covered by finitely many balls with radius ε.

(iii) From every sequence of continuous functions $f_k \in \mathcal{F}$ one can extract a subsequence converging to some function f, uniformly on E.

Theorem 3.12 (Ascoli). *Let E be a compact metric space. Let $\mathcal{F} \subset \mathcal{C}(E)$ be an equicontinuous family of functions, such that*

$$\sup_{f \in \mathcal{F}} |f(x)| < \infty \qquad \textit{for every } x \in E. \tag{3.11}$$

Then $\mathcal{F}$ is a relatively compact subset of $\mathcal{C}(E)$.

Proof. It suffices to prove that $\mathcal{F}$ is precompact.

1. Let $\varepsilon > 0$ be given. By Lemma 3.11, there exists $\delta > 0$ such that (3.10) holds. Since E is compact, it can be covered by finitely many balls of radius

δ, say

$$E \subseteq \bigcup_{i=1}^{n} B(x_i, \delta).$$

By the assumption (3.11),

$$M \doteq \max_{i\in\{1,\ldots,n\}} \sup_{f\in\mathcal{F}} |f(x_i)| < \infty.$$

We now choose finitely many numbers $\alpha_1, \ldots, \alpha_m$ such that the balls $B(\alpha_j, \varepsilon)$ cover the compact interval $[-M, M]$.

2. Consider the set Θ of all maps $\theta : \{x_1, \ldots, x_n\} \mapsto \{\alpha_1, \ldots, \alpha_m\}$. This is a finite set. Indeed, there are exactly m^n such maps. For every $\theta \in \Theta$, define the family of continuous functions

$$\mathcal{F}_\theta \doteq \Big\{f \in \mathcal{F};\ f(x_i) \in B(\theta(x_i), \varepsilon) \text{ for all } i = 1, \ldots, n\Big\}.$$

Since the interval $[-M, M]$ is covered by the balls $B(\alpha_j, \varepsilon)$, we clearly have $\bigcup_{\theta\in\Theta} \mathcal{F}_\theta = \mathcal{F}$.

3. We claim that each set $\mathcal{F}_\theta$ has diameter $\leq 4\varepsilon$. Indeed, assume $f, g \in \mathcal{F}_\theta$. For any $x \in E$, choose an index i such that $x \in B(x_i, \delta)$. We then have

$$\begin{aligned} &|f(x) - g(x)| \\ &\quad\leq |f(x) - f(x_i)| + |f(x_i) - \theta(x_i)| + |\theta(x_i) - g(x_i)| + |g(x_i) - g(x)| \\ &\quad\leq \varepsilon + \varepsilon + \varepsilon + \varepsilon\,. \end{aligned}$$

Therefore

$$\|f - g\|_{\mathcal{C}} \doteq \max_{x\in E}|f(x) - g(x)| \leq 4\varepsilon\,.$$

The above arguments show that, for any $\varepsilon > 0$, the set $\mathcal{F}$ can be covered with finitely many sets having diameter $\leq 4\varepsilon$. Hence it can also be covered by finitely many closed balls of radius 4ε. Since ε is arbitrary, this proves that $\mathcal{F}$ is precompact. □

The above theorem can be easily extended to functions taking values in the complex domain $\mathbb{C}$ or in a Euclidean space $\mathbb{R}^n$.

Corollary 3.13. *Let E be a compact metric space. Let $(f_k)_{k\geq 1}$ be a sequence of continuous functions from E into $\mathbb{R}^n$ such that*

(i) *the family $\{f_k\}$ is equicontinuous;*

(ii) $\sup_{k\geq 1} |f_k(x)| < \infty$ *for every $x \in E$.*

Then the sequence $(f_k)_{k\geq 1}$ admits a uniformly convergent subsequence.

Proof. Let us write out the components:

$$f_k(x) \;=\; (f_{k,1}(x), f_{k,2}(x), \ldots, f_{k,n}(x)) \;\in\; \mathbb{R}^n.$$

Applying Theorem 3.12 to the sequence of scalar functions $(f_{k,1})_{k\geq 1}$, we can extract a subsequence such that these first components converge uniformly on E. From this subsequence we can extract a further subsequence such that the second components converge uniformly, etc. After n steps we obtain a subsequence where all components converge, uniformly on E. □

3.4. Spaces of Hölder continuous functions

Let $\Omega \subset \mathbb{R}^n$ be an open set, and $0 < \gamma \leq 1$. We say that a function $f : \Omega \mapsto \mathbb{R}$ is **Hölder continuous** with exponent γ if there exists a constant C such that

$$|f(x) - f(y)| \;\leq\; C\,|x-y|^\gamma \qquad \text{for all } x, y \in \Omega\,.$$

We denote by $\mathcal{C}^{0,\gamma}(\Omega)$ the space of all bounded Hölder continuous functions on Ω, with norm

$$\|f\|_{\mathcal{C}^{0,\gamma}(\Omega)} \;\doteq\; \sup_{x\in\Omega} |f(x)| \;+\; \sup_{x,y\in\Omega,\ x\neq y} \frac{|f(x)-f(y)|}{|x-y|^\gamma}\,. \tag{3.12}$$

More generally, given an integer $k \geq 0$, we denote by $\mathcal{C}^{k,\gamma}(\Omega)$ the space of all continuous functions with Hölder continuous partial derivatives up to order k. This space is endowed with the norm[1]

$$\|f\|_{\mathcal{C}^{k,\gamma}(\Omega)} \;\doteq\; \sum_{|\alpha|\leq k} \Big(\sup_{x\in\Omega} |D^\alpha f(x)|\Big) + \sum_{|\alpha|=k} \left(\sup_{x,y\in\Omega,\ x\neq y} \frac{|D^\alpha f(x) - D^\alpha f(y)|}{|x-y|^\gamma}\right). \tag{3.13}$$

Theorem 3.14 (Hölder spaces are complete). *Let $\Omega \subseteq \mathbb{R}^n$ be an open set. For every integer $k \geq 0$ and any $0 < \gamma \leq 1$, the space $\mathcal{C}^{k,\gamma}(\Omega)$ is a Banach space.*

Proof. The fact that (3.13) defines a norm is clear. To prove that the space $\mathcal{C}^{k,\gamma}(\Omega)$ is complete, let $(f_m)_{m\geq 1}$ be a Cauchy sequence with respect to the norm (3.13). Then, for every $x \in \Omega$, the sequence $f_m(x)$ is Cauchy and converges to some value $f(x)$ uniformly on Ω.

The assumption also imply that, for every $|\alpha| \leq k$, the sequence of partial derivatives $D^\alpha f_m$ is Cauchy; hence it converges to some continuous function $v_\alpha(x) = D^\alpha f(x)$ uniformly on Ω.

---

[1]Given a multi-index $\alpha = (\alpha_1, \ldots, \alpha_n)$, we use the notation $D^\alpha f = \left(\frac{\partial}{\partial x_1}\right)^{\alpha_1} \cdots \left(\frac{\partial}{\partial x_n}\right)^{\alpha_n} f$ to denote a partial derivative of f of order $|\alpha| = \alpha_1 + \cdots + \alpha_n$.

It remains to prove that the convergence $f_m \to f$ takes place also with respect to the norm of $\mathcal{C}^{m,\gamma}(\Omega)$. In other words, for $|\alpha| = k$ we need to show that

$$\lim_{m\to\infty} \sup_{x,y\in\Omega,\ x\neq y} \frac{\left|D^\alpha(f_m - f)(x) - D^\alpha(f_m - f)(y)\right|}{|x-y|^\gamma} = 0. \tag{3.14}$$

By assumption,

$$\lim_{m,n\to\infty} \sup_{x,y\in\Omega,\ x\neq y} \frac{\left|D^\alpha(f_m - f_n)(x) - D^\alpha(f_m - f_n)(y)\right|}{|x-y|^\gamma} = 0.$$

Hence, for any $\varepsilon > 0$, there exists N large enough so that

$$\sup_{x,y\in\Omega,\ x\neq y} \frac{\left|D^\alpha(f_m - f_n)(x) - D^\alpha(f_m - f_n)(y)\right|}{|x-y|^\gamma} \leq \varepsilon \quad \text{for all } m,n \geq N.$$

Keeping m fixed and letting $n \to \infty$ we obtain

$$\sup_{x,y\in\Omega,\ x\neq y} \frac{\left|D^\alpha(f_m - f)(x) - D^\alpha(f_m - f)(y)\right|}{|x-y|^\gamma} \leq \varepsilon \quad \text{for all } m \geq N.$$

Since $\varepsilon > 0$ was arbitrary, this proves (3.14). □

3.5. Problems

1. Let E be a compact metric space. Assume that a family of real-valued continuous functions $\mathcal{F} \subset \mathcal{C}(E)$ satisfies the following two conditions:

(i) For every $x, y \in E$ and $a, b \in \mathbb{R}$, there exists a function $f \in \mathcal{F}$ such that $f(x) = a$ and $f(y) = b$.

(ii) If $f, g \in \mathcal{F}$, then the functions $\max\{f, g\}$ and $\min\{f, g\}$ lie in the closure $\overline{\mathcal{F}}$.

Prove that $\mathcal{F}$ is dense in $\mathcal{C}(E)$.

2. Prove or disprove the following statements.

(i) Given any continuous function $f : \mathbb{R} \mapsto \mathbb{R}$ (possibly unbounded), there exists a sequence of polynomials $(p_n)_{n\geq 1}$ such that $p_n(x) \to f(x)$ uniformly on every bounded interval $[a, b]$. (Note: Here the sequence of polynomials should be independent of the interval $[a, b]$.)

(ii) There exists a sequence of polynomials p_n that converges to the function $f(x) = e^{-x^2}$ uniformly on $\mathbb{R}$.

3. Let $g : [0, \pi] \mapsto \mathbb{R}$ be a continuous function.

(i) Prove that, for any $\varepsilon > 0$, there exists a trigonometric polynomial of the form

$$p(x) = \sum_{k=0}^{N} b_k \cos kx$$

such that $|p(x) - g(x)| < \varepsilon$ for every $x \in [0, \pi]$.

(ii) If $g(0) = g(\pi) = 0$, prove that, for any $\varepsilon > 0$, there exists a trigonometric polynomial of the form

$$p(x) = \sum_{k=1}^{N} a_k \sin kx$$

such that $|p(x) - g(x)| < \varepsilon$ for every $x \in [0, \pi]$.

4. Let $f : \mathbb{R} \mapsto \mathbb{R}$ be continuously differentiable. Show that, for every bounded interval $[a, b]$, there exists a sequence of polynomials p_n such that $p_n \to f$ and $p'_n \to f'$, uniformly on $[a, b]$.

5. Let E be the unit circle $\{x^2 + y^2 = 1\}$ in $\mathbb{R}^2$. Points on E will be parameterized with the angle $\theta \in [0, 2\pi]$. Let $\mathcal{A}$ be the family of all complex trigonometric polynomials of the form

$$p(\theta) = \sum_{n=0}^{N} c_n e^{in\theta},$$

where $N \geq 0$ is any integer and the c_n are complex-valued coefficients.

(i) Is $\mathcal{A}$ an algebra?

(ii) Does $\mathcal{A}$ contain the constant functions and separate points?

(iii) Is $\mathcal{A}$ dense on the space of all continuous, complex-valued functions $f : E \mapsto \mathbb{C}$?

6. Let $\Omega \subset \mathbb{R}^n$ be a bounded open set. Prove that the family of all polynomials $p = p(x_1, x_2, \ldots, x_n)$ in n variables is dense on the space $\mathbf{L}^p(\Omega)$, for every $1 \leq p < \infty$.

7. Consider the rectangle $Q = \{(x, y)\,;\ x \in [0, a]\,,\ y \in [0, b]\}$.

(i) Given any continuous function $f = f(x, y)$ on Q and any $\varepsilon > 0$, construct a finite number of continuous functions $g_1, \ldots, g_N : [0, a] \mapsto \mathbb{R}$ and $h_1, \ldots, h_n : [0, b] \mapsto \mathbb{R}$ such that

$$\left| f(x, y) - \sum_{i=1}^{N} g_i(x) h_i(y) \right| \leq \varepsilon \quad \text{for all } (x, y) \in Q.$$

(ii) Given any continuous function $f : Q \mapsto \mathbb{R}$ that vanishes on the boundary of Q, show that one can choose an integer $M \geq 1$ and finitely many coefficients c_{mn}, $1 \leq m, n \leq M$, such that

$$\left| f(x,y) - \sum_{m,n=1}^{M} c_{mn} \sin \frac{\pi m\, x}{a} \sin \frac{\pi n\, y}{b} \right| \leq \varepsilon \quad \text{for all } (x,y) \in Q.$$

8. Let $\Omega \subset \mathbb{R}^n$ be a bounded open set, and $0 < \gamma \leq 1$. Prove that the embedding $\mathcal{C}^{0,\gamma}(\Omega) \subset\subset \mathcal{C}^0(\Omega)$ is compact. In other words, if $(f_n)_{n\geq 1}$ is a bounded sequence in $\mathcal{C}^{0,\gamma}(\Omega)$, then it admits a subsequence that converges in $\mathcal{C}^0(\Omega)$.

9. Decide for which values of $0 < \gamma \leq 1$ the following functions lie in the Hölder space $\mathcal{C}^{0,\gamma}(\,]0,1[\,)$:

$$f_1(x) \;=\; x^{1/3}, \qquad f_2(x) \;=\; \sqrt{x} \cdot \sin \frac{1}{x}, \qquad f_3(x) \;=\; x\,|\ln x|\,.$$

10. Explain what is wrong with the following argument.

"Let $0 < \gamma \leq 1$ and let $(f_n)_{n\geq 1}$ be a sequence of functions in the Hölder space $\mathcal{C}^{0,\gamma}([0,1])$, with $\|f_n\|_{\mathcal{C}^{0,\gamma}} \leq 1$ for every n. Since the functions f_n are uniformly bounded and equicontinuous, by Ascoli's theorem we can extract a subsequence converging to some limit function f, uniformly for $x \in [0,1]$. It is now easy to check that $\|f\|_{\mathcal{C}^{0,\gamma}} \leq 1$. This shows that the closed unit ball in $\mathcal{C}^{0,\gamma}$ is compact, and hence by Theorem 2.22 in Chapter 2 the space $\mathcal{C}^{0,\gamma}$ is finite-dimensional."

11. Let $\varphi : \mathbb{R}_+ \mapsto \mathbb{R}_+$ be a smooth function such that

$$\varphi(0) \;=\; 0, \qquad \varphi'(s) > 0, \quad \varphi''(s) \leq 0 \quad \text{for all } s > 0\,.$$

Given an open set $\Omega \subseteq \mathbb{R}^n$, consider the space of continuous functions

$$\mathcal{C}^{\varphi}(\Omega) \;\doteq\; \left\{ f : \Omega \mapsto \mathbb{R};\; \|f\|_{\varphi} \doteq \sup_x |f(x)| \;+\; \sup_{x \neq y} \frac{|f(x) - f(y)|}{\varphi(|x-y|)} \;<\; \infty \right\}.$$

Prove that $\mathcal{C}^{\varphi}(\Omega)$ is a Banach space. (Note: Here the function φ can be regarded as a modulus of continuity. In the case $\varphi(s) = s^{\gamma}$ with $0 < \gamma \leq 1$ one has $\mathcal{C}^{\varphi}(\Omega) = \mathcal{C}^{0,\gamma}(\Omega)$.)

12. Prove the following more general version of Ascoli's theorem. Let E be a compact metric space and let K be a compact subset of a normed space X. Assume that a sequence of maps $\phi_n : E \mapsto K$ is equicontinuous. Namely, for every $x \in E$ and $\varepsilon > 0$, there exists $\delta > 0$ such that

$$d(y,x) < \delta \quad \text{implies} \quad \|\phi_n(y) - \phi_n(x)\| \;<\; \varepsilon \quad \text{for all } n \geq 1\,.$$

Then the sequence $(\phi_n)_{n\geq 1}$ admits a uniformly convergent subsequence.

Chapter 4

Bounded Linear Operators

In this chapter we look in more detail at linear operators in Banach spaces. As in Chapter 2, $\mathcal{B}(X,Y)$ will denote the space of bounded linear operators $\Lambda : X \mapsto Y$, with norm

$$\|\Lambda\| \doteq \sup_{\|x\|\leq 1} \|\Lambda x\| .$$

We begin with some results based on the Baire category theorem.

4.1. The uniform boundedness principle

Theorem 4.1 (Banach-Steinhaus uniform boundedness principle). *Let X,Y be Banach spaces. Let $\mathcal{F} \subset \mathcal{B}(X,Y)$ be any family of bounded linear operators. Then either $\mathcal{F}$ is uniformly bounded, so that*

$$\sup_{\Lambda\in\mathcal{F}} \|\Lambda\| < \infty,$$

or else there exists a dense set $S \subset X$ such that

$$\sup_{\Lambda\in\mathcal{F}} \|\Lambda x\| = \infty \qquad \text{for all } x \in S. \tag{4.1}$$

Proof. For every integer n, consider the open set

$$S_n \doteq \{x \in X\,;\ \|\Lambda x\| > n \text{ for some } \Lambda \in \mathcal{F}\}. \tag{4.2}$$

If one of these sets, say S_k, is not dense in X, then there exists $x_0 \in X$ and a radius $r > 0$ such that the closed ball $\overline{B}(x_0, r)$ does not intersect S_k. In other words,

$$\|\Lambda x\| \leq k \quad \text{for all } \Lambda \in \mathcal{F}, \quad x \in \overline{B}(x_0, r).$$

If now $\|x\| \leq r$, then

$$\|\Lambda x\| = \|\Lambda(x_0+x) - \Lambda x_0\| \leq \|\Lambda(x_0+x)\| + \|\Lambda x_0\| \leq 2k \quad \text{for all } \Lambda \in \mathcal{F}.$$

Therefore

$$\|\Lambda\| \doteq \sup_{\|x\|\leq 1} \|\Lambda x\| = \frac{1}{r} \sup_{\|x\|\leq r} \|\Lambda x\| \leq \frac{2k}{r}$$

for every $\Lambda \in \mathcal{F}$. In this case, the family of operators $\mathcal{F}$ is uniformly bounded.

The other possibility is that the open sets S_n in (4.2) are all dense in X. By Baire's category theorem, the intersection $S \doteq \bigcap_{n\geq 1} S_n$ is dense in X. By construction, for each $x \in S$ and $n \geq 1$ there exists an operator $\Lambda \in \mathcal{F}$ such that $\|\Lambda x\| > n$. Hence (4.1) holds. □

Remark 4.2. From the above theorem it follows that, if a family of operators $\Lambda \in \mathcal{B}(X, Y)$ is pointwise bounded on the unit ball, then it is uniformly bounded. In other words, the condition

$$\sup_{\Lambda\in\mathcal{F}} \|\Lambda x\| < \infty \quad \text{for each } x \in X \text{ with } \|x\| \leq 1$$

implies

$$\sup_{\Lambda\in\mathcal{F}} \sup_{\|x\|\leq 1} \|\Lambda x\| < \infty .$$

This justifies the name "uniform boundedness principle".

Corollary 4.3 (Continuity of the pointwise limit). *Let X, Y be Banach spaces. Let $(\Lambda_n)_{n\geq 1}$ be a sequence of bounded linear operators in $\mathcal{B}(X; Y)$. Assume that the pointwise limit*

$$\Lambda x \doteq \lim_{n\to\infty} \Lambda_n x \tag{4.3}$$

exists for every $x \in X$. Then the map Λ defined by (4.3) is a bounded linear operator.

Proof. For every $x \in X$, the sequence $(\Lambda_n x)_{n\geq 1}$ is bounded. Hence by the previous theorem the sequence of operators Λ_n is uniformly bounded. This implies

$$\|\Lambda\| \doteq \sup_{\|x\|\leq 1} \|\Lambda x\| = \sup_{\|x\|\leq 1} \left(\lim_{n\to\infty} \|\Lambda_n x\|\right) \leq \sup_{n\geq 1} \|\Lambda_n\| < \infty ,$$

showing that the linear operator Λ is bounded. □

4.2. The open mapping theorem

If X, Y are metric spaces, we say that a map $f : X \mapsto Y$ is **open** if, for every open subset $U \subseteq X$, the image $f(U)$ is an open subset of Y. This is the case if and only if, for every $x \in X$ and $r > 0$, the image $f(B(x, r))$ of the open ball centered at x with radius $r > 0$ contains a ball centered at $f(x)$.

Theorem 4.4 (Open mapping). *Let X, Y be Banach spaces. Let $\Lambda : X \mapsto Y$ be a bounded, surjective linear operator. Then Λ is open.*

Proof. 1. By linearity, the image of an open ball $B(x, r)$ can be written as

$$\Lambda(B(x,r)) \;=\; \Lambda(x) + \Lambda(B(0,r)) \;=\; \Lambda(x) + r\,\Lambda(B(0,1)).$$

Calling $B_r \doteq B(0, r)$ the open ball centered at the origin with radius r, to prove the theorem it thus suffices to show that the image $\Lambda(B_1)$ contains an open ball around the origin in Y.

2. Since Λ is surjective, $Y = \bigcup_{n=1}^{\infty} \Lambda(B_n)$. Recalling that Y is a complete metric space, by Baire's theorem at least one of the closures $\overline{\Lambda(B_n)} \subset Y$ has nonempty interior.

By a rescaling argument, $\overline{\Lambda(B_1)} = n^{-1}\overline{\Lambda(B_n)}$ must also have nonempty interior. Namely, there exist $y_0 \in Y$ and $r > 0$ such that $B(y_0,\, r) \subset \overline{\Lambda(B_1)}$.

Since the open unit ball B_1 is convex and symmetric, the same is true of its image $\Lambda(B_1)$ and of the closure $\overline{\Lambda(B_1)}$. In particular, by symmetry $B(-y_0, r) \subseteq \overline{\Lambda(B_1)}$, while convexity implies that the ball $B(0, r) \subset Y$ satisfies

$$B(0,r) \;=\; \frac{1}{2}B(y_0,r) + \frac{1}{2}B(-y_0,r) \;\subseteq\; \overline{\Lambda(B_1)}.$$

Using again the linearity of Λ, by rescaling we obtain

$$B(0, 2^{-n}r) \;\subseteq\; \overline{\Lambda(B_{2^{-n}})} \quad \text{for all } n \geq 1\,. \tag{4.4}$$

3. We conclude the proof by showing that $B(0, r/2) \subseteq \Lambda(B_1)$. Indeed, consider any point $y \in B(0, r/2)$. We proceed by induction.

- By density, we can find $x_1 \in B_{2^{-1}}$ such that $\|y - \Lambda x_1\| < 2^{-2}r$.
- In turn, we can find $x_2 \in B_{2^{-2}}$ such that $\|(y - \Lambda x_1) - \Lambda x_2\| < 2^{-3}r$.
- Continuing by induction, for each n we have

$$y - \sum_{j=1}^{n-1} \Lambda x_j \;\in\; B(0, 2^{-n}r) \;\subseteq\; \overline{\Lambda(B_{2^{-n}})}.$$

Therefore we can select a point $x_n \in B_{2^{-n}}$ such that

$$\left\| \left(y - \sum_{j=1}^{n-1} \Lambda x_j\right) - \Lambda x_n \right\| < 2^{-n-1} r\,.$$

Since X is a Banach space and $\sum_n \|x_n\| < \infty$, the series $\sum_n x_n$ converges, say $\sum_{n=1}^{\infty} x_n = x$. We observe that

$$\|x\| \leq \sum_{n=1}^{\infty} \|x_n\| < \sum_{n=1}^{\infty} 2^{-n} = 1, \qquad \Lambda x = \lim_{n\to\infty} \sum_{j=1}^{n} \Lambda x_j = y\,.$$

Hence the image $\Lambda(B_1)$ contains all points $y \in Y$ with $\|y\| < r/2$. □

Corollary 4.5. *If X, Y are Banach spaces and $\Lambda : X \mapsto Y$ is a continuous bijection, then its inverse $\Lambda^{-1} : Y \mapsto X$ is continuous as well.*

Proof. Since Λ is a bijection, Λ is open if and only if the inverse mapping Λ^{-1} is continuous. □

4.3. The closed graph theorem

Let X, Y be Banach spaces. The product space $X \times Y$ is the set of all ordered couples (x, y) with $x \in X$ and $y \in Y$. This is a Banach space with norm

$$\|(x,y)\| \doteq \|x\| + \|y\|\,. \tag{4.5}$$

Next, let Λ be a (possibly unbounded) linear operator, with domain $\mathrm{Dom}(\Lambda) \subseteq X$ and values in Y. We say that Λ is **closed** if its graph

$$\mathrm{Graph}(\Lambda) \doteq \Big\{(x,y)\,;\; x \in \mathrm{Dom}(\Lambda) \subseteq X\,,\; y = \Lambda x\Big\} \subset X \times Y$$

is a closed subset of the product space $X \times Y$. In other words, the linear operator Λ is closed provided that the following holds:

Given two sequences of points $x_n \in \mathrm{Dom}(\Lambda)$ and $y_n = \Lambda x_n \in Y$, if $x_n \to x$ and $y_n \to y$, then $x \in \mathrm{Dom}(\Lambda)$ and $\Lambda x = y$.

Notice that every continuous linear operator $\Lambda : X \mapsto Y$ is closed. The next result shows that a converse is also true, provided that the domain *Dom*(Λ) is the entire space X.

Theorem 4.6 (Closed graph). *Let X, Y be Banach spaces, and let $\Lambda : X \mapsto Y$ be a closed linear operator defined on the entire space X. Then Λ is continuous.*

Proof. Call $\Gamma \doteq \text{Graph}(\Lambda)$. By assumption, Γ is a closed subspace of the Banach space $X \times Y$, hence it is a Banach space as well.

Consider the projections $\pi_1 : \Gamma \mapsto X$ and $\pi_2 : \Gamma \mapsto Y$, defined as

$$\pi_1(x, \Lambda x) = x\,, \qquad \pi_2(x, \Lambda x) = \Lambda x\,.$$

The map π_1 is a bijection between Γ and X, hence by Corollary 4.2, its inverse π_1^{-1} is continuous. Therefore $\Lambda = \pi_2 \circ \pi_1^{-1}$ is the composition of two continuous maps, hence continuous. □

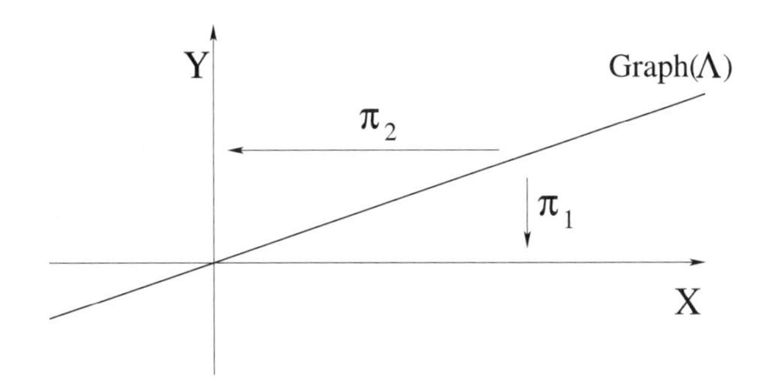

Figure 4.3.1. Proving the closed graph theorem.

Example 4.7. Consider the space $X = \mathcal{C}^0(\mathbb{R})$ of all bounded continuous functions $f : \mathbb{R} \mapsto \mathbb{R}$, with norm $\|f\|_{\mathcal{C}^0} \doteq \sup_x |f(x)|$. Let Λ be the differentiation operator defined by $\Lambda f = f'$. Its domain is the subspace

$$\text{Dom}(\Lambda) \;=\; \Big\{ f \in \mathcal{C}^0(\mathbb{R})\,;\; f' \in \mathcal{C}^0(\mathbb{R}) \Big\} \;=\; \mathcal{C}^1(\mathbb{R})$$

consisting of all continuously differentiable functions with bounded derivative.

Observe that the linear operator Λ is not bounded (hence not continuous). For example, the functions $f_n(x) = \sin nx$ are uniformly bounded: $\|f_n\|_{\mathcal{C}^0} = 1$ for all $n \geq 1$. However, the sequence of derivatives $\Lambda f_n = f_n'$ is unbounded, because $f_n'(x) = n \cos nx$ and hence $\|f_n'\|_{\mathcal{C}^0} = n$.

On the other hand, the linear operator $\Lambda : f \mapsto f'$ has closed graph. To see this, consider a sequence $f_n \in \text{Dom}(\Lambda)$ such that, for some functions $f, g \in \mathcal{C}^0$, one has

$$\|f_n - f\|_{\mathcal{C}^0} \to 0\,, \qquad \|f_n' - g\|_{\mathcal{C}^0} \to 0\,. \tag{4.6}$$

If (4.6) holds, then f is continuously differentiable and $f' = g$. Hence the point $(f, g) \in X \times X$ lies in the graph of Λ.

Notice that this example does not contradict the closed graph theorem, because Λ is not defined on the entire space X.

4.4. Adjoint operators

Let X be a Banach space on the field $\mathbb{K}$. By definition, its **dual** is the space X^* of all bounded (hence continuous) linear functionals $x^* : X \mapsto \mathbb{K}$, with norm

$$\|x^*\| \doteq \sup_{\|x\|\leq 1} |x^*(x)|.$$

In turn, every element $x \in X$ induces a bounded linear functional on X^*, namely $x^* \mapsto x^*(x) \in \mathbb{K}$. Using this identification, we can thus write $X \subseteq (X^*)^*$. Since the spaces X and X^* often play a symmetric role, it will be convenient to use the notation

$$\langle x^*, x\rangle \doteq x^*(x).$$

Now let X, Y be Banach spaces over the field $\mathbb{K}$, and let X^*, Y^* be their duals. Let $\Lambda : X \mapsto Y$ be a bounded linear operator. Then, for every bounded linear functional $y^* : Y \mapsto \mathbb{K}$, the composed map $x^* : X \mapsto \mathbb{K}$ defined as $x^*(x) = y^*(\Lambda x)$ is a bounded linear functional on X. The map $y^* \mapsto \Lambda^* y^* \doteq y^* \circ \Lambda$ is a bounded linear operator from Y^* into X^*, which we call the **adjoint** of Λ. By definition (see Figure 4.4.1),

$$\langle \Lambda^* y^*, x\rangle = \langle y^*, \Lambda x\rangle \quad \text{for all } x \in X.$$

In the following, given a subset $V \subseteq X$, we define its orthogonal set as

$$V^\perp \doteq \{x^* \in X^*;\ \langle x^*, x\rangle = 0 \text{ for all } x \in V\}.$$

Similarly, if $W \subseteq X^*$, we define

$$W^\perp \doteq \{x \in X;\ \langle x^*, x\rangle = 0 \text{ for all } x^* \in W\}.$$

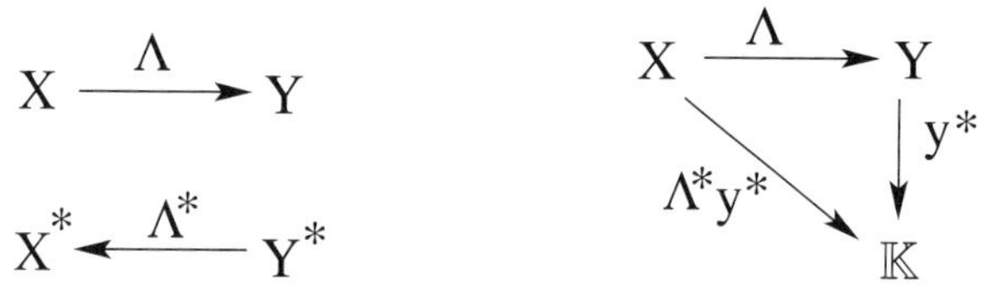

Figure 4.4.1. The maps involved in the definition of adjoint operators.

Theorem 4.8 (Properties of adjoint operators). *Let $\Lambda : X \mapsto Y$ be a bounded linear operator, and let $\Lambda^* : Y^* \mapsto X^*$ be its adjoint operator. Then:*

(i) $\|\Lambda^*\| = \|\Lambda\|$.

(ii) $\mathrm{Ker}(\Lambda) = [\mathrm{Range}(\Lambda^*)]^\perp$ *and* $\mathrm{Ker}(\Lambda^*) = [\mathrm{Range}(\Lambda)]^\perp$.

Proof. The statement (i) follows from

$$\begin{aligned}
\|\Lambda\| &= \sup\Big\{\|\Lambda x\|\,;\ \|x\|\leq 1\Big\}\\
&= \sup\Big\{|\langle y^*,\,\Lambda x\rangle|\,;\ \|x\|\leq 1\,,\ \|y^*\|\leq 1\Big\}\\
&= \sup\Big\{|\langle \Lambda^* y^*,\,x\rangle|\,;\ \|x\|\leq 1\,,\ \|y^*\|\leq 1\Big\}\\
&= \sup\Big\{\|\Lambda^* y^*\|\,;\ \|y^*\|\leq 1\Big\} \;=\; \|\Lambda^*\|\,.
\end{aligned}$$

To prove (ii), we observe that the following statements are all equivalent:

$$\begin{aligned}
x &\in \mathrm{Ker}(\Lambda),\\
\Lambda x &= 0,\\
\langle y^*,\,\Lambda x\rangle &= 0 \text{ for all } y^*\in Y^*,\\
\langle \Lambda^* y^*,\,x\rangle &= 0 \text{ for all } y^*\in Y^*,\\
x &\in [\mathrm{Range}(\Lambda^*)]^{\perp}.
\end{aligned}$$

Similarly, the following statements are all equivalent:

$$\begin{aligned}
y^* &\in \mathrm{Ker}(\Lambda^*),\\
\Lambda^* y^* &= 0,\\
\langle \Lambda^* y^*,\,x\rangle &= 0 \text{ for all } x\in X,\\
\langle y^*,\,\Lambda x\rangle &= 0 \text{ for all } x\in X,\\
y^* &\in [\mathrm{Range}(\Lambda)]^{\perp}.
\end{aligned}$$

□

We conclude this section by showing a useful application of the uniform boundedness principle.

Corollary 4.9 (Weakly convergent sequences are bounded). *Let X be a Banach space. Any sequence $x_n \in X$ which converges weakly to some $x \in X$ is necessarily bounded.*

Proof. Let X^* be the dual space of X. Each x_n determines a linear functional ψ_n on X^*, namely

$$x^* \;\mapsto\; \psi_n(x^*) \;\doteq\; \langle x^*, x_n\rangle.$$

Here $\langle\cdot,\cdot\rangle$ denotes the duality between X and X^*. By assumption, as $n\to\infty$ we have the pointwise convergence

$$\psi_n(x^*) \;\doteq\; \langle x^*, x_n\rangle \;\to\; \langle x^*, x\rangle$$

for each $x^* \in X^*$. This implies

$$\sup_n |\psi_n(x^*)| < \infty \quad \text{for every } x^* \in X^*.$$

Using the uniform boundedness principle, we conclude that the family of linear functionals $\{\psi_n \,;\, n \geq 1\}$ is uniformly bounded. Since

$$\|\psi_n\| = \sup_{\|x^*\| \leq 1} |\psi_n(x^*)| = \|x_n\|,$$

this completes the proof. □

4.5. Compact operators

Let X, Y be Banach spaces. A bounded linear operator $\Lambda : X \mapsto Y$ is **compact** if, for every bounded sequence $(x_n)_{n\geq 1}$ of points in X, there exists a subsequence $(x_{n_j})_{j\geq 1}$ such that Λx_{n_j} converges. Equivalently, Λ is compact if and only if, for any bounded set $U \subset X$, the image $\Lambda(U) \subset Y$ has compact closure.

Theorem 4.10 (**Examples of compact operators**).

(i) *Let X, Y be Banach spaces, and let $\Lambda : X \mapsto Y$ be a bounded linear operator. If the range of Λ is finite-dimensional, then Λ is compact.*

(ii) *Let $\Lambda_n : X \mapsto Y$ be a compact operator, for each $n \geq 1$. Assume $\lim_{n\to\infty} \|\Lambda_n - \Lambda\| = 0$. Then the operator Λ is compact as well.*

Proof. 1. A bounded linear operator $\Lambda : X \mapsto Y$ is compact if and only the unit ball $B_1 \subset X$ has image $\Lambda(B_1) \subset Y$ whose closure is compact. If Range(Λ) is finite-dimensional and Λ is bounded, then the closure $\overline{\Lambda(B_1)}$ is a closed bounded subset of a finite-dimensional space, hence compact. This proves (i).

2. To prove (ii) we observe that, since Y is complete, the closure $\overline{\Lambda(B_1)}$ is compact if and only if $\Lambda(B_1)$ is precompact. This means: for every $\varepsilon > 0$, the set $\Lambda(B_1)$ can be covered by finitely many balls of radius ε.

Under the assumptions (ii), let $\varepsilon > 0$ be given. Choose k such that $\|\Lambda - \Lambda_k\| < \varepsilon/2$. Since Λ_k is compact, we can select finitely many elements $y_1, \ldots, y_N \in Y$ such that

$$\Lambda_k(B_1) \subseteq \bigcup_{i=1}^{N} B\Big(y_i, \frac{\varepsilon}{2}\Big). \tag{4.7}$$

If $\|x\| \leq 1$, then $\|\Lambda x - \Lambda_k x\| < \varepsilon/2$. By (4.7) there exists a point y_i with $\|\Lambda_k x - y_i\| < \varepsilon/2$. By the triangle inequality, $\|\Lambda x - y_i\| < \varepsilon$. This proves that the finitely many balls $B(y_i, \varepsilon)$ cover $\Lambda(B_1)$. □

Theorem 4.11 (Adjoint of a compact operator). *Let X, Y be Banach spaces, and let $\Lambda : X \mapsto Y$ be a bounded linear operator. Then Λ is compact if and only if its adjoint $\Lambda^* : Y^* \mapsto X^*$ is compact.*

Proof. 1. Assume that Λ is compact. Let $(y_n^*)_{n\geq 1}$ be a sequence in Y^*, with $\|y_n^*\| \leq 1$ for every n. To prove that Λ^* is compact we need to show that the sequence $(\Lambda^* y_n^*)_{n\geq 1}$ admits a convergent subsequence.

Let $B_1 \doteq \{x \in X\,;\ \|x\| \leq 1\}$ be the closed unit ball in X. By assumption, the image $\Lambda(B_1)$ has compact closure, which we denote by $E \doteq \overline{\Lambda(B_1)} \subset Y$.

2. By definition, each $y_n^* \in Y^*$ is a linear map from Y into the field $\mathbb{K}$. Let $f_n : E \mapsto \mathbb{K}$ be the restriction of y_n^* to the compact set E. We claim that the family of functions $\{f_n\,;\ n \geq 1\}$ satisfies the assumptions of Ascoli's theorem. Indeed, all these functions are uniformly Lipschitz continous, because

$$|f_n(y) - f_n(y')| \ \leq\ \|y_n^*\|\,\|y - y'\| \ \leq\ \|y - y'\| \quad \text{for all } y, y' \in E\,.$$

Moreover, observing that

$$\sup_{y\in E} \|y\| \ =\ \sup_{\|x\|\leq 1} \|\Lambda x\| \ =\ \|\Lambda\|\,,$$

we obtain

$$|f_n(y)| \ \leq\ \|y_n^*\|\,\|y\| \ \leq\ 1\cdot\|\Lambda\| \quad \text{for all } y \in E\,.$$

Hence all functions $f_n : E \mapsto \mathbb{K}$ are uniformly bounded.

By Theorem 3.12, there exists a subsequence $(f_{n_j})_{j\geq 1}$ which converges to a function f uniformly on the compact set $E = \overline{\Lambda(B_1)}$.

3. We now observe that

$$\begin{aligned}\|\Lambda^* y_{n_i}^* - \Lambda^* y_{n_j}^*\| &= \sup_{\|x\|\leq 1} |\langle \Lambda^* y_{n_i}^* - \Lambda^* y_{n_j}^*\,,\ x\rangle| \\ &= \sup_{\|x\|\leq 1} |\langle y_{n_i}^* - y_{n_j}^*\,,\ \Lambda x\rangle| \ =\ \sup_{\|x\|\leq 1} |f_{n_i}(\Lambda x) - f_{n_j}(\Lambda x)|\,,\end{aligned}$$

where the right-hand side approaches zero as $i, j \to \infty$. This shows that the subsequence $(\Lambda^* y_{n_j}^*)_{j\geq 1}$ is Cauchy, hence it converges to a limit $x^* \in X^*$. Therefore Λ^* is compact.

The converse implication can be proved by the same arguments. □

4.5.1. Integral operators. Compact operators often arise in the form of integral operators. To see an example, consider the Banach space $\mathcal{C}([a,b])$ of all continuous, real-valued functions defined on the closed interval $[a,b]$.

Theorem 4.12 (Compactness of an integral operator). *Let $K : [a,b] \times [a,b] \mapsto \mathbb{R}$ be a continuous map. Then the integral operator*

$$(\Lambda f)(x) \doteq \int_a^b K(x,y)\, f(y)\, dy \tag{4.8}$$

is a compact linear operator from $\mathcal{C}([a,b])$ into itself.

Proof. Consider a bounded sequence of continuous functions $f_n \in \mathcal{C}([a,b])$. We need to prove that the sequence Λf_n admits a uniformly convergent subsequence. By Ascoli's compactness theorem, it suffices to show that the functions Λf_n are uniformly bounded and equicontinuous.

1. Since K is continuous on the compact set $[a,b] \times [a,b]$, it is bounded and uniformly continuous. Namely, there exists a constant κ such that

$$|K(x,y)| \leq \kappa \qquad \text{for all } x,y.$$

Moreover, for every $\varepsilon > 0$ there exists $\delta > 0$ such that

$$|K(x,y) - K(\tilde{x},y)| \leq \varepsilon \qquad \text{whenever } |x - \tilde{x}| \leq \delta\,, \quad x, \tilde{x}, y \in [a,b]\,. \tag{4.9}$$

2. By assumption, there exists a constant M such that

$$\|f_n\| = \max_{x \in [a,b]} |f_n(x)| \leq M \qquad \text{for all } n \geq 1\,.$$

This implies

$$\left|(\Lambda f_n)(x)\right| \leq \int_a^b |K(x,y)|\,|f_n(y)|\, dy \leq (b-a)\kappa M\,,$$

proving that the functions Λf_n are uniformly bounded.

3. Next, let $\varepsilon > 0$ be given. Choose $\delta > 0$ such that (4.9) holds. If $|x - \tilde{x}| \leq \delta$, then for any $n \geq 1$ we have

$$\left|(\Lambda f_n)(x) - (\Lambda f_n)(\tilde{x})\right| \leq \int_a^b |K(x,y) - K(\tilde{x},y)|\; |f_n(y)|\, dy \leq (b-a)\varepsilon\, M\,.$$

Since $\varepsilon > 0$ was arbitrary, this proves the equicontinuity of the sequence $(\Lambda f_n)_{n \geq 1}$. An application of Ascoli's theorem completes the proof. □

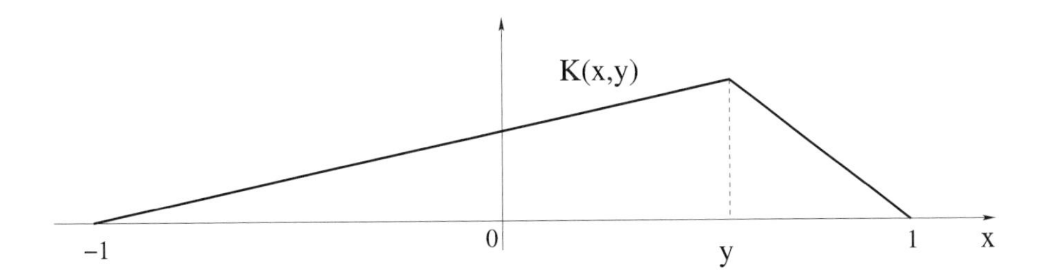

Figure 4.5.1. The kernel function K in (4.11).

Example 4.13. For any continuous function $f \in \mathcal{C}([-1,1])$, let $u = \Lambda f$ be the solution to the two-point boundary value problem

$$u''(x) + f(x) = 0, \qquad u(-1) = u(1) = 0. \tag{4.10}$$

Observe that the solution must be unique. Indeed, if u_1, u_2 are solutions, then the function $w = u_1 - u_2$ satisfies

$$w''(x) = 0, \qquad w(-1) = w(1) = 0,$$

hence $w(x) \equiv 0$. A direct computation shows that

$$u(x) = \int_{-1}^{x} \frac{(1+y)(1-x)}{2} f(y)\,dy + \int_{x}^{1} \frac{(1-y)(1+x)}{2} f(y)\,dy$$

provides a solution to (4.10). The solution operator $\Lambda : f \mapsto u = \Lambda f$ is thus a linear, compact operator on $\mathcal{C}([-1,1])$. It can be written in the form (4.9), with

$$K(x,y) \doteq \begin{cases} \frac{(1-y)(1+x)}{2} & \text{if } -1 \le x \le y, \\[2mm] \frac{(1+y)(1-x)}{2} & \text{if } y \le x \le 1. \end{cases} \tag{4.11}$$

Referring to Figure 4.5.1, the piecewise affine map $x \mapsto K(x,y)$ for a fixed y can be uniquely determined by the three equations

$$K(-1,y) = K(1,y) = 0, \qquad K_x(y+,\, y) - K_x(y-,\, y) = -1.$$

4.6. Problems

1. On the Banach space $X = \mathcal{C}([0,1])$, decide whether the following operators are (i) linear, (ii) bounded, (iii) compact:

(1) $(\Lambda f)(x) = f(\sin x)$.

(2) $(\Lambda f)(x) = \sin(f(x))$.

(3) $(\Lambda f)(x) = x f(x)$.

(4) $(\Lambda f)(x) = x\,f(0) + \int_0^1 f(s)\,ds.$

(5) $(\Lambda f)(x) = y(x)$, where $y(\cdot)$ is the solution to the Cauchy problem

$$y'(x) + y(x) = f(x), \qquad y(0) = 0.$$

2. Prove that $\mathbf{L}^2([0,1])$ is a vector subspace of $\mathbf{L}^1([0,1])$ of first category. Indeed, for each $n \geq 1$, the set $E_n \doteq \{f : [0,1] \mapsto \mathbb{R};\ \int_0^1 |f|^2\,dx \leq n\}$ is a closed subset of $\mathbf{L}^1$ with empty interior.

3. Let X be a Banach space and let $\Lambda : X \mapsto \ell^\infty$ be a linear operator, so that $\Lambda(x) = (\Lambda_1(x), \Lambda_2(x), \ldots)$ is a bounded sequence of real numbers, for every $x \in X$. Prove that the operator Λ is bounded if and only if each linear functional Λ_n is bounded.

4. Let $1 \leq p \leq \infty$. Consider a linear operator $\Lambda : \mathbf{L}^p([0,1]) \mapsto \mathbf{L}^p([0,1])$ (defined on the entire space $\mathbf{L}^p$) which has the following property. If a sequence of functions $f_n \in \mathbf{L}^p$ converges pointwise a.e. to some $f \in \mathbf{L}^p$, then the sequence $(\Lambda f_n)(x)$ converges to $(\Lambda f)(x)$ for a.e. $x \in [0,1]$. Prove that Λ is continuous.

5. Let X be an infinite-dimensional Banach space. Let $K : X \mapsto X$ be a compact linear operator. Show that, if K is one-to-one, then the range of K cannot be closed.

6. Let X, Y, Z be Banach spaces and consider two linear operators $\Lambda_1 : X \mapsto Y$, $\Lambda_2 : Y \mapsto Z$. Assume that one of the two operators is continuous while the other is compact. Prove that the composition $\Lambda_2 \circ \Lambda_1 : X \mapsto Z$ is a compact linear operator.

7. Let X be a Banach space and let $\Lambda : X \mapsto X$ be a compact linear operator such that $\Lambda = \Lambda^2$. Namely, $\Lambda(x) = \Lambda(\Lambda(x))$ for all $x \in X$. Prove that Range(Λ) is finite-dimensional.

8. Let X be a Banach space. Let $U \subset X$ be a closed, convex set such that $\bigcup_{n\geq 1} nU = X$. Prove that U contains a neighborhood of the origin.

9. Let X, Y be infinite-dimensional Banach spaces. If $K : X \mapsto Y$ is a compact linear operator, prove that $K(X) \neq Y$, i.e., K cannot be surjective.

As an example, consider the map $K : \ell^1 \mapsto \ell^1$ defined as follows. If $\mathbf{x} = (x_1, x_2, x_3, \ldots)$ with $\|\mathbf{x}\|_{\ell^1} \doteq \sum_{n\geq 1} |x_n| < \infty$, then $K\mathbf{x} = \left(\frac{x_1}{1}, \frac{x_2}{2}, \frac{x_3}{3}, \ldots, \frac{x_n}{n}, \ldots\right)$. Find a point $\mathbf{y} \in \ell^1 \setminus K(\ell^1)$.

10. Let $f : \mathbb{R} \mapsto \mathbb{R}$ be a bounded function with closed graph. Prove that f is continuous.

On the other hand, construct a function $g : \mathbb{R} \mapsto \mathbb{R}$ which is one-to-one and onto, has closed graph, but is not continuous.

11. Let X, Y be Banach spaces. Prove that every continuous function $f : X \mapsto Y$ (not necessarily linear) has closed graph.

Construct a bounded function $g : \mathbb{R} \mapsto \mathbf{L}^1(\mathbb{R})$ which has closed graph but is not continuous.

12. Let $1 \leq p < \infty$, and consider the Banach space ℓ^p of all sequences $\mathbf{x} = (x_1, x_2, x_3, \ldots)$ of real numbers such that $\|\mathbf{x}\|_p \doteq \left(\sum_{k\geq 1} |x_k|^p\right)^{1/p} < \infty$. Let $(\lambda_1, \lambda_2, \lambda_3, \ldots)$ be a bounded sequence of real numbers, and define the bounded linear operator $\Lambda : \ell^p \mapsto \ell^p$ by setting

$$\Lambda(x_1, x_2, x_3, \ldots) \doteq (\lambda_1 x_1, \lambda_2 x_2, \lambda_3 x_3, \ldots).$$

Prove that Λ is compact if and only if $\lim_{k\to\infty} \lambda_k = 0$.

13. Let X, Y, Z be Banach spaces. Let $B : X \times Y \mapsto Z$ be a bilinear map.[1] Assume that B is continuous at the origin. Prove that B is bounded; namely there exists a constant C such that

$$\|B(x, y)\| \leq C\,\|x\|\,\|y\| \qquad \text{for all } x \in X\,,\ y \in Y.$$

14. Let X be the vector space of all polynomials in one real variable, with norm

$$\|p\| \doteq \int_0^1 |p(t)|\,dt\,.$$

Consider the bilinear functional $B : X \times X \mapsto \mathbb{R}$ defined as

$$B(p, q) = \int_0^1 p(t)q(t)\,dt.$$

Show that, for each fixed $p \in X$, the map $q \mapsto B(p, q)$ is a continuous linear functional on X. Similarly, for $q \in X$ fixed, the map $p \mapsto B(p, q)$ is a continuous linear functional. However, prove that B is not continuous from the product space $X \times X$ into $\mathbb{R}$. We recall that $X \times X$ has norm $\|(p, q)\| \doteq \max\Big\{ \int_0^1 |p(x)|dx\,,\ \int_0^1 |q(x)|dx \Big\}$.

15. Let S be a closed bounded subset of a Banach space X. Assume that for every $\varepsilon > 0$ there exists a finite-dimensional subspace $Y_\varepsilon \subseteq X$ such that

$$d(S, Y_\varepsilon) \doteq \sup_{x\in S} d(x, Y_\varepsilon) \leq \varepsilon.$$

Prove that the set S is compact.

16. Let A be a bounded linear operator on the Banach space X. Assuming that $AK = KA$ for every compact operator K, prove that A is a scalar multiple of the identity, i.e., there exists a number λ such that $A = \lambda I$.

[1] Saying that B is bilinear means that $x \mapsto B(x, y)$ is a linear map for every given $y \in Y$, and $y \mapsto B(x, y)$ is a linear map for every fixed $x \in X$.

17. Let X be a Banach space and let $\Lambda : X \mapsto X$ be a bounded linear operator. Assume that there exist $\varepsilon > 0$ and an infinite-dimensional subspace $V \subseteq X$ such that $\|\Lambda x\| \geq \varepsilon \|x\|$ for every $x \in V$. Prove that the operator Λ cannot be compact.

18. Let X be a Banach space and let X^* be its dual space. Consider a sequence of elements $x_k \in X$ with the property that the series $\sum_{k\geq 1}\langle x^*, x_k\rangle$ converges for every $x^* \in X^*$. Prove that

$$x^* \mapsto \varphi(x^*) \doteq \sum_{k\geq 1}\langle x^*, x_k\rangle$$

is a bounded linear functional on X^*.

19. On the Banach space $X = \mathcal{C}([0,1])$ consider the linear operator $\Lambda : X \mapsto X$ defined by

$$(\Lambda f)(0) = f(0), \qquad (\Lambda f)(t) = \frac{1}{t}\int_0^t f(s)\,ds \quad \text{for } t > 0.$$

(i) Prove that Λ is continuous.

(ii) Prove that Λ is one-to-one but not onto.

(iii) Show that Λ is not compact.

20. Let $\Omega \subset \mathbb{R}^n$ be an open set and let $g : \Omega \mapsto \mathbb{R}$ be a bounded, measurable function. As in Example 2.18, for any $1 \leq p \leq \infty$, on the space $\mathbf{L}^p(\Omega)$ consider the multiplication operator $(M_g f)(x) \doteq g(x)\,f(x)$.

(i) Determine for which functions g the operator M_g is one-to-one.

(ii) Determine for which functions g the operator M_g has closed range.

(iii) Determine for which functions g the operator M_g is compact.

21. Find an example showing that, for an infinite-dimensional Banach space X, the set of bounded linear operators $\Lambda : X \mapsto X$ which are one-to-one may not be an open subset of the space $\mathcal{B}(X;\, X)$ of all bounded linear operators.

Similarly, show that the set of bounded linear operators whose range is dense may not be open in $\mathcal{B}(X;\, X)$.

22. Let X, Y be Banach spaces, and let $\Lambda : X \mapsto Y$ be a linear continuous bijection.

(i) Prove that there exists $\beta > 0$ such that

$$\|\Lambda x\| \geq \beta\,\|x\| \quad \text{for all } x \in X.$$

(ii) Let $\Psi \in \mathcal{B}(X;\, Y)$ be any bounded linear operator with norm $\|\Psi\| < \beta$. Using the contraction mapping theorem, prove that, for any $f \in Y$, the equation

$$u = \Lambda^{-1}(f - \Psi u)$$

has a unique solution.

(iii) Prove that, within the Banach space $\mathcal{B}(X;\, Y)$, the set of all bijective operators is open.

23. Recalling the spaces introduced in Examples 2.6 and 2.7 of Chapter 2, define the operator $\Lambda : \ell^\infty \mapsto \ell^\infty$ by setting $\Lambda x = y = (y_1, y_2, y_3, \ldots)$, where

$$y_n \doteq \frac{1}{n}\sum_{i=1}^{n} x_i\,. \tag{4.12}$$

(i) Prove that Λ is a bounded linear operator and compute its norm. Is Λ a compact operator?

(ii) Consider the operator Λ defined as in (4.12), but on the space ℓ^1 of absolutely summable sequences. Is Λ a bounded linear operator? Is it compact?

24. Let X, Y be vector spaces, with $Y \subseteq X$. Assume that there exist norms such that $(X\,;\ \|\cdot\|_X)$ and $(Y\,;\ \|\cdot\|_Y)$ are both Banach spaces, with

$$\|y\|_X \;\leq\; C\cdot\|y\|_Y \quad \text{for all } y \in Y\,.$$

(i) If $Y = X$, prove that the two norms $\|\cdot\|_X$ and $\|\cdot\|_Y$ are equivalent, namely $\|x\|_Y \;\leq\; C'\cdot\|x\|_X$ for some constant C'.

(ii) If $Y \neq X$, prove that Y is a subset of first category in X. Namely, each set $S_n \;\doteq\; \{y \in Y\,;\ \|y\|_Y \leq n\}$ is a closed, nowhere dense subset of X.

Chapter 5

Hilbert Spaces

The Euclidean space $\mathbb{R}^n$ is equipped with a natural inner product $\langle \cdot, \cdot \rangle$. This inner product is useful in many ways:

- It defines the Euclidean norm $\|\mathbf{x}\| \doteq \sqrt{\langle \mathbf{x}, \mathbf{x} \rangle}$.
- It determines perpendicular spaces and perpendicular projections, and it allows us to construct bases of mutually orthogonal vectors $\{\mathbf{v}_1, \ldots, \mathbf{v}_n\}$. Thanks to the inner product, computing the components of a vector $\mathbf{v} \in \mathbb{R}^n$ with respect to an orthogonal basis is an easy matter.
- Every linear functional $\varphi : \mathbb{R}^n \mapsto \mathbb{R}$ can be represented as an inner product: $\varphi(\mathbf{x}) = \langle \mathbf{w}, \mathbf{x} \rangle$ for a suitable vector $\mathbf{w} \in \mathbb{R}^n$.
- Having an inner product, one can define a class of symmetric operators, with many useful properties. We recall that $A : \mathbb{R}^n \mapsto \mathbb{R}^n$ is *symmetric* if $\langle A\mathbf{x}, \mathbf{y} \rangle = \langle \mathbf{x}, A\mathbf{y} \rangle$ for all $x, y \in \mathbb{R}^n$. In the standard basis of $\mathbb{R}^n$, symmetric operators correspond to symmetric $n \times n$ matrices.
- Always relying on the inner product, one can define a class of positive operators. We recall that $A : \mathbb{R}^n \mapsto \mathbb{R}^n$ is *strictly positive definite* if $\langle A\mathbf{x}, \mathbf{x} \rangle > 0$ for all $\mathbf{x} \in \mathbb{R}^n$, $\mathbf{x} \neq 0$. In this case, the map $\mathbf{x} \mapsto \langle A\mathbf{x}, \mathbf{x} \rangle$ is a positive definite quadratic form.

The goal of this chapter is to show how the definition and properties of the Euclidean inner product can be extended to infinite-dimensional vector spaces.

5.1. Spaces with an inner product

Let H be a vector space over the field $\mathbb{K}$ of real or complex numbers. An **inner product** on H is a map $(\cdot,\cdot)$ that, to each couple of elements $x, y \in H$, associates a number $(x, y) \in \mathbb{K}$ with the following properties. For every $x, y, z \in H$ and $\alpha \in \mathbb{K}$ one has

(i) $(x, y) = \overline{(y, x)}$, where the upper bar denotes complex conjugation;

(ii) $(x + y,\, z) = (x, z) + (y, z)$;

(iii) $(\alpha x, z) = \alpha(x, z)$;

(iv) $(x, x) \geq 0$, with equality holding if and only if $x = 0$.

Notice that the above properties also imply

$$(x,\, y + z) \;=\; (x, y) + (x, z), \qquad (x,\, \alpha y) \;=\; \bar{\alpha}(x, y). \tag{5.1}$$

In the case where $\mathbb{K} = \mathbb{R}$, the properties (i)–(iii) simply say that an inner product on a real vector space is a symmetric bilinear mapping. In connection with the above inner product, we also define

$$\|x\| \;\doteq\; \sqrt{(x, x)}\,. \tag{5.2}$$

The Minkowski inequality, proved below, shows that this is indeed a norm on the vector space H.

Theorem 5.1 (Two basic inequalities). *Let H be a vector space with inner product $(\cdot,\cdot)$. Then*

(i) $|(x, y)| \;\leq\; \|x\|\,\|y\|$ *(Cauchy-Schwarz inequality);*

(ii) $\|x + y\| \;\leq\; \|x\| + \|y\|$ *(Minkowski inequality).*

Proof. (i) If $y = 0$, the first inequality is trivial. To cover the general case, set

$$a \doteq (x, x), \qquad b \doteq (x, y), \qquad c \doteq (y, y).$$

For every scalar $\lambda \in \mathbb{K}$ one has

$$0 \;\leq\; (x + \lambda y,\, x + \lambda y) \;=\; a + b\bar{\lambda} + \bar{b}\lambda + c\lambda\bar{\lambda}.$$

Choosing $\lambda = -b/c$, we obtain

$$0 \;\leq\; a - \frac{b\bar{b}}{c}\,.$$

Since $c = (y, y) > 0$, multiplying both sides by c we obtain

$$0 \leq\; ac - |b|^2 \;=\; \|x\|^2\,\|y\|^2 - |(x, y)|^2,$$

proving (i).

(ii) By the Schwarz inequality,

$$\mathrm{Re}(x,y) \leq |(x,y)| \leq \|x\|\,\|y\|.$$

Therefore

$$\begin{aligned}\|x+y\|^2 &= (x+y,\, x+y) = \|x\|^2 + \|y\|^2 + 2\,\mathrm{Re}(x,y)\\ &\leq \|x\|^2 + \|y\|^2 + 2\|x\|\,\|y\| = \Big(\|x\| + \|y\|\Big)^2.\end{aligned}$$

Taking square roots we obtain (ii). □

A vector space H with an inner product $(\cdot,\cdot)$, which is complete with respect to the norm $\|x\| \doteq \sqrt{(x,x)}$, is called a **Hilbert space**.

Example 5.2. The Euclidean space $\mathbb{R}^n$ with inner product $(x,y) = x_1y_1 + \cdots + x_ny_n$ is a Hilbert space over the real numbers.

Example 5.3. The space ℓ^2 of all sequences of complex numbers $x = (x_1, x_2, \ldots)$ such that

$$\|x\| = \left(\sum_{k=1}^{\infty} |x_k|^2\right)^{1/2} < \infty$$

is a Hilbert space over the complex numbers, with inner product

$$(x,y) \doteq \sum_{k=1}^{\infty} x_k\,\overline{y_k}\,.$$

Example 5.4. Let $\Omega \subseteq \mathbb{R}^n$ be any open set. The space $\mathbf{L}^2(\Omega;\,\mathbb{R})$ of square summable maps $f : \Omega \mapsto \mathbb{R}$ is a Hilbert space, with

$$(f,g) \doteq \int_\Omega f(x)\,g(x)\,dx\,, \qquad \|f\|_{\mathbf{L}^2} \doteq \left(\int_\Omega |f(x)|^2\,dx\right)^{1/2}.$$

5.2. Orthogonal projections

Given a subset $S \subseteq H$, by $\mathrm{span}(S)$ we denote the set of all finite linear combinations of elements of S, namely

$$\mathrm{span}(S) \doteq \left\{\sum_{i=1}^{N} c_i x_i\,;\ N \geq 1,\ c_i \in \mathbb{K},\ x_i \in S\right\}. \tag{5.3}$$

In general, $\mathrm{span}(S)$ is a subspace of H, possibly not closed. The closure $V \doteq \overline{\mathrm{span}(S)}$ is called the **space generated by** S. We say that the set S

is **total** if it generates the whole space H. In other words, S is total if, for every $x \in H$, there exists a sequence of elements $x_n \in \text{span}(S)$ such that $\|x_n - x\| \to 0$ as $n \to \infty$.

Two elements x, y in a Hilbert space H are said to be **orthogonal** if $(x, y) = 0$.

Given any subset $S \subseteq H$, its **orthogonal subspace** is defined as

$$S^\perp \doteq \{y \in H\,;\ (y, x) = 0 \text{ for all } x \in S\}.$$

Notice that $S^\perp$ is always a closed subspace of H.

Theorem 5.5 (Perpendicular projections). *Let H be a Hilbert space and let $V \subseteq H$ be a closed subspace. Then*

(i) *$H = V \oplus V^\perp$, in the sense that each $x \in H$ can be uniquely written as $x = y + z$, where $y \in V$ and $z \in V^\perp$.*

(ii) *$y \doteq P_V(x)$ is the unique point in V having minimal distance from x, while $z \doteq P_{V^\perp}(x)$ is the unique point in $V^\perp$ having minimal distance from x.*

(iii) *The perpendicular projections $x \mapsto y \doteq P_V(x)$ and $x \mapsto z \doteq P_{V^\perp}(x)$ are linear continuous operators, with norm ≤ 1.*

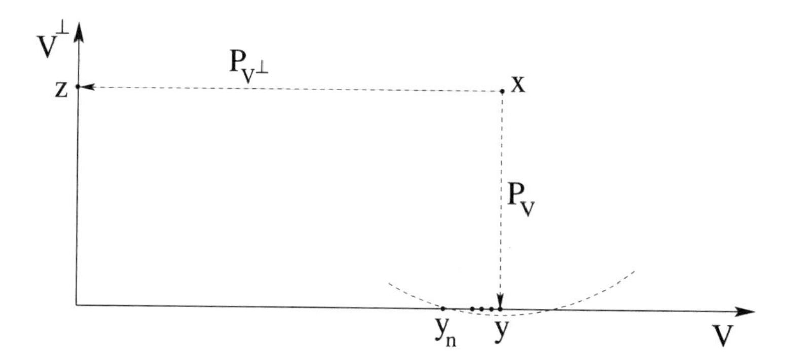

Figure 5.2.1. Constructing the perpendicular projections on the subspace V and on the orthogonal subspace $V^\perp$.

Proof. 1. Given $x \in H$, we begin by showing that there exists a unique point $y \in V$ having minimal distance from x. Let

$$\alpha \doteq d(x, V) \doteq \inf_{y \in V} \|x - y\|.$$

Then there exists a sequence of points $(y_n)_{n\geq 1}$ such that $\lim_{n\to\infty} \|x - y_n\| = \alpha$. We claim that $(y_n)_{n\geq 1}$ is a Cauchy sequence. Indeed, for any two points $u, v \in H$ one has

$$\|u + v\|^2 + \|u - v\|^2 = 2\|u\|^2 + 2\|v\|^2.$$

Applying this equality to $u = x - y_m$ and $v = x - y_n$ we obtain

$$\|y_m - y_n\|^2 \;=\; 2\,\|x - y_m\|^2 + 2\,\|x - y_n\|^2 - 4\left\|x - \frac{y_m + y_n}{2}\right\|^2. \tag{5.4}$$

Since $\frac{y_m+y_n}{2} \in V$, we have $\left\|x - \frac{y_m+y_n}{2}\right\|^2 \geq \alpha^2$. Therefore

$$\begin{aligned}\limsup_{m,n\to\infty} \|y_m - y_n\|^2 \;&\leq\; 2\,\limsup_{m\to\infty} \|x - y_m\|^2 + 2\,\limsup_{n\to\infty} \|x - y_n\|^2 \\ &\qquad -4\,\liminf_{m,n\to\infty} \left\|x - \tfrac{y_m+y_n}{2}\right\|^2 \\ &\leq\; 2\alpha^2 + 2\alpha^2 - 4\alpha^2 \;=\; 0,\end{aligned}$$

proving our claim.

2. Since V is closed (and hence complete), the sequence $(y_n)_{n\geq 1}$ converges to a unique limit y such that $\|x - y\| = d(x, V)$. We claim that this point is unique: if $\|x - y'\| = d(x, V)$ for some other point y', then the same argument used in (5.4) yields

$$\|y - y'\|^2 \;=\; 2\,\|x - y\|^2 + 2\,\|x - y'\|^2 - 4\left\|x - \frac{y + y'}{2}\right\|^2 \;\leq\; 2\alpha^2 + 2\alpha^2 - 4\alpha^2,$$

because $\frac{y+y'}{2} \in V$. Hence $y' = y$. This proves that the map $x \mapsto P_V(x)$ is well defined.

3. We now show that $P_V(x)$ can also be characterized as the unique point $y \in V$ such that

$$x - y \;\in\; V^\perp. \tag{5.5}$$

Choose any vector $v \in V$. By step **1**, for $\lambda \in \mathbb{R}$, the real-valued map

$$\lambda \;\mapsto\; \|x - (y + \lambda v)\|^2 \;=\; \|x - y\|^2 + |\lambda|^2\|v\|^2 + 2\,\mathrm{Re}(x - y,\, \lambda v)$$

attains its unique global minimum at $\lambda = 0$. Therefore, its derivative at $\lambda = 0$ must be zero. This already proves that $\mathrm{Re}(x - y,\, v) = 0$ for every $v \in V$. If H is a complex Hilbert space, we can replace v with $-iv$ and conclude that $\mathrm{Im}(x - y,\, v) = \mathrm{Re}(x - y,\, -iv) = 0$ as well. This proves (5.5).

4. Next, we prove that the point $y \in V$ such that $x - y \in V^\perp$ is unique. Indeed, if $y' \in V$ is another point such that $x - y' \in V^\perp$, then

$$\|y - y'\|^2 \;=\; (y - y',\, y - y') \;=\; (y - y',\, x - y') - (y - y',\, x - y) \;=\; 0$$

because $y - y' \in V$ while $x - y' \in V^\perp$ and $x - y \in V^\perp$.

5. Finally, we show that the perpendicular projection $P_V : H \mapsto V$ is a linear operator with norm $\|P_V\| \leq 1$.

If $y = P_V(x)$ and $y' = P_V(x')$, then for any scalars $\alpha, \alpha' \in \mathbb{K}$ one has

$$\alpha y + \alpha' y' \in V, \qquad \alpha x + \alpha' x' - (\alpha y + \alpha' y') \in V^{\perp}.$$

By step **3**, this suffices to conclude that $P_V(\alpha x + \alpha' x') = \alpha y + \alpha' y'$, proving that the projection operator P_V is linear. Similarly, the operator $P_{V^\perp} = I - P_V$ is linear.

Since the vectors $P_V(x)$ and $P_{V^\perp}(x) = x - P_V(x)$ are perpendicular, by Pythagoras' theorem we have

$$\|P_V(x)\|^2 + \|x - P_V(x)\|^2 = \|x\|^2.$$

This proves that $\|P_V\| \leq 1$ and $\|P_{V^\perp}\| \leq 1$. □

Remark 5.6. In the previous theorem, if $V \neq \{0\}$, then $\|P_V\| = 1$. If $V \neq H$, then $\|P_{V^\perp}\| = 1$.

5.3. Linear functionals on a Hilbert space

The next theorem shows that a Hilbert space can be identified with its dual. In other words, every element $x \in H$ determines a bounded linear functional $\phi^x : H \mapsto \mathbb{K}$ defined as $\phi^x(y) = (y, x)$. Conversely, every bounded linear functional on H is of the form $y \mapsto (y, x)$, for some $x \in H$.

Theorem 5.7 (Riesz representation of linear functionals). *Let H be a Hilbert space.*

(i) *For every $x \in H$, the map $y \mapsto (y, x)$ is a continuous linear functional on H.*

(ii) *Let $y \mapsto Ay$ be a continuous linear functional on H. Then there exists a unique element $\mathbf{a} \in H$ such that $Ay = (y, \mathbf{a})$ for every $y \in H$.*

Proof. (i) Let $x \in H$ be given. By the definition of inner product, the map ϕ^x defined as $\phi^x(y) \doteq (y, x)$ is linear. The boundedness of this linear map is a consequence of the Cauchy-Schwarz inequality:

$$\|\phi^x\| \doteq \sup_{\|y\| \leq 1} |(y, x)| \leq \sup_{\|y\| \leq 1} \|y\| \, \|x\| = \|x\| .$$

(ii) Conversely, let a linear continuous functional $y \mapsto Ay$ be given. If $Ay \equiv 0$ for all $y \in H$, then the conclusion clearly holds with $\mathbf{a} = \mathbf{0}$ (the zero vector in H).

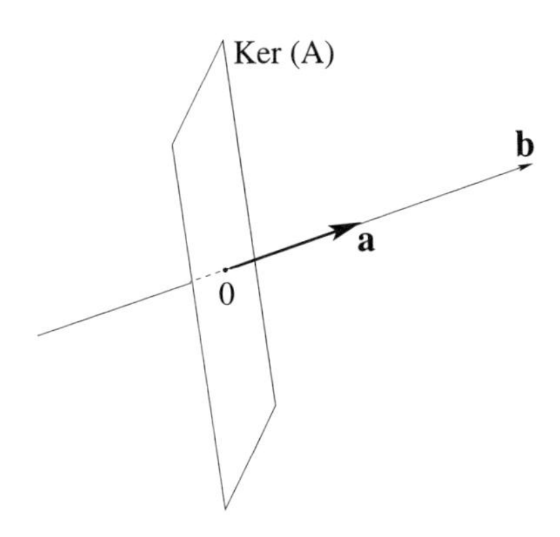

Figure 5.3.1. If $Ay = (y, \mathbf{a})$ for every y, the vector $\mathbf{a}$ must be perpendicular to $\mathrm{Ker}(A)$. Since $[\mathrm{Ker}(A)]^\perp$ is one-dimensional, given any nonzero vector $\mathbf{b} \in [\mathrm{Ker}(A)]^\perp$, we must have $\mathbf{a} = t\mathbf{b}$, where the number t is determined by the identity $A\mathbf{b} = (\mathbf{b},\, t\mathbf{b})$.

Otherwise, $V \doteq \mathrm{Ker}(A)$ is a closed hyperplane. The orthogonal complement $V^\perp$ is a subspace of dimension one. Choose any nonzero vector $\mathbf{b} \in V^\perp$ and define $\kappa = A\mathbf{b}/\|\mathbf{b}\|^2$, $\mathbf{a} \doteq \bar{\kappa}\mathbf{b}$. This choice yields

$$0 \neq A\mathbf{b} \;=\; \kappa(\mathbf{b}, \mathbf{b}) \;=\; (\mathbf{b}, \bar{\kappa}\mathbf{b}) \;=\; (\mathbf{b}, \mathbf{a}). \tag{5.6}$$

Given any vector $y \in H$, we can decompose y as a sum of a vector in V and a vector in $V^\perp$:

$$y \;=\; P_V(y) + \alpha\mathbf{b}$$

for some $\alpha \in \mathbb{K}$. By (5.6) it now follows that

$$\begin{aligned} Ay &= A(P_V(y)) + A(\alpha\mathbf{b}) = 0 + \alpha A\mathbf{b} = 0 + \alpha(\mathbf{b}, \mathbf{a}) \\ &= (P_V(y),\, \mathbf{a}) + (\alpha\mathbf{b}, \mathbf{a}) = (y, \mathbf{a}). \end{aligned}$$

□

Remark 5.8. Let H be a Hilbert space over the reals. By the previous theorem, the map $x \mapsto \phi^x$ is an isometric isomorphism between H and its dual space H^* (= the space of all bounded linear functionals on H). We can thus identify the two spaces H and H^*.

If now $\Lambda : H \mapsto H$ is a bounded linear operator, its adjoint $\Lambda^* : H^* \mapsto H^*$ can be identified with an operator $\Lambda^* : H \mapsto H$. This adjoint is characterized by the identities

$$(x\,,\ \Lambda^* y) \;=\; (\Lambda x\,,\ y) \qquad \text{for all } \ x, y \in H\,.$$

5.4. Gram-Schmidt orthogonalization

Let H be a Hilbert space. We say that a vector $x \in H$ is **normalized** if $\|x\| = 1$. A subset $E \subset H$ is **orthonormal** if every vector in E has unit norm and any two vectors in E are orthogonal to each other.

Next, consider a set $S = \{v_1, v_2, \ldots, v_n\}$ of finitely many linearly independent vectors. Assume that $x \in \text{span}\{v_1, \ldots, v_n\}$, so that

$$x = \sum_{k=1}^{n} \theta_k v_k$$

for some coefficients θ_k. To actually compute these coefficients, we observe that, for every $j = 1, \ldots, n$, one must have

$$(x, v_j) = \sum_{k=1}^{n} \theta_k (v_k, v_j).$$

Therefore the numbers $\theta_1, \ldots, \theta_n$ are obtained by solving the system of n linear equations

$$\begin{pmatrix} (v_1, v_1) & \cdots & (v_n, v_1) \\ \vdots & \ddots & \vdots \\ (v_1, v_n) & \cdots & (v_n, v_n) \end{pmatrix} \begin{pmatrix} \theta_1 \\ \vdots \\ \theta_n \end{pmatrix} = \begin{pmatrix} (x, v_1) \\ \vdots \\ (x, v_n) \end{pmatrix}. \tag{5.7}$$

This system is much easier to solve when the matrix is diagonal, namely $(v_i, v_j) = 0$ for $i \neq j$. This happens precisely when the vectors $v_1, v_2, \ldots, v_n$ are orthogonal to each other. The explicit solution is then computed as

$$\theta_k = \frac{(x, v_k)}{(v_k, v_k)}.$$

In the special case where the set $\{v_1, \ldots, v_n\}$ is orthonormal, so that $(v_j, v_j) = 1$ for every $j = 1, \ldots, n$, the above formula simplifies further to

$$\theta_k = (x, v_k).$$

The previous analysis shows that, if we have an orthonormal basis at our disposal, computations become much easier. Given a (finite or countable) linearly independent set $S = \{v_1, v_2, \ldots\}$, we now describe a general procedure to construct an orthonormal set $\{e_1, e_2, \ldots\}$ in such a way that

$$E_n \doteq \text{span}\{e_1, \ldots, e_n\} = \text{span}\{v_1, \ldots, v_n\}$$

for every $n \geq 1$. This will be achieved by induction.

Gram-Schmidt orthogonalization algorithm:

(i) *Start by defining* $e_1 \doteq \frac{v_1}{\|v_1\|}$.

(ii) *If* $e_1, \ldots, e_{n-1}$ *have been constructed, let* $\tilde{v}_n$ *be the perpendicular projection of* v_n *on the subspace* $E_{n-1} = \text{span}\{v_1, \ldots, v_{n-1}\} = \text{span}\{e_1, \ldots, e_{n-1}\}$. *Then define*

$$e_n \doteq \frac{v_n - \tilde{v}_n}{\|v_n - \tilde{v}_n\|}. \tag{5.8}$$

Observe that $v_n \neq \tilde{v}_n$ because $v_n \notin \text{span}\{v_1, \ldots, v_{n-1}\}$. Hence e_n is well defined and has norm one. Moreover, e_n is perpendicular to all vectors $e_1, \ldots, e_{n-1}$.

Notice that the projection of v_n on the subspace E_{n-1} is computed by

$$\tilde{v}_n = \sum_{k=1}^{n-1} (v_n, e_k) e_k .$$

By (5.8), this yields the explicit formula

$$e_n = \frac{v_n - \sum_{k=1}^{n-1}(v_n, e_k)e_k}{\left\| v_n - \sum_{k=1}^{n-1}(v_n, e_k)e_k \right\|}.$$

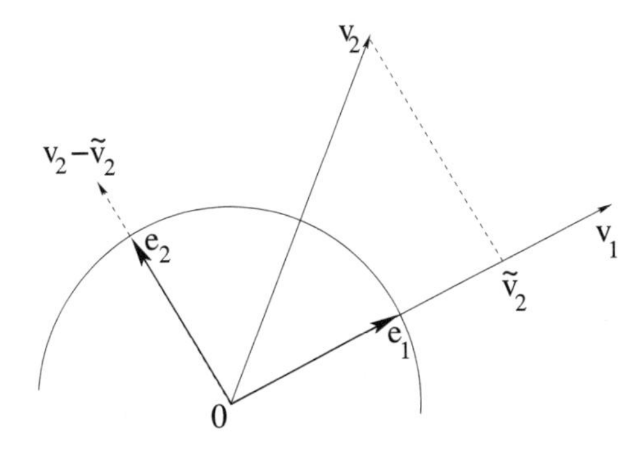

Figure 5.4.1. The Gram-Schmidt orthonormalization procedure, applied to the two vectors v_1, v_2.

5.5. Orthonormal sets

If $\{e_1, e_2, \ldots, e_n\}$ is an orthonormal basis of $\mathbb{R}^n$, then every vector $x \in \mathbb{R}^n$ can be uniquely written as a linear combination

$$x = \sum_{k=1}^{n} (x, e_k) e_k .$$

Notice that each term $(x, e_k)e_k$ represents the perpendicular projection of the vector x into the one-dimensional subspace spanned by e_k.

In an infinite-dimensional Hilbert space H, the above finite sum should be replaced by an infinite series. For many applications, it is important to understand in which cases the corresponding series converges and when we achieve the equality

$$x = \sum_{k\geq 1} (x, e_k) e_k \,. \tag{5.9}$$

Lemma 5.9 (Properties of the orthogonal subspace). *Let H be a Hilbert space. For any subset $S \subseteq H$, the orthogonal subspace $S^\perp$ is a closed subspace of H. Moreover, the following are equivalent:*

(i) $\mathrm{span}(S)$ *is dense in* H.

(ii) $S^\perp = \{0\}$.

Proof. 1. The fact that $S^\perp$ is a subspace of H is clear. Now assume that $x_n \in S^\perp$ and $x_n \to x$ as $n \to \infty$. Then for every $a \in S$ we have

$$(x, a) = \lim_{n\to\infty} (x_n, a) = 0.$$

Hence $x \in S^\perp$ as well, showing that this subspace is closed.

2. To prove the implication (i) $\Longrightarrow$ (ii), assume that $\mathrm{span}(S)$ is dense in H and let $x \in S^\perp$, so that $(x, a) = 0$ for every $a \in S$. Then there exists a sequence of linear combinations, say

$$x_n = \sum_{k=1}^{N_n} \theta_{n,k} a_{n,k}$$

with $a_{n,k} \in S$, such that $x_n \to x$ as $n \to \infty$. This implies

$$\begin{aligned}(x, x) &= \lim_{n\to\infty} (x, x_n) = \lim_{n\to\infty} \left(x\,, \sum_{k=1}^{N_n} \theta_{n,k} a_{n,k} \right) \\ &= \lim_{n\to\infty} \sum_{k=1}^{N_n} \theta_{n,k} \, (x, a_{n,k}) = 0\,,\end{aligned}$$

showing that $x = 0$.

3. To prove the implication (ii) $\Longrightarrow$ (i), let V be the closure of $\mathrm{span}(S)$. If (i) fails, then $V \neq H$ and there exists an element $y \notin V$. Consider the perpendicular projection $P_V(y)$. Then $w \doteq y - P_V(y)$ is a nonzero vector perpendicular to V. Since $w \in S^\perp$, this is a contradiction with (ii). □

Given an orthonormal sequence, the next theorem provides the convergence of the series (5.9) and characterizes its sum.

Theorem 5.10 (Sum of an orthogonal series). *Let $S = \{e_1, e_2, \ldots\}$ be a (finite or countable) orthonormal set in a Hilbert space H. Let V be the closed subspace generated by S and call $P_V : H \mapsto V$ the perpendicular projection. Then, for every $x \in H$, one has the Bessel inequality*

$$\sum_{k\geq 1} |(x, e_k)|^2 \;=\; \|P_V x\|^2 \;\leq\; \|x\|^2. \tag{5.10}$$

Moreover,

$$\sum_{k\geq 1} (x, e_k) e_k \;=\; P_V x\,. \tag{5.11}$$

Proof. 1. For any $n \geq 1$, call $V_n \doteq \operatorname{span}\{e_1, \ldots, e_n\}$. Then

$$P_{V_n} x \;=\; \sum_{k=1}^{n} (x, e_k) e_k\,,$$

and hence, by orthonormality,

$$\|P_{V_n} x\|^2 \;=\; \Big(\sum_{j=1}^{n} (x, e_j) e_j\,,\, \sum_{k=1}^{n} (x, e_k) e_k\Big) \;=\; \sum_{k,j=1}^{n} (x, e_j)\overline{(x, e_k)}(e_j, e_k)$$

$$=\; \sum_{k=1}^{n} |(x, e_k)|^2.$$

Since $\|P_{V_n} x\| \leq \|x\|$ for every n, the inequality in (5.10) is proved.

2. If the set S is finite, the identity (5.11) is clear. To cover the case where S is countable, we first show that the sequence of partial sums

$$x_n \;\doteq\; \sum_{k=1}^{n} (x, e_k) e_k$$

is Cauchy. Indeed, all terms in the series (5.11) are orthogonal to each other. For $m < n$, using Pythagoras' theorem and (5.10) we obtain

$$\|x_n - x_m\|^2 \;=\; \sum_{k=m+1}^{n} |(x, e_k)|^2 \;\to\; 0 \qquad \text{as } m, n \to \infty.$$

Since H is complete, we thus have the convergence $x_n \to \tilde{x}$ for some $\tilde{x}$ which provides the sum of the series in (5.11).

3. To complete the proof, we need to show that $\tilde{x} = P_V x$.

Since $x_n \in V$ for every $n \geq 1$ and $x_n \to \tilde{x}$, it is clear that $\tilde{x}$ lies in the closed subspace V. Moreover,

$$(x - \tilde{x}\,,\, e_k) \;=\; \lim_{n\to\infty} (x - x_n, e_k) = 0$$

because $(x - x_n, e_k) = 0$ as soon as $n \geq k$. This proves that $x - \tilde{x}$ is perpendicular to all vectors e_k, and hence to every linear combination of these vectors. Therefore, $x - \tilde{x}$ is perpendicular to every vector $v \in V$. The two properties $\tilde{x} \in V$ and $x - \tilde{x} \in V^\perp$ together imply $\tilde{x} = P_V x$. □

We say that an orthonormal set $S = \{e_1, e_2, \ldots\} \subset H$ is an **orthonormal basis** if span(S) is dense in H. In this case, the closed subspace generated by S is $V = H$. Hence $P_V x = x$ for every $x \in H$, and (5.10)–(5.11) yield

$$\sum_{k \geq 1} |(x, e_k)|^2 = \|x\|^2, \qquad \sum_{k \geq 1} (x, e_k) e_k = x. \tag{5.12}$$

5.5.1. Fourier series. As an application of the previous theory, consider the Hilbert space of complex-valued functions $\mathbf{L}^2([-\pi, \pi]\,;\ \mathbb{C})$, with inner product

$$(f, g) \doteq \int_{-\pi}^{\pi} f(x)\,\overline{g(x)}\,dx. \tag{5.13}$$

Within this space, the set of functions

$$\varphi_n(x) \doteq \frac{e^{inx}}{\sqrt{2\pi}}, \qquad n \in \mathbb{Z},$$

is orthonormal. Indeed,

$$\int_{-\pi}^{\pi} e^{imx}\,\overline{e^{inx}}\,dx = \int_{-\pi}^{\pi} e^{i(m-n)x}\,dx = \begin{cases} 0 & \text{if } m \neq n, \\ 2\pi & \text{if } m = n. \end{cases}$$

We claim that the countable set $S \doteq \{e^{inx}\,;\ n \in \mathbb{Z}\}$ is an orthonormal basis of $\mathbf{L}^2([-\pi,\ \pi])$.

To show that S is dense, consider any function $f \in \mathbf{L}^2([-\pi, \pi]\,;\ \mathbb{C})$. For any $\varepsilon > 0$ we can find a continuous function $f_\varepsilon : [-\pi, \pi] \mapsto \mathbb{C}$ such that

$$\|f_\varepsilon - f\|_{\mathbf{L}^2([-\pi,\pi])} < \varepsilon, \qquad f_\varepsilon(-\pi) = f_\varepsilon(\pi). \tag{5.14}$$

In turn, as shown by Example 3.8 in Chapter 3, we can find a complex trigonometric polynomial of the form

$$p(x) = \sum_{k=-N}^{N} \alpha_k e^{ikx}$$

such that

$$\|f_\varepsilon - p\|_{\mathcal{C}^0([-\pi,\ \pi])} \doteq \max_{x \in [-\pi,\pi]} |f_\varepsilon(x) - p(x)| < \varepsilon. \tag{5.15}$$

Observing that

$$\|f_\varepsilon - p\|_{\mathbf{L}^2} = \left(\int_{-\pi}^{\pi} |f_\varepsilon(x) - p(x)|^2\,dx \right)^{1/2} \leq \sqrt{2\pi}\,\|f_\varepsilon - p\|_{\mathcal{C}^0},$$

it is clear that f_ε can be approximated by trigonometric polynomials $p(\cdot)$ also with respect to the $\mathbf{L}^2$ norm. Using (5.14) together with (5.15) we conclude that $\overline{\mathrm{span}(S)} = \mathbf{L}^2([-\pi,\,\pi])$.

Now consider the complex trigonometric series

$$\sum_{k=-\infty}^{\infty} a_k \frac{e^{ikx}}{\sqrt{2\pi}}, \qquad a_k \doteq (f, \varphi_k) = \int_{-\pi}^{\pi} f(x)\, \frac{e^{-ikx}}{\sqrt{2\pi}}\, dx\,. \tag{5.16}$$

By the previous theorems, this series converges to f in $\mathbf{L}^2([-\pi,\,\pi])$, namely

$$\lim_{N\to\infty} \left\| f - \sum_{k=-N}^{N} a_k \varphi_k \right\|_{\mathbf{L}^2([-\pi,\,\pi])} = 0\,. \tag{5.17}$$

This result can be restated as

Corollary 5.11. *Let $f \in \mathbf{L}^2([-\pi,\pi]\,;\mathbb{C})$ be a complex-valued, square summable function. Defining the coefficients*

$$c_k \doteq \frac{1}{2\pi} \int_{-\pi}^{\pi} f(y)\, e^{-iky}\, dy\,,$$

one has the convergence

$$\lim_{N\to\infty} \int_{-\pi}^{\pi} \left| f(x) - \sum_{k=-N}^{N} c_k e^{ikx} \right|^2 dx = 0.$$

5.6. Positive definite operators

A basic problem of linear algebra is to solve the system of linear equations

$$A\mathbf{x} = \mathbf{b}\,,$$

where A is an $n \times n$ matrix and $\mathbf{b}$ is a vector in $\mathbb{R}^n$. One condition which guarantees the existence and uniqueness of solutions is that A be strictly positive definite. Indeed if the inner product satisfies $\langle A\mathbf{x}, \mathbf{x}\rangle > 0$ for every $\mathbf{x} \neq 0$, then A must have full rank. Hence the above system of linear equations will have a unique solution. In this section we show that this result remains valid also in an infinite-dimensional Hilbert space.

Let H be a Hilbert space over the real numbers. We say that a linear operator $A : H \mapsto H$ is **strictly positive definite**[1] if there exists $\beta > 0$ such that

$$(Au, u) \geq \beta\,\|u\|^2 \qquad \text{for all } \ u \in H. \tag{5.18}$$

[1] In the literature, operators satisfying the inequality (5.18) are usually called *strictly monotone*. Here we prefer to call them *strictly positive definite*, to stress the relationship with positive definite matrices in linear algebra.

Theorem 5.12 (Inverse of a positive definite operator). *Let H be a real Hilbert space. Let $A : H \mapsto H$ be a bounded linear operator which is strictly positive definite, so that (5.18) holds. Then, for every $f \in H$, there exists a unique $u \doteq A^{-1}f \in H$ such that*

$$Au = f. \tag{5.19}$$

The inverse operator A^{-1} satisfies

$$\|A^{-1}\| \leq \frac{1}{\beta}. \tag{5.20}$$

Proof. We need to show that, under the assumption (5.18), the continuous map A is one-to-one and onto.

1. From (5.18) it follows that

$$\beta\|u\|^2 \leq (Au, u) \leq \|Au\|\,\|u\|.$$

Hence

$$\beta\|u\| \leq \|Au\|. \tag{5.21}$$

If $Au = 0$, then $u = 0$, proving that $\mathrm{Ker}(A) = \{0\}$ and A is one-to-one.

2. Next, we claim that Range(A) is closed. Consider any sequence of points $v_n \in$ Range(A), such that $v_n \to v$. We need to show that $v = Au$ for some $u \in H$.

By assumption, $v_n = Au_n$ for some $u_n \in H$. Using (5.21) we obtain

$$\limsup_{m,n\to\infty} \|u_m - u_n\| \leq \limsup_{m,n\to\infty} \frac{1}{\beta}\|Au_m - Au_n\| = \limsup_{m,n\to\infty} \frac{1}{\beta}\|v_m - v_n\| = 0.$$

Hence the sequence $(u_n)_{n\geq 1}$ is Cauchy and converges to some limit $u \in H$. By continuity, $Au = v$, proving our claim.

3. We now claim that Range$(A) = H$. If not, since Range(A) is closed, we could find a nonzero vector $w \in [\mathrm{Range}(A)]^{\perp}$. But this would imply

$$\beta\|w\|^2 \leq (Aw, w) = 0,$$

reaching a contradiction.

4. By the previous steps, $A : H \mapsto H$ is a bijection, hence the equation (5.19) has a unique solution $u \doteq A^{-1}f$. By (5.21) it follows that

$$\|A^{-1}f\| = \|u\| \leq \frac{\|f\|}{\beta} \qquad \text{for all } f \in H,$$

proving (5.20). □

The previous result can also be conveniently formulated in terms of bilinear forms.

Theorem 5.13 (Lax-Milgram). *Let H be a Hilbert space over the reals and let $B : H \times H \mapsto \mathbb{R}$ be a continuous bilinear functional. This means that*

$$\begin{aligned} B[au + bu',\ v] &= aB[u,v] + bB[u',v], \\ B[u,\ av + bv'] &= aB[u,v] + bB[u,v'], \\ |B[u,v]| &\leq C\,\|u\|\,\|v\|, \end{aligned}$$

for some constant C and all $u, u', v, v' \in H$, $a, b \in \mathbb{R}$. In addition, assume that B is strictly positive definite, i.e., there exists a constant $\beta > 0$ such that

$$B[u,u] \geq \beta\,\|u\|^2 \qquad \text{for all } u \in H\,.$$

Then, for every $f \in H$, there exists a unique $u \in H$ such that

$$B[u,v] = (f,v) \qquad \text{for all } v \in H\,. \tag{5.22}$$

Moreover,

$$\|u\| \leq \beta^{-1}\|f\|\,. \tag{5.23}$$

Proof. For every fixed $u \in H$ the map $v \mapsto B[u,v]$ is a continuous linear functional on H. By the Riesz representation theorem, there exists a unique vector, which we call $Au \in H$, such that

$$B[u,v] = (Au,v) \qquad \text{for all } v \in H\,.$$

We claim that A is a bounded, positive definite linear operator.

The linearity of A is easy to check. To prove that A is bounded we observe that, for every $u \in H$,

$$\|Au\| \doteq \sup_{\|v\|=1} |(Au,v)| = \sup_{\|v\|=1} |B[u,v]| \leq C\,\|u\|\,.$$

Hence $\|A\| \leq C$.

Moreover,

$$(Au,u) \doteq B[u,u] \geq \beta\,\|u\|^2,$$

proving that A is strictly positive definite.

We can now apply Theorem 5.12 and conclude that the equation $Au = f$ has a unique solution $u = A^{-1}f$, satisfying $\|u\| \leq \beta^{-1}\|f\|$. By the definition of A, this provides a solution to (5.22). $\square$

5.7. Weak convergence

Let H be a Hilbert space. We say that a sequence of points $x_n \in H$ **converges weakly** to a point $x \in H$, and write $x_n \rightharpoonup x$, if

$$\lim_{n\to\infty} (y, x_n) \;=\; (y, x) \qquad \text{for every } y \in H. \tag{5.24}$$

If a weak limit exists, then it is necessarily unique. Indeed, assume that $x_n \rightharpoonup x$ and $x_n \rightharpoonup \tilde{x}$. Choosing $y = x - \tilde{x}$ in (5.24), one obtains

$$0 \;=\; \lim_{n\to\infty} (x-\tilde{x},\, x_n) - \lim_{n\to\infty} (x-\tilde{x},\, x_n) \;=\; (x-\tilde{x},\, x) - (x-\tilde{x},\, \tilde{x}) \;=\; \|x-\tilde{x}\|^2.$$

Hence $x = \tilde{x}$.

We recall that, if H is infinite-dimensional, then the closed unit ball $B_1 \doteq \{x \in H\,;\ \|x\| \le 1\}$ is not compact (with respect to the topology determined by the norm). In particular, B_1 contains an orthonormal sequence $\{e_1, e_2, \ldots\}$, which does not admit any convergent subsequence. On the other hand, replacing strong convergence by weak convergence, one can still prove a useful compactness property.

Theorem 5.14 (Weakly convergent sequences). *Let H be a Hilbert space.*

- (i) *Every weakly convergent sequence is bounded. Namely, if $x_n \rightharpoonup x$, then $\|x_n\| \le C$ for some constant C and all $n \ge 1$.*
- (ii) *Every bounded sequence of points $x_n \in H$ admits a weakly convergent subsequence: $x_{n_j} \rightharpoonup x$ for some $x \in H$.*

Proof. 1. Every x_n can be regarded as a continuous linear map, namely $y \mapsto (y, x_n)$ from H into $\mathbb{K}$ (the field of reals, or of complex numbers). The Hilbert norm of x_n coincides with the norm of x_n as a linear functional on H, namely

$$\|x_n\| \;=\; \sup_{\|y\|\le 1} |(y, x_n)|\,.$$

By assumption, for every $y \in H$, the set $\{(y, x_n)\,;\ n \ge 1\}$ is bounded. Hence by Remark 4.2 in Chapter 4 (the uniform boundedness principle), the countable set of linear functionals $\{x_n\,;\ n \ge 1\}$ is uniformly bounded. This establishes (i).

2. To prove (ii), consider the vector space $X \doteq \overline{\text{span}\{x_n\,;\ n \ge 1\}}$, i.e., the closure of the set of all linear combinations of the points $x_1, x_2, \ldots$. We observe that X is separable. Indeed, consider the set of all finite linear combinations

$$\sum_{i=1}^{N} \theta_i x_i$$

with $N \geq 1$ and $\theta_1, \ldots, \theta_N$ rational. This is a countable set dense in X.

Since $X \subseteq H$ is itself a Hilbert space, by Riesz's theorem each x_n can be identified with a bounded linear functional on X. By Theorem 2.34 in Chapter 2 (Banach-Alaoglu), the sequence $(x_n)_{n\geq 1}$ admits a weak-star convergent subsequence, say $(x_{n_j})_{j\geq 1}$. This means that

$$\lim_{j\to\infty} (y, x_{n_j}) = \varphi(y) \qquad \text{for all } y \in X,$$

for some bounded linear functional $\varphi : X \mapsto \mathbb{K}$.

3. By Riesz's representation theorem there exists a unique element $x \in X$ such that $\varphi(y) = (y, x)$ for all $y \in X$. We conclude the proof by showing that the subsequence x_{n_j} converges weakly to x in the entire Hilbert space H. Indeed, consider any $y \in H$. Denoting by $P_X : H \mapsto X$ the perpendicular projection, one has

$$\lim_{j\to\infty} (y, x_{n_j}) = \lim_{j\to\infty} (P_X y, x_{n_j}) = (y, x). \qquad \square$$

We conclude this section by showing that a compact operator maps weakly convergent sequences into strongly convergent ones.

Theorem 5.15. *In a Hilbert space H, consider a weakly convergent sequence $x_n \rightharpoonup x$. Let $\Lambda : H \mapsto H$ be a compact operator. Then one has the strong convergence*

$$\|\Lambda x_n - \Lambda x\| \to 0. \tag{5.25}$$

Proof. To prove (5.25), it suffices to show that, from any subsequence $(x_n)_{n\in I_1}$ one can extract a further subsequence $(x_n)_{n\in I_2}$, with $I_2 \subseteq I_1$, such that

$$\lim_{n\to\infty,\, n\in I_2} \|\Lambda x_n - \Lambda x\| = 0.$$

Let a subsequence $(x_n)_{n\in I_1}$ be given. Since this sequence is weakly convergent, by the previous theorem it is globally bounded, say $\|x_n\| \leq C$ for every $n \in I_1$.

Since Λ is compact, from this bounded sequence we can extract a further subsequence $(x_n)_{n\in I_2}$, with $I_2 \subseteq I_1$, such that the images converge strongly:

$$\lim_{n\to\infty,\, n\in I_2} \|\Lambda x_n - y\| = 0$$

for some $y \in H$. It remains to show that $y = \Lambda x$.

Calling Λ^* the adjoint operator, one has

$$(v\,,\ \Lambda x_n - \Lambda x) \;=\; (\Lambda^* v\,,\ x_n - x) \;\to\; 0 \qquad \text{for all } \ v \in H,$$

proving the weak convergence $\Lambda x_n \rightharpoonup \Lambda x$. Since the weak limit is unique, this implies $\Lambda x = y$, completing the proof. □

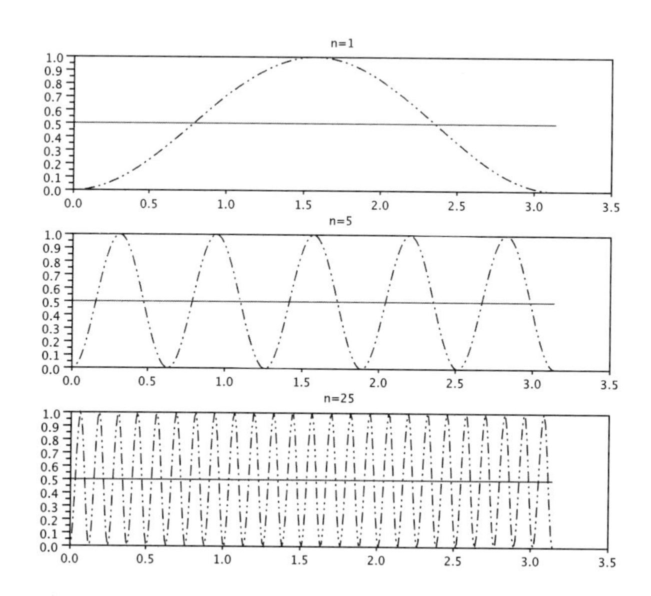

Figure 5.7.1. A plot of the functions $f_n(x) = \sin^2 nx$, for $n = 1, 5, 25$. As $n \to \infty$, the functions f_n do not converge pointwise, or in the $\mathbf{L}^2$ norm. However, we have the weak convergence $f_n \rightharpoonup f \equiv 1/2$, because the average value of f_n on every interval $[a, b]$ converges to the constant $1/2$.

Example 5.16. In the space $\mathbf{L}^2([0, \pi])$, consider the sequence of functions $f_n(x) = \sin^2 nx$ (see Figure 5.7.1). We claim that this sequence converges weakly to the constant function $f(x) \equiv 1/2$. Indeed, consider any $g \in \mathbf{L}^2([0, \pi])$. We need to prove that

$$\lim_{n\to\infty} \int_0^\pi g(x)\, \sin^2 nx\, dx \;=\; \int_0^\pi \frac{g(x)}{2}\, dx\,. \tag{5.26}$$

In the special case where

$$g(x) \;=\; \begin{cases} 1 & \text{if } x \in [0, b]\,, \\ 0 & \text{if } x \in \,]b, \pi]\,, \end{cases} \tag{5.27}$$

the result is clear, because

$$\lim_{n\to\infty} \int_0^b \sin^2 nx\, dx \;=\; \lim_{n\to\infty} \left(\frac{b}{2} - \frac{\sin 2nb}{4n} \right) \;=\; \frac{b}{2} \;=\; \int_0^\pi \frac{g(x)}{2}\, dx\,.$$

By linearity, (5.26) remains true for every linear combination of functions of the form (5.27), i.e., for every piecewise constant function.

Now consider any function $g \in \mathbf{L}^2$. Given any $\varepsilon > 0$, one can find a piecewise constant function $\tilde{g}$ such that $\|g - \tilde{g}\|_{\mathbf{L}^2} < \varepsilon$. This yields

$$\begin{aligned}
&\int_0^\pi g(x) \left(\sin^2 nx - \frac{1}{2}\right) dx \\
&= \int_0^\pi \tilde{g}(x) \left(\sin^2 nx - \frac{1}{2}\right) dx + \int_0^\pi [\tilde{g}(x) - g(x)] \cdot \left(\sin^2 nx - \frac{1}{2}\right) dx \\
&\doteq A_n + B_n .
\end{aligned}$$

Since $\tilde{g}$ is piecewise constant, we already know that $A_n \to 0$ as $n \to \infty$. On the other hand, by Cauchy's inequality,

$$|B_n| \leq \|\tilde{g} - g\|_{\mathbf{L}^2} \cdot \left(\int_0^\pi \left(\sin^2 nx - \frac{1}{2}\right)^2 dx \right)^{1/2} < \varepsilon \cdot \left(\frac{\pi}{4}\right)^{1/2} .$$

Since ε can be taken arbitrarily small, this proves the weak convergence $f_n \rightharpoonup f$.

Notice that strong convergence does not hold. Indeed,

$$\lim_{n\to\infty} \|f_n - f\|^2_{\mathbf{L}^2} = \lim_{n\to\infty} \int_0^\pi \left(\sin^2 nx - \frac{1}{2}\right)^2 dx = \frac{1}{8} .$$

Next, consider the compact operator Λ from $\mathbf{L}^2([0, \pi])$ into itself defined by

$$(\Lambda g)(x) \doteq \int_0^x g(y)\, dy .$$

This integral operator maps the sequence $f_n(x) = \sin^2 nx$ into the sequence

$$(\Lambda f_n)(x) = \int_0^x \sin^2 ny\, dy = \frac{x}{2} - \frac{\sin 2nx}{4n} .$$

Moreover, $(\Lambda f)(x) = x/2$. Observe that Λf_n converges to Λf uniformly on $[0, \pi]$. In particular, we have the strong convergence $\|\Lambda f_n - \Lambda f\|_{\mathbf{L}^2} \to 0$. This provides an illustration of Theorem 5.15, stating that a compact operator maps weakly convergent sequences into strongly convergent ones.

5.8. Problems

1. Let H be a Hilbert space. Using the definition of inner product, prove the two identities in (5.1). Moreover, prove Pythagoras' theorem: if two vectors $x, y \in H$ are orthogonal to each other, then $\|x\|^2 + \|y\|^2 = \|x + y\|^2$.

2. Let H be a vector space with inner product $(\cdot,\cdot)$. Prove that the inner product is a continuous map from the product space $H\times H$ into $\mathbb{K}$. In other words, if $x_k\to x$ and $y_k\to y$ in the norm (5.2), then the sequence of inner products (x_k,y_k) converges to (x,y). On $H\times H$, consider the norm $\|(x,y)\|\doteq(\|x\|^2+\|y\|^2)^{1/2}$.

3. Let X be a Banach space over the reals, with norm $\|\cdot\|$. Does there exist an inner product $(\cdot,\cdot)$ on X such that $\|x\|=\sqrt{(x,x)}$ for every $x\in X$? To answer, follow the two steps below.

(i) Let H be a real Hilbert space. Prove that its norm $\|x\|=\sqrt{(x,x)}$ satisfies the parallelogram identity

$$\|x+y\|^2+\|x-y\|^2 = 2\|x\|^2+2\|y\|^2 \qquad \text{for all } x,y\in X. \tag{5.28}$$

(ii) Conversely, let X be a Banach space over the real numbers, whose norm satisfies the parallelogram identity (5.28). Show that

$$(x,y)\doteq\frac{1}{2}\Big(\|x+y\|^2-\|x\|^2-\|y\|^2\Big) = \frac{1}{4}\Big(\|x+y\|^2-\|x-y\|^2\Big)$$

is an inner product on X, which yields exactly the same norm $\|\cdot\|$. Hints: To prove that $(x+x',y)=(x,y)+(x',y)$, first establish the identity

$$\|x+x'+y\|^2 = \|x\|^2+\|x'\|^2+\|x+y\|^2+\|x'+y\|^2-\frac{1}{2}\|x+y-x'\|^2-\frac{1}{2}\|x'+y-x\|^2.$$

To prove that $(\lambda x,y)=\lambda(x,y)$, first show that this identity holds when λ is an integer, then for λ rational, and finally by continuity for every $\lambda\in\mathbb{R}$.

4. Using the above problem, prove that:

(i) On the vector space $\mathbb{R}^2$, for all $p\ge 1$ with $p\neq 2$, the norm $\|\mathbf{x}\|_p\doteq(|x_1|^p+|x_2|^p)^{1/p}$ is not generated by an inner product.

(ii) On the space of real-valued continuous functions $\mathcal{C}([0,1])$, the norm $\|f\|\doteq\max_{x\in[0,1]}|f(x)|$ is not generated by an inner product.

5. Consider the Hilbert space $H=\mathbf{L}^2([-1,1])$, with inner product $(f,g)=\int_{-1}^{1}fg\,dx$. Show that the sequence of polynomials $\{1,x,x^2,x^3,\ldots\}$ is linearly independent, and its span is dense in H. Applying the Gram-Schmidt orthogonalization procedure, construct an orthonormal sequence of polynomials $\{p_0,p_1,p_2,p_3,\ldots\}$ (proportional to the Legendre polynomials). Explicitly compute the first three polynomials.

6. Let $(e_n)_{n\ge1}$ be an orthonormal sequence in a Hilbert space H. Prove the weak convergence $e_n\rightharpoonup 0$. In other words show that, for every $x\in H$, one has $\lim_{n\to\infty}(x,e_n)=0$.

7. Let H be an infinite-dimensional Hilbert space and let any vector $x\in H$ be given, with $\|x\|\le 1$. Construct a sequence of vectors x_n with $\|x_n\|=1$ for every $n\ge1$, such that the weak convergence holds: $x_n\rightharpoonup x$.

8. Let $(v_k)_{k\geq 1}$ be a sequence of unit vectors in a real Hilbert space H. Let $(\alpha_k)_{k\geq 1}$ be a sequence of real numbers.

(i) If $\sum_{k=1}^{\infty} |\alpha_k| < \infty$, prove that the series $\sum_{k=1}^{\infty} \alpha_k v_k$ converges.

(ii) Assume that the sequence $(v_k)_{k\geq 1}$ is orthonormal. In this case, prove that the series $\sum_{k=1}^{\infty} \alpha_k v_k$ converges if and only if $\sum_{k=1}^{\infty} |\alpha_k|^2 < \infty$.

9. Let $\phi : \mathbb{R}^n \mapsto \mathbb{R}^n$ be a smooth bijection. Assume that the determinant of the Jacobian matrix satisfies $\det D\phi(x) = 1$ for all $x \in \mathbb{R}^n$, so that ϕ is volume-preserving. On the space $\mathbf{L}^2(\mathbb{R}^n)$ consider the linear operator $(\Lambda f)(x) \doteq f(\phi(x))$. Prove that

$$\|\Lambda\| = 1, \qquad \Lambda^* = \Lambda^{-1}.$$

Observe that Λ can be regarded as a rotation in the space $\mathbf{L}^2(\mathbb{R}^n)$, because the adjoint operator coincides with the inverse.

10. Consider the Hilbert space ℓ^2 of all sequences of real numbers $\mathbf{x} = (x_1, x_2, \ldots)$ such that $\sum_{k=1}^{\infty} |x_k|^2 < \infty$, with inner product $(\mathbf{x}, \mathbf{y}) \doteq \sum_{k=1}^{\infty} x_k y_k$. Define the operator $\Lambda : H \mapsto H$ by setting $\Lambda(x_1, x_2, x_3, \ldots) \doteq (x_2, x_3, x_4, \ldots)$. Compute the adjoint operator Λ^*. Are the operators Λ, Λ^* surjective? One-to-one?

11. Let S be a convex set. We say that $x \in S$ is an **extreme point** of S if x cannot be expressed as a convex combination of distinct points of S. In other words,

$$x \neq \theta x_1 + (1-\theta)x_2 \qquad \text{whenever } 0 < \theta < 1, \qquad x_1, x_2 \in S, \qquad x_1 \neq x_2 .$$

Prove the following.

(i) If S is the closed unit ball in a Hilbert space H, then every point $x \in S$ with $\|x\| = 1$ is an extreme point of S. This is true, in particular, for the space $H = \mathbf{L}^2([0,1])$.

(ii) On the other hand, consider the unit ball in $\mathbf{L}^1([0,1])$, i.e.,

$$B \doteq \left\{ f : [0,1] \mapsto \mathbb{R}\,; \int_0^1 |f(t)|\, dt \leq 1 \right\}.$$

Prove that B does not contain any extreme point.

12. Let $(x_n)_{n\geq 1}$ be a sequence of points in a Hilbert space H such that $C \doteq \liminf_{n\to\infty} \|x_n\| < \infty$. Prove that there exists a weakly convergent subsequence $x_{n_j} \rightharpoonup x$, for some point $x \in H$ satisfying $\|x\| \leq C$.

13. Let H be an infinite-dimensional, separable Hilbert space over the reals, and let ℓ^2 be the space of all sequences of real numbers $\mathbf{a} = (a_1, a_2, \ldots)$ such that $\|\mathbf{a}\|_{\ell^2} \doteq \left(\sum_{k=1}^{\infty} a_k^2\right)^{1/2} < \infty$. Construct a linear bijection $\Lambda : H \mapsto \ell^2$ which preserves distances, i.e., such that

$$\|\Lambda x\|_{\ell^2} = \|x\|_H \qquad \text{for all } x \in H.$$

14. Given n vectors $v_1, v_2, \ldots, v_n$ in a real Hilbert space H, define their *Gram determinant* as the determinant of the $n \times n$ symmetric matrix in (5.7):

$$G(v_1, \ldots, v_m) \doteq \det \begin{pmatrix} (v_1, v_1) & \cdots & (v_n, v_1) \\ \vdots & \ddots & \vdots \\ (v_1, v_n) & \cdots & (v_n, v_n) \end{pmatrix}.$$

(i) Using (5.7) with $x = 0$, prove that the vectors $v_1, \ldots, v_n$ are linearly independent if and only if $G(v_1, \ldots, v_n) \neq 0$.

(ii) Assuming that the vectors $v_1, \ldots, v_n$ are linearly independent, consider the subspace $V \doteq \text{span}\{v_1, \ldots, v_n\}$. For any vector $x \in H$, show that the distance of x to V is

$$d(x, V) = \|x - P_V(x)\| = \sqrt{\frac{G(x, v_1, v_2, \ldots, v_n)}{G(v_1, v_2, \ldots, v_n)}}.$$

(iii) As shown in Figure 5.8.1, prove that the n-dimensional volume of the parallelepiped with edges $v_1, \ldots, v_n$ can be expressed as

$$\|v_1\| \cdot d\Big(v_2\,;\ \text{span}\{v_1\}\Big) \cdot d\Big(v_3\,;\ \text{span}\{v_1, v_2\}\Big) \ \cdots\ d\Big(v_n\,;\ \text{span}\{v_1, \ldots, v_{n-1}\}\Big).$$

Using (ii), show that the volume of this parallelepiped is $\sqrt{G(v_1, v_2, \ldots, v_n)}$.

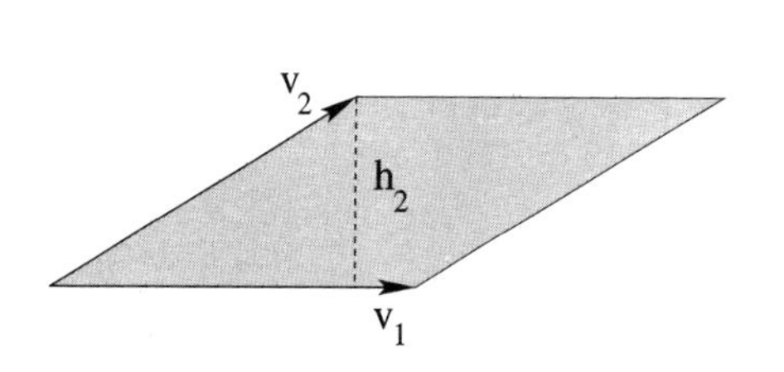

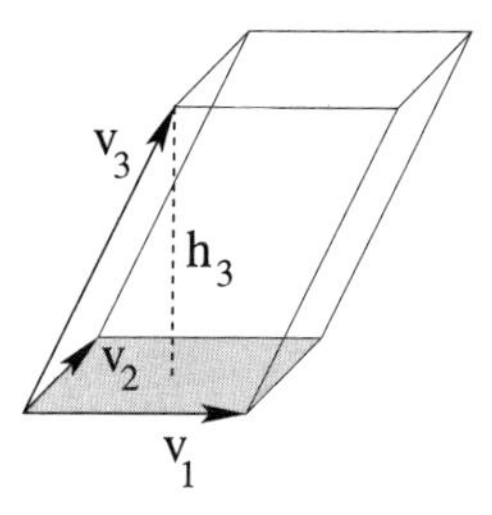

Figure 5.8.1. Computing the area of a parallelogram and the volume of a parallelepiped. Here $h_2 = d(v_2\,,\ \text{span}\{v_1\})$, while $h_3 = d(v_3\,,\ \text{span}\{v_1, v_2\})$.

15. Let $f \in \mathbf{L}^2(\mathbb{R})$. Prove that there exists a unique even[2] function g_0 such that

$$\|f - g_0\|_{\mathbf{L}^2} = \min_{g \in \mathbf{L}^2(\mathbb{R}),\ g \text{ even}} \|f - g\|_{\mathbf{L}^2}.$$

Explicitly determine the function g_0.

16. Let $K \in \mathcal{B}(X; H)$ be a bounded linear operator from a Banach space X into a Hilbert space H. Prove that the following conditions are equivalent.

(i) K is compact.

(ii) For every $\varepsilon > 0$ there exists an operator $K_\varepsilon : X \mapsto H$ with finite-dimensional range, such that $\|K_\varepsilon - K\| < \varepsilon$.

[2]We recall that g is an even function if $g(x) = g(-x)$ for all $x \in \mathbb{R}$.

17. Let $\{u_n\}_{n\geq 1}$ and $\{v_n\}_{n\geq 1}$ be two orthonormal sets in a Hilbert space H. Assume that

$$\sum_{n=1}^{\infty} \|u_n - v_n\| < 1.$$

Show that if one set is complete, then the other set is also complete.

18. Let $\{u_n\,;\; n \geq 1\}$ be a countable set of linearly independent unit vectors in a Hilbert space H. Consider the vector $v \doteq \sum_{n=1}^{\infty} 2^{-n} u_n$.

(i) Assuming that all vectors u_n are mutually orthogonal, prove that the set $S \doteq \{v, u_1, u_2, u_3, \ldots\}$ is linearly independent.

(ii) Show by an example that, if the vectors u_n are not mutually orthogonal, the above set S can be linearly dependent.

19. Given a sequence $(x_n)_{n\geq 1}$ in a Hilbert space H, show that the strong convergence $\|x_n - x\| \to 0$ holds if and only if

$$\|x_n\| \to \|x\| \qquad \text{and} \qquad x_n \rightharpoonup x \text{ (weak convergence)}.$$

20. Let $Q = [0,1] \times [0,1]$ be the unit square. Within the space $\mathbf{L}^2(Q)$, consider the subspace of all functions depending only on the variable y:

$$U = \Big\{ u \in \mathbf{L}^2(Q)\,;\; u(x,y) = \varphi(y) \text{ for some function } \varphi : [0,1] \mapsto \mathbb{R} \text{ and a.e. } (x,y) \in Q \Big\}.$$

(i) Find the orthogonal subspace $W = U^{\perp}$.

(ii) Given any $f \in \mathbf{L}^2(Q)$, determine the function $g \in U$ such that

$$\|f - g\|_{\mathbf{L}^2(Q)} = \min_{u\in U} \|f - u\|_{\mathbf{L}^2(Q)}.$$

21. Let H be a Hilbert space, and let $\Omega \subset H$ be a closed, convex subset.

(i) Prove that, for any $x \in H$, there exists a unique point $y \in \Omega$ such that $\|y - x\| = \min_{\omega\in\Omega} \|\omega - x\|$. This point of minimum distance $y = \pi_\Omega(x)$ is called the *perpendicular projection* of x into Ω.

(ii) Show that $y = \pi_\Omega(x)$ if and only if $\langle \omega - y\,,\; y - x\rangle \geq 0$ for all $\omega \in \Omega$.

22. On the Hilbert space $H = \mathbf{L}^2(\mathbb{R})$, consider the subset

$$\Omega \doteq \Big\{ f\,;\; f(x) \leq e^x \text{ for a.e. } x \in \mathbb{R} \Big\}.$$

(i) Prove that Ω is a closed, convex subset of H.

(ii) Prove that the perpendicular projection $\pi : H \mapsto \Omega$ is the map defined as $(\pi f)(x) = \min\{f(x),\, e^x\}$.

23. On the Hilbert space $H = \mathbf{L}^2(\mathbb{R})$, consider the linear operator $\Lambda : H \mapsto H$ defined by $(\Lambda f)(x) = f(|x|)$.

(i) Find the operator norm of Λ.

(ii) Find the kernel and the range of Λ.

(iii) Compute the adjoint operator Λ^*.

24. On the Hilbert space $H = \mathbf{L}^2([0, \infty[)$, consider the operator $(\Lambda f)(x) = f(e^x)$.

(i) Compute the norm of linear operator Λ.

(ii) Find the kernel and the range of Λ.

(iii) Compute the adjoint operator Λ^*.

25. Consider a bounded sequence of functions $f_n \in \mathbf{L}^2([0,T])$. As $n \to \infty$, show that the weak convergence $f_n \rightharpoonup f$ holds if and only if

$$\lim_{n\to\infty} \int_0^b f_n(x)\,dx = \int_0^b f(x)\,dx \qquad \text{for every } b \in [0,T].$$

26. On the space $\mathbf{L}^2([0,1])$, consider the two sequences of functions

$$f_n(x) = \sqrt{n}\cdot\cos nx, \qquad f_n(x) = \begin{cases} n^{2/3} & \text{if } x \in [0, n^{-1}], \\ 0 & \text{if } x > n^{-1}. \end{cases}$$

(i) In both cases, prove that $\lim_{n\to\infty} \int_0^b f_n(x)\,dx = 0$ for every $b \in [0,1]$.

(ii) By taking linear combinations, show that $\lim_{n\to\infty} \int_0^1 f_n\, g\,dx = 0$ for every piecewise constant function g.

(iii) Is it true that $f_n \rightharpoonup 0$?

27. Given a sequence $(x_n)_{n\geq 1}$ in a Hilbert space H, prove that the following statements are equivalent.

(i) The weak convergence holds: $x_n \rightharpoonup x$.

(ii) The sequence (x_n) is bounded and $(y, x_n) \to (y, x)$ for all y in a subset $S \subseteq H$ whose closure has nonempty interior.

28. Let $\Omega \subseteq \mathbb{R}^N$ be an open set. Prove that a sequence of functions $f_n \in \mathbf{L}^2(\Omega)$ converges weakly to f if and only if there exists a constant C such that $\|f_n\|_{\mathbf{L}^2} \leq C$ for every $n \geq 1$ and moreover

$$\lim_{n\to\infty} \int_Q f_n\,dx = \int_Q f\,dx$$

for every box $Q = [a_1, b_1] \times [a_2, b_2] \times \cdots \times [a_N, b_N]$ entirely contained in Ω.

29. In a real Hilbert space H, consider a weakly convergent sequence: $x_n \rightharpoonup y$. Let $S \doteq \overline{\text{co}}\{x_n\,;\ n \geq 1\}$ be the smallest closed convex set containing all points x_n. Prove that $y \in S$.

Chapter 6

Compact Operators on a Hilbert Space

For a linear operator $A : \mathbb{R}^n \mapsto \mathbb{R}^n$ in a finite-dimensional space, several results are known concerning its kernel, range, eigenvalues, and eigenvectors. In particular:

(i) A is one-to-one if and only if A is onto. Indeed, the subspaces $\mathrm{Ker}(A)$ and $[\mathrm{Range}(A)]^\perp$ have the same dimension.

(ii) If A is symmetric, then its eigenvalues are real. Moreover, the space $\mathbb{R}^n$ admits an orthonormal basis consisting of eigenvectors of A.

In this chapter we shall prove similar results, valid for operators Λ on an infinite-dimensional Hilbert space H. Indeed, (i) remains valid for operators of the form $\Lambda = I - K$, where I is the identity and K is a compact operator. Moreover, the statements in (ii) can be extended to any compact, selfadjoint operator $\Lambda : H \mapsto H$.

6.1. Fredholm theory

Let H be a Hilbert space. We recall that a bounded linear operator $K : H \mapsto H$ is **compact** if for every bounded sequence of points $u_n \in H$ one can extract a subsequence $(u_{n_j})_{j\geq 1}$ such that the images converge: $Ku_{n_j} \to v$ for some $v \in H$.

The next theorem describes various relations between the kernel and range of an operator having the form $I - K$ and of its adjoint.

Theorem 6.1 (Fredholm). *Let H be a Hilbert space over the reals and let $K : H \mapsto H$ be a compact linear operator. Then*

(i) $\text{Ker}(I - K)$ *is finite-dimensional;*

(ii) $\text{Range}(I - K)$ *is closed;*

(iii) $\text{Range}(I - K) = Ker(I - K^*)^\perp$*;*

(iv) $\text{Ker}(I - K) = \{0\}$ *if and only if* $\text{Range}(I - K) = H$*;*

(v) $\text{Ker}(I - K)$ *and* $\text{Ker}(I - K^*)$ *have the same dimension.*

Proof. 1. If the kernel of $(I - K)$ is infinite-dimensional, one can find an orthonormal sequence $(e_n)_{n\geq 1}$ contained in $\text{Ker}(I - K)$. In this case $Ke_n = e_n$ for every n. Moreover, by Pithagoras' theorem, for $m \neq n$ one has

$$\|e_m - e_n\|^2 = \|e_m\|^2 + \|e_n\|^2 = 2.$$

Therefore $\|Ke_m - Ke_n\| = \|e_m - e_n\| = \sqrt{2}$ for every $m \neq n$. Hence from the sequence $(Ke_n)_{n\geq 1}$ one cannot extract any convergent subsequence. This contradiction establishes (i).

2. Toward the proof of (ii), we first show that there exists $\beta > 0$ such that

$$\|u - Ku\| \geq \beta\|u\| \qquad \text{for all } u \in \text{Ker}(I - K)^\perp. \tag{6.1}$$

Indeed, if (6.1) fails, we could find a sequence of points $u_n \in \text{Ker}(I - K)^\perp$ such that $\|u_n\| = 1$ and $\|u_n - Ku_n\| < 1/n$.

Since the sequence $(u_n)_{n\geq 1}$ is bounded, by extracting a subsequence and relabeling, we can assume that this sequence converges weakly, say $u_n \rightharpoonup u$ for some $u \in H$.

Since K is a compact operator, by Theorem 5.15 this implies the strong convergence $Ku_n \to Ku$. We now have

$$\|u_n - Ku\| \leq \|u_n - Ku_n\| + \|Ku_n - Ku\| \to 0 \qquad \text{as } n \to \infty.$$

This yields the strong convergence $u_n \to Ku$. Recalling the weak convergence $u_n \rightharpoonup u$, we conclude that $u = Ku$ and $u_n \to u$ strongly.

By construction, we now have

$$\|u\| = \lim_{n\to\infty} \|u_n\| = 1, \qquad u \in \text{Ker}(I - K)^\perp,$$

while, at the same time, $u - Ku = 0$, hence $u \in \text{Ker}(I - K)$. We thus reached a contradiction, proving (6.1).

3. We now prove that $\text{Range}(I - K)$ is closed. Consider a sequence of points $v_n \in \text{Range}(I - K)$, with $v_n \to v$ as $n \to \infty$. We need to find some u such that $v = u - Ku$. By assumption, for each $n \geq 1$, there exists u_n such that

$v_n = u_n - Ku_n$. Notice that, if $u_n \to u$ for some $u \in H$, then by continuity we could immediately conclude that $v = u - Ku$. In general, however, there is no guarantee that the sequence $(u_n)_{n\geq 1}$ converges.

To overcome this difficulty, let $\tilde{u}_n$ be the perpendicular projection of u_n on $\mathrm{Ker}(I-K)$, and let $z_n \doteq u_n - \tilde{u}_n$. These definitions yield

$$z_n \doteq u_n - \tilde{u}_n \in \mathrm{Ker}(I-K)^\perp, \qquad v_n = u_n - Ku_n = z_n - Kz_n .$$

Using (6.1), for every pair of indices m, n we obtain

$$\|v_m - v_n\| \geq \beta\|z_m - z_n\|.$$

Since the sequence $(v_n)_{n\geq 1}$ is Cauchy, this proves that the sequence $(z_n)_{n\geq 1}$ is a Cauchy sequence as well. Therefore there exists $u \in H$ such that $z_n \to u$, and hence

$$u - Ku = \lim_{n\to\infty} z_n - Kz_n = \lim_{n\to\infty} v_n = v.$$

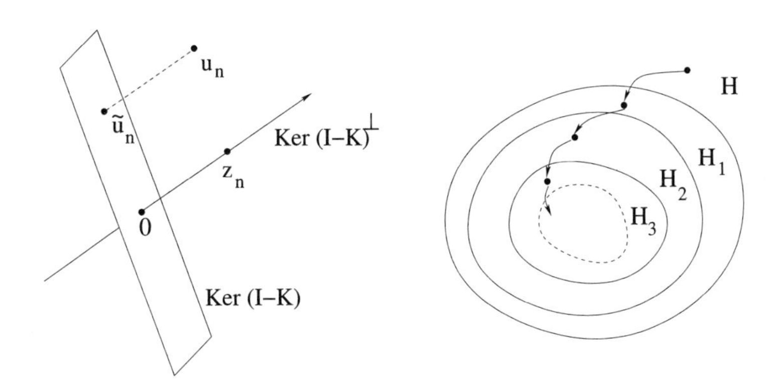

Figure 6.1.1. Left: by choosing $z_n \in \mathrm{Ker}(I-K)^\perp$ we achieve the inequality $\beta\|z_n\| \leq \|v_n\|$, used in the proof of (ii). Right: if $I-K$ is one-to-one but not onto, the sequence of subspaces $H_n \doteq (I-K)^n(H)$ is strictly decreasing. This leads to a contradiction, used in the proof of (iv).

4. Since $\mathrm{Range}(I-K)$ and $\mathrm{Ker}(I-K^*)^\perp$ are closed subspaces, the assertion (iii) holds if and only if

$$\mathrm{Range}(I-K)^\perp = \mathrm{Ker}(I-K^*). \tag{6.2}$$

This is proved by observing that the following statements are all equivalent:

$$\begin{aligned} & x \in \mathrm{Ker}(I-K^*), \\ & (I-K^*)x = 0, \\ & (y\,,\,(I-K^*)x) = 0 \qquad \text{for all } y \in H, \\ & ((I-K)y\,,\,x) = 0 \qquad \text{for all } y \in H, \\ & x \in \mathrm{Range}(I-K)^\perp. \end{aligned}$$

5. Toward a proof of (iv), assume that $\mathrm{Ker}(I - K) = \{0\}$, so that the operator $I - K$ is one-to-one. If $\mathrm{Range}(I - K) \neq H$, we shall derive a contradiction.

Indeed, assume that this range is not the entire space: $H_1 \doteq (I - K)(H) \neq H$. By (ii), H_1 is a closed subspace of H. Since $I - K$ is one-to-one, we must have

$$H_2 \doteq (I - K)(H_1) \subset H_1 .$$

We can continue this process by induction: for every n, we set

$$H_n \doteq (I - K)^n(H).$$

By induction, we see that each H_n is a closed subspace of H, and

$$H \supset H_1 \supset H_2 \supset \cdots .$$

For each $n \geq 1$ we now choose a vector $e_n \in H_n \cap H_{n+1}^{\perp}$ with $\|e_n\| = 1$. Observe that, if $m < n$, then

$$Ke_m - Ke_n = -(e_m - Ke_m) + (e_n - Ke_n) + (e_m - e_n) = e_m + z_m$$

with $z_m \doteq -(e_m - Ke_m) + (e_n - Ke_n) - e_n \in H_{m+1}$.

Since $e_m \in H_{m+1}^{\perp}$, by Pythagoras' theorem this implies

$$\|Ke_m - Ke_n\| \geq \|e_m\| = 1.$$

Therefore, the sequence $(Ke_n)_{n\geq 1}$ cannot have any strongly convergent subsequence, contradicting the compactness of K.

6. To prove the converse implication in (iv), we use a duality argument. Assume that $\mathrm{Range}(I - K) = H$. By Theorem 4.8, $\mathrm{Ker}(I - K^*) = \mathrm{Range}(I - K)^{\perp} = H^{\perp} = \{0\}$. Since K^* is compact, by the previous step we have $\mathrm{Range}(I - K^*) = H$. Using again Theorem 4.8, we obtain $\mathrm{Ker}(I - K) = \mathrm{Range}(I - K^*)^{\perp} = H^{\perp} = \{0\}$, as claimed.

7. Toward a proof of (v), we first show that the dimension of $\mathrm{Ker}(I - K)$ is greater than or equal to the dimension of $\mathrm{Range}(I - K)^{\perp}$. Indeed, suppose on the contrary that

$$\dim \mathrm{Ker}(I - K) < \dim \mathrm{Range}(I - K)^{\perp}. \tag{6.3}$$

Then there exists a linear map $A : \mathrm{Ker}(I - K) \mapsto \mathrm{Range}(I - K)^{\perp}$ which is one-to-one but not onto. We extend A to a linear map $A : H \mapsto \mathrm{Range}(I - K)^{\perp}$ defined on the whole space H, by requiring that $Au = 0$ if $u \in \mathrm{Ker}(I - K)^{\perp}$. Since the range of A is finite-dimensional, the operator A is compact, and so is $K + A$.

We claim that $\text{Ker}(I - (K + A)) = \{0\}$. Indeed, consider any vector $u \in H$ and write

$$u = u_1 + u_2\,, \qquad u_1 \in \text{Ker}(I-K), \quad u_2 \in \text{Ker}(I-K)^\perp.$$

Then

$$(I-K-A)(u_1+u_2) \;=\; (I-K)u_2 - Au_1 \;\in\; \text{Range}(I-K)\oplus\text{Range}(I-K)^\perp. \tag{6.4}$$

Since $(I-K)u_2$ is orthogonal to Au_1, the sum $(I-K)u_2 + Au_1$ can vanish only if $(I-K)u_2 = 0$ and $Au_1 = 0$. Recalling that the operator $I - K$ is one-to-one on $\text{Ker}(I-K)^\perp$ and A is one-to-one on $\text{Ker}(I-K)$, we conclude that $u_1 = u_2 = 0$.

Applying (iv) to the compact operator $K + A$, we obtain $\text{Range}(I - (K + A)) = H$. However, this is impossible: by construction, there exists a vector $v \in \text{Range}(I-K)^\perp$ with $v \notin \text{Range}(A)$. By (6.4), the equation

$$u - Ku - Au \;=\; v$$

has no solution. This contradiction shows that (6.3) cannot hold.

8. Recalling that $\text{Range}(I-K^*)^\perp \;=\; \text{Ker}(I-K)$, from the previous step we deduce that

$$\dim\text{Ker}(I-K^*) \;\geq\; \dim\text{Range}(I-K^*)^\perp \;=\; \dim\text{Ker}(I-K).$$

Interchanging the roles of K and K^* we obtain the opposite inequality. □

Remark 6.2. When K is a compact operator, the above theorem provides information about the existence and uniqueness of solutions to the linear equation

$$u - Ku \;=\; f. \tag{6.5}$$

Namely, two cases can arise.

CASE 1: $\text{Ker}(I-K) = \{0\}$. Then the operator $I-K$ is one-to-one and onto. For every $f \in H$ the equation (6.5) has exactly one solution.

CASE 2: $\text{Ker}(I-K) \neq \{0\}$. This means that the homogeneous equation $u - Ku = 0$ has a nontrivial solution. In this case, the equation (6.5) has solutions if and only if $f \in \text{Ker}(I-K^*)^\perp$, i.e., if and only if

$$(f,u) \;=\; 0 \qquad \text{for every } u \in H \text{ such that } u - K^*u = 0. \tag{6.6}$$

The above dichotomy is known as the **Fredholm alternative**.

6.2. Spectrum of a compact operator

Let H be a Hilbert space over the reals, and let $\Lambda : H \mapsto H$ be a bounded linear operator.

• The **resolvent set** of Λ, denoted as $\rho(\Lambda)$, is the set of numbers $\eta \in \mathbb{R}$ such that $\eta I - \Lambda$ is a bijection (i.e., one-to-one and onto). Notice that in this case, by the open mapping theorem, the inverse operator $(\eta I - \Lambda)^{-1}$ is continuous.

• The complement of the resolvent set: $\sigma(\Lambda) \doteq \mathbb{R} \setminus \rho(\Lambda)$ is called the **spectrum** of Λ.

• The **point spectrum** of Λ, denoted as $\sigma_p(\Lambda)$, is the set of numbers $\eta \in \mathbb{R}$ such that $\eta I - \Lambda$ is not one-to-one. Equivalently, $\eta \in \sigma_p(\Lambda)$ if there exists a nonzero vector $w \in H$ such that

$$\Lambda w = \eta w.$$

In this case, η is called an **eigenvalue** of Λ and w is an associated **eigenvector**.

• The **essential spectrum** of Λ, denoted as $\sigma_e(\Lambda) = \sigma(\Lambda) \setminus \sigma_p(\Lambda)$, is the set of numbers $\eta \in \mathbb{R}$ such that $\eta I - \Lambda$ is one-to-one but not onto.

Theorem 6.3 (Spectrum of a compact operator). *Let H be an infinite-dimensional Hilbert space, and let $K : H \mapsto H$ be a compact linear operator. Then*

(i) $0 \in \sigma(K)$.

(ii) $\sigma(K) = \sigma_p(K) \cup \{0\}$.

(iii) *Either $\sigma_p(K)$ is finite, or else $\sigma_p(K) = \{\lambda_k ;\ k \geq 1\}$, where the eigenvalues satisfy* $\lim_{k\to\infty} \lambda_k = 0$.

Proof. 1. To prove (i) we argue by contradiction. If $0 \notin \sigma(K)$, then K has a continuous inverse $K^{-1} : H \mapsto H$. We thus have $I = K \circ K^{-1}$. Since the composition of a continuous operator with a compact operator is compact, this implies that the identity is a compact operator. But this is false, because H is an infinite-dimensional space and the closed unit ball in H is not compact.

2. To prove (ii), assume that $\lambda \in \sigma(K)$, with $\lambda \neq 0$. If $\text{Ker}(\lambda I - K) = \{0\}$, the Fredholm alternative would imply $\text{Range}(\lambda I - K) = H$. By the open mapping theorem, $(\lambda I - K)$ would have a bounded inverse, against the assumptions. This contradiction proves that $\lambda \in \sigma_p(K)$.

3. To prove (iii), assume that $(\lambda_n)_{n\geq 1}$ is a sequence of distinct eigenvalues of K, with $\lambda_n \to \lambda$. We claim that $\lambda = 0$.

Indeed, since $\lambda_n \in \sigma_p(K)$, for each $n \geq 1$ there exists an eigenvector w_n such that $Kw_n = \lambda_n w_n$. Call $H_n \doteq \text{span}\{w_1, \ldots, w_n\}$. Since eigenvectors corresponding to distinct eigenvalues are linearly independent, $H_n \subset H_{n+1}$.

Observe that, for every $n \geq 2$, one has $(K - \lambda_n I)H_n \subseteq H_{n-1}$. For each n we can thus choose an element $e_n \in H_n \cap H_{n-1}^{\perp}$ with $\|e_n\| = 1$. If $m < n$, then

$$(Ke_n - \lambda_n e_n) \in H_{n-1}\,, \qquad (Ke_m - \lambda_m e_m) \in H_{m-1} \subset H_{n-1}\,,$$

and $e_m \in H_m \subseteq H_{n-1}$ while $e_n \in H_{n-1}^{\perp}$. Hence

$$\begin{aligned} \|Ke_n - Ke_m\| &= \Big\|(Ke_n - \lambda_n e_n) - (Ke_m - \lambda_m e_m) + \lambda_n e_n - \lambda_m e_m\Big\| \\ &\geq \|\lambda_n e_n\| = |\lambda_n|\,. \end{aligned}$$

Therefore

$$\liminf_{m,n\to\infty} \|Ke_n - Ke_m\| \;\geq\; \lim_{n\to\infty} |\lambda_n| \;=\; |\lambda|\,.$$

If $|\lambda| > 0$, then the sequence $(Ke_n)_{n\geq 1}$ cannot have any convergent subsequence, contradicting the assumption that K is compact. □

6.3. Selfadjoint operators

Let $\Lambda : H \mapsto H$ be a bounded linear operator on a real Hilbert space H. We say that Λ is **symmetric** if

$$(\Lambda \mathbf{x},\, \mathbf{y}) \;=\; (\mathbf{x},\, \Lambda \mathbf{y}) \qquad \text{for all } \mathbf{x}, \mathbf{y} \in H\,.$$

Notice that this is equivalent to saying that Λ is selfadjoint.

Example 6.4. Let $A = (a_{ij})_{i,j=1,\ldots,n}$ be a symmetric $n \times n$ matrix. Then A determines a symmetric linear operator $\mathbf{x} \mapsto A\mathbf{x}$ from $\mathbb{R}^n$ into $\mathbb{R}^n$. It also determines the quadratic form

$$\mathbf{x} \;\mapsto\; \langle \mathbf{x}, A\mathbf{x}\rangle \;=\; \sum_{i,j=1}^{n} a_{ij}\, x_i x_j\,.$$

The quantities

$$m \;\doteq\; \min_{|\mathbf{x}|=1} \langle \mathbf{x},\, A\mathbf{x}\rangle, \qquad\qquad M \;\doteq\; \max_{|\mathbf{x}|=1} \langle \mathbf{x},\, A\mathbf{x}\rangle$$

provide the smallest and the largest eigenvalue of A, respectively.

The theory of symmetric linear operators on a Hilbert space extends many well-known properties of symmetric matrices to an infinite-dimensional setting.

Lemma 6.5 (Bounds on the spectrum of a symmetric operator). *Let $\Lambda : H \mapsto H$ be a bounded linear selfadjoint operator on a real Hilbert space H. Define the upper and lower bounds*

$$m \doteq \inf_{u\in H\,,\ \|u\|=1} (\Lambda u, u), \qquad M \doteq \sup_{u\in H\,,\ \|u\|=1} (\Lambda u, u).$$

Then

(i) *The spectrum $\sigma(\Lambda)$ is contained in the interval $[m, M]$.*

(ii) $m, M \in \sigma(\Lambda)$.

(iii) $\|\Lambda\| = \max\{-m, M\}$.

Proof. 1. Let $\eta > M$. Then

$$(\eta u - \Lambda u\,,\ u) \ \geq\ (\eta - M)\|u\|^2 \qquad \text{for all} \ \ u \in H.$$

By the Lax-Milgram theorem, the linear continuous operator $\eta I - \Lambda$ is one-to-one and onto. By the open mapping theorem, it has a continuous inverse. This proves that every $\eta > M$ is in the resolvent set of Λ. Similarly, replacing Λ with $-\Lambda$, we see that every $\eta < m$ lies in the resolvent set. This proves (i).

2. From now on, to fix the ideas, we assume $|m| \leq M$. The opposite case, where $M < -m$, can be handled by entirely similar arguments, replacing Λ by $-\Lambda$.

For every $u, v \in H$ we have

$$\begin{aligned} 4(\Lambda u, v) &= (\Lambda(u+v),\ u+v) - (\Lambda(u-v),\ u-v) \\ &\leq M\Big(\|u+v\|^2 + \|u-v\|^2\Big) \\ &= 2M(\|u\|^2 + \|v\|^2). \end{aligned}$$

If $\Lambda u \neq 0$, setting $v \doteq (\|u\|/\|\Lambda u\|)\Lambda u$ we obtain

$$2\|u\|\,\|\Lambda u\| \ = \ 2(\Lambda u, v) \ \leq\ M(\|u\|^2 + \|v\|^2) \ = \ 2M\,\|u\|^2.$$

Therefore

$$\|\Lambda u\| \ \leq\ M\|u\| \qquad\qquad \text{for all } u \in H\,. \tag{6.7}$$

Indeed, (6.7) trivially holds also if $\Lambda u = 0$.

Since $\|\Lambda\| \geq \sup_{\|u\|=1}(\Lambda u, u) = M$, from (6.7) it follows that $\|\Lambda\| = M$, proving (iii).

3. Next, we claim that $M \in \sigma(\Lambda)$. Choose a sequence $(u_n)_{n\geq 1}$ with

$$(\Lambda u_n, u_n) \ \to\ M\,, \qquad\qquad \|u_n\| = 1 \quad \text{for all } n.$$

Then

$$\begin{aligned}\|\Lambda u_n - M u_n\|^2 &= \|\Lambda u_n\|^2 - 2M(\Lambda u_n, u_n) + M^2\|u_n\|^2\\ &\leq 2M^2 - 2M(\Lambda u_n, u_n) \to 0.\end{aligned}$$

Therefore the operator $\Lambda - MI$ cannot have a bounded inverse. □

According to a classical theorem in linear algebra, every symmetric $n \times n$ matrix A can be reduced to diagonal form by an orthogonal transformation. This can be achieved by choosing an orthonormal basis of $\mathbb{R}^n$ consisting of eigenvectors of A. The following theorem shows that the result remains valid for compact symmetric operators.

Theorem 6.6 (Hilbert-Schmidt; eigenvectors of a compact symmetric operator). *Let H be a separable real Hilbert space, and let $K : H \mapsto H$ be a compact symmetric linear operator. Then there exists a countable orthonormal basis of H consisting of eigenvectors of K.*

Proof. 1. If $H = \mathbb{R}^n$, this is a classical result in linear algebra. We thus assume that H is infinite-dimensional. Let $\eta_0 = 0$ and let $\{\eta_1, \eta_2, \ldots\}$ be the set of all nonzero eigenvalues of K. Consider the eigenspaces

$$H_0 \doteq \mathrm{Ker}(K),\ \ H_1 \doteq \mathrm{Ker}(K - \eta_1 I),\ \ H_2 \doteq \mathrm{Ker}(K - \eta_2 I),\ \ldots.$$

Observe that $0 \leq \dim(H_0) \leq \infty$, while $0 < \dim(H_k) < \infty$ for every $k \geq 1$.

2. We claim that, for $m \neq n$, the subspaces H_m and H_n are orthogonal. Indeed, assume $u \in H_m$, $v \in H_n$. Then

$$\eta_m(u, v) = (Ku, v) = (u, Kv) = \eta_n(u, v).$$

Since $\eta_m \neq \eta_n$, this implies $(u, v) = 0$.

3. Next, we show that the subspaces H_k generate the entire space H. More precisely, consider the set of all linear combinations

$$\widetilde{H} \doteq \Big\{\sum_{k=1}^{N} \alpha_k u_k\,;\ N \geq 1,\ u_k \in H_k\,,\ \alpha_k \in \mathbb{R}\Big\}.$$

We claim that

$$\widetilde{H}^{\perp} \subseteq \mathrm{Ker}(K) \doteq H_0\,. \tag{6.8}$$

Indeed, $K(\widetilde{H}) \subseteq \widetilde{H}$. Moreover if $u \in \widetilde{H}^{\perp}$ and $v \in \widetilde{H}$, then $Kv \in \widetilde{H}$ and hence

$$(Ku, v) = (u, Kv) = 0.$$

This shows that $K(\widetilde{H}^{\perp}) \subseteq \widetilde{H}^{\perp}$.

Let $\widetilde{K}$ be the restriction of K to the subspace $\widetilde{H}^\perp$. Clearly $\widetilde{K}$ is a compact, symmetric operator. By Lemma 6.5 we have

$$\|\widetilde{K}\| = \sup_{u\in\widetilde{H}^\perp,\ \|u\|=1} |(\widetilde{K}u,u)| \doteq M .$$

If $M \neq 0$, then by the lemma either $\lambda = M$ or $\lambda = -M$ is in the spectrum of $\widetilde{K}$. In this case, since $\widetilde{K}$ is compact, λ is in the point spectrum, and there exists an eigenvector $w \in \widetilde{H}^\perp$ such that

$$\widetilde{K}w = Kw = \lambda w .$$

But this is impossible, because all eigenvectors of K corresponding to a nonzero eigenvalue are already contained in the union of the subspaces H_k, $k \geq 1$. We thus conclude that $\|\widetilde{K}\| = 0$, proving (6.8).

In turn, (6.8) yields

$$\widetilde{H}^\perp \subseteq H_0^\perp \cap H_0 = \{0\},$$

proving that $\widetilde{H}$ is dense in H.

4. For each $k \geq 1$, the finite-dimensional subspace H_k admits an orthonormal basis $\mathcal{B}_k = \{e_{k,1}, e_{k,2}, \ldots, e_{k,N(k)}\}$. Moreover, since H is separable, the closed subspace $H_0 = \mathrm{Ker}(K)$ admits a countable orthonormal basis $\mathcal{B}_0 = \{e_{0,1}, e_{0,2}, \ldots\}$. Hence the union $\mathcal{B} \doteq \bigcup_{k\geq 0}\mathcal{B}_k$ is an orthonormal basis of H. □

Remark 6.7. Let $\{\mathbf{w}_1, \mathbf{w}_2, \ldots\}$ be an orthonormal basis of a real Hilbert space H, consisting of eigenvectors of a linear, compact, selfadjoint operator K. Let $\lambda_1, \lambda_2, \ldots$ be the corresponding eigenvalues. For a given $f \in H$, consider the equation

$$u - Ku = f . \tag{6.9}$$

If $1 \notin \sigma(K)$, then (6.9) admits a unique solution. Writing

$$u = \sum_{k=1}^{\infty} c_k \mathbf{w}_k , \qquad f = \sum_{k=1}^{\infty} b_k \mathbf{w}_k ,$$

this solution can be computed as

$$u = \sum_{k=1}^{\infty} \frac{b_k}{1-\lambda_k}\mathbf{w}_k = \sum_{k=1}^{\infty} \frac{(f,\mathbf{w}_k)}{1-\lambda_k}\mathbf{w}_k . \tag{6.10}$$

Remark 6.8. Given a countable set $S = \{u_1, u_2, \ldots\}$ in a Banach space X over the reals, a basic problem is to decide whether span(S) is dense on X. Positive answers can be provided in two important cases.

(I) $X = \mathcal{C}(E)$ is the space of all continuous real-valued functions on a compact metric space E. In this case, if span(S) is an algebra that contains the constant functions and separates points, then the Stone-Weierstrass theorem yields $\overline{\text{span}(S)} = X$.

(II) X is a separable Hilbert space, and there exists a compact selfadjoint operator $\Lambda : X \mapsto X$ such that
(i) span(S) contains all eigenvectors of Λ, and
(ii) span(S) contains the kernel of Λ.
In this case, the Hilbert-Schmidt theorem yields $\overline{\text{span}(S)} = X$.

6.4. Problems

1. (i) Find two bounded linear operators Λ_1, Λ_2 from $\mathbf{L}^2([0,\infty[)$ into itself such that $\Lambda_1 \circ \Lambda_2 = I$ (the identity operator), but $\Lambda_2 \circ \Lambda_1 \neq I$.

(ii) Let H be a real Hilbert space, and let $\Lambda, K : H \mapsto H$ be bounded linear operators, with K compact. Show that in this case

$$\Lambda(I - K) = I \qquad \text{if and only if} \qquad (I - K)\Lambda = I\,.$$

2. On the Hilbert space $\mathbf{L}^2([0,\infty[)$, consider the linear operator defined by

$$(\Lambda f)(x) \;=\; 2f(x+1), \qquad x \geq 0\,.$$

(i) Is Λ a bounded operator? Is it compact?
(ii) Explicitly determine the adjoint operator Λ^*.
(iii) Describe Ker(Λ) and Ker(Λ^*).

3. Let H be a real Hilbert space, and let $\Lambda : H \mapsto H$ be a bounded linear operator, with norm $\|\Lambda\| = M$. Give a direct proof that $\sigma(\Lambda) \subseteq [-M, M]$.

4. On the Hilbert space $H = \mathbf{L}^2([0,1])$, consider the linear operator

$$(\Lambda f)(x) \;=\; \int_0^1 K(x,y)\, f(y)\, dy\,, \tag{6.11}$$

where $K : [0,1] \times [0,1] \mapsto \mathbb{R}$ is a continuous map, satisfying

$$K(x,y) = K(y,x)\,, \qquad \text{for all } x, y \in [0,1]. \tag{6.12}$$

(i) Prove that Λ is a compact selfadjoint linear operator.
(ii) As a special case, note that the assumption (6.12) is satisfied by the integral kernel

$$K(x,y) \;\doteq\; \begin{cases} (1-y)x & \text{if } 0 \leq x \leq y\,, \\ (1-x)y & \text{if } y \leq x \leq 1\,. \end{cases}$$

Show that, when f is continuous, the function $u(x) = (\Lambda f)(x)$ provides a solution to the boundary value problem

$$u''(x) + f(x) = 0, \qquad u(0) = u(1) = 0.$$

5. On the Hilbert space $H \doteq \mathbf{L}^2([0,\pi])$, consider the multiplication operator $(\Lambda f)(x) \doteq f(x) \cdot \sin x$.

(i) Check whether Λ is selfadjoint.

(ii) Compute the operator norm $\|\Lambda\|$ and check whether $\|\Lambda\| \in \sigma_p(\Lambda)$.

(iii) Is Λ compact?

6. On the space $\mathbf{L}^2([0,1])$, consider the integral operator $(\Lambda u)(t) \doteq \int_0^t u(s)\, ds$.

(i) Prove that for every $u \in \mathbf{L}^2$ the function Λu is Hölder continuous, namely $\Lambda u \in \mathcal{C}^{0,1/2}([0,1])$.

(ii) Prove that the operator Λ is compact.

(iii) Compute the adjoint operator Λ^*.

(iv) Given a function $g \in \mathbf{L}^2$, does the equation $u - Ku = g$ have a unique solution? Assuming that g is continuously differentiable, write the ODE satisfied by this solution.

7. As in Remark 6.7, let $\{\mathbf{w}_1, \mathbf{w}_2, \ldots\}$ be an orthonormal basis of the Hilbert space H, consisting of eigenvectors of the linear, compact, selfadjoint operator K. Assume that $1 \in \sigma(K)$. Give a necessary and sufficient condition for the equation (6.9) to have solutions. Write a formula, similar to (6.10), describing all such solutions.

8. Let A, B be continuous, selfadjoint linear operators on a Hilbert space H. Prove that the composition AB is selfadjoint if and only if $AB = BA$.

9. Let $\Lambda : H \mapsto H$ be a bounded linear operator. Prove that the following are equivalent.

(i) Λ is compact.

(ii) $\lim_{n\to\infty} \Lambda \mathbf{v}_n = 0$ for *every* orthonormal sequence $(\mathbf{v}_n)_{n\geq 1}$ of vectors in H.

On the other hand, let H be a separable Hilbert space with orthonormal basis $\{\mathbf{e}_1, \mathbf{e}_2, \ldots\}$. Show that there exists a bounded linear operator $\Lambda \in \mathcal{B}(H)$ which satisfies $\lim_{n\to\infty} \Lambda \mathbf{e}_n = 0$ but is not compact.

10. Let K be a compact linear operator on the Hilbert space H. Prove that, for every closed subspace $V \subset H$, the image $(I - K)(V) = \{x - Kx\, ;\ x \in V\}$ is a closed subspace of H.

11. In the setting of Remark 6.7, consider the linear ODE on the Hilbert space H:

$$\frac{d}{dt}u(t) = Ku(t), \qquad u(0) = f, \tag{6.13}$$

for some $f \in H$. Write the solution in the form

$$u(t) = \sum_{k=1}^{\infty} c_k(t)\mathbf{w}_k, \tag{6.14}$$

computing the time-dependent coefficients $c_k(\cdot)$.

12. Extend the result of Problem 11 to equations of the more general form

$$\frac{d}{dt}u(t) = Ku(t) + g(t), \qquad u(0) = f,$$

where $f \in H$ and $t \mapsto g(t) \in H$ is a continuous function.

13. In the setting of Remark 6.7, consider the second-order linear ODE on the Hilbert space H:

$$\frac{d^2}{dt^2}u(t) = Ku(t), \qquad u(0) = f, \quad \frac{d}{dt}u(0) = g, \tag{6.15}$$

for some $f, g \in H$. Write the solution in the form (6.14), computing the coefficients $c_k(\cdot)$.

14. Let $\Lambda : H \mapsto H$ be a bounded linear operator. Using Problem 22 in Chapter 4, show that the resolvent set $\rho(\Lambda) \doteq \{\eta \in \mathbb{R};\ \eta I - \Lambda \text{ is a bijection}\}$ is open.

Chapter 7

Semigroups of Linear Operators

7.1. Ordinary differential equations in a Banach space

The classical existence-uniqueness theory for ODEs with Lipschitz continuous right-hand side can be extended to Banach spaces without any substantial change. Let X be a Banach space, and let $F : X \mapsto X$ be a Lipschitz continuous map, so that

$$\|F(x) - F(y)\| \leq L \|x - y\| \tag{7.1}$$

for some Lipschitz constant L and every $x, y \in X$.

Given an initial point $\bar{x} \in X$, consider the Cauchy problem

$$\dot{x}(t) = F(x(t)), \qquad x(0) = \bar{x}. \tag{7.2}$$

Here and throughout the sequel, the upper dot denotes a derivative with respect to time. As in the finite-dimensional case, the global existence and uniqueness of a solution can be proved using the contraction mapping theorem.

Theorem 7.1 (Existence-uniqueness of solutions to a Cauchy problem for an ODE with Lipschitz continuous right-hand side). *Let X be a Banach space and assume that the vector field $F : X \mapsto X$ satisfies the Lipschitz condition* (7.1). *Then for every $\bar{x} \in X$ the Cauchy problem* (7.2) *has a unique solution $t \mapsto x(t)$, defined for all $t \in \mathbb{R}$.*

Proof. Fix any $T > 0$ and consider the Banach space $\mathcal{C}([0,T];\, X)$ of all continuous mappings $w : [0,T] \mapsto X$, with the equivalent norm

$$\|w\|_\dagger \doteq \max_{t\in[0,T]} e^{-2Lt}\|w(t)\|\,. \tag{7.3}$$

Observe that a function $x : [0,T] \mapsto X$ provides a solution to the Cauchy problem (7.2) if and only if $x(\cdot)$ is a fixed point of the Picard operator

$$\Phi(w)(t) \doteq \bar{x} + \int_0^t F(w(s))\,ds, \qquad t \in [0,T]. \tag{7.4}$$

We claim that Φ is a strict contraction, with respect to the equivalent norm (7.3). Indeed, given any $u, v \in \mathcal{C}([0,T];\, X)$, set $\delta \doteq \|u - v\|_\dagger$. By the definition (7.3), this implies

$$\|u(s) - v(s)\| \leq \delta e^{2Ls} \qquad \text{for all } s \in [0,T]\,.$$

For every $t \in [0,T]$, the assumption of Lipschitz continuity in (7.1) implies

$$\begin{aligned} e^{-2Lt}\Big\|\Phi(u)(t) - \Phi(v)(t)\Big\| &= e^{-2Lt}\left\|\int_0^t (F(u(s)) - F(v(s)))\,ds\right\| \\ &\leq e^{-2Lt}\int_0^t \Big\|F(u(s)) - F(v(s))\Big\|\,ds \leq e^{-2Lt}\int_0^t L\|u(s) - v(s)\|\,ds \\ &\leq e^{-2Lt}\int_0^t L\delta e^{2Ls}\,ds < \frac{\delta}{2}\,. \end{aligned}$$

Therefore

$$\|\Phi(u) - \Phi(v)\|_\dagger \leq \frac{1}{2}\,\|u - v\|_\dagger\,. \tag{7.5}$$

We can now apply the contraction mapping theorem, obtaining the existence of a unique fixed point for Φ, i.e., a continuous mapping $x(\cdot)$ such that

$$x(t) = \bar{x} + \int_0^t F(x(s))\,ds \qquad \text{for all } t \in [0,T].$$

This function $x(\cdot)$ provides the unique solution to the Cauchy problem. By reversing time, we can construct a unique solution on any time interval of the form $[-T, 0]$. □

There are two well-known methods for constructing approximate solutions to the Cauchy problem (7.2), shown in Figure 7.1.1. In the following we fix a time step $h > 0$ and define the times $t_j = j \cdot h$, $j = 0, 1, 2, \ldots$.

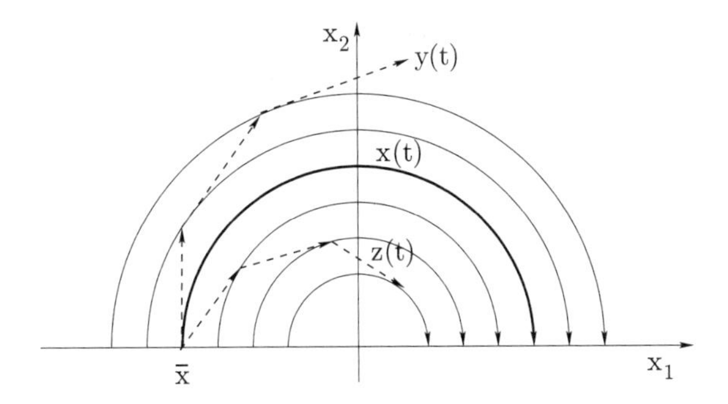

Figure 7.1.1. Euler approximations to the system $\dot{x}_1 = x_2$, $\dot{x}_2 = -x_1$. Here $x(\cdot)$ is an exact solution, $y(\cdot)$ is a piecewise affine approximation obtained by the forward Euler scheme, while $z(\cdot)$ is obtained by the backward Euler scheme.

- **Forward Euler approximations.** The values of the approximate solution at the times t_j are defined by induction, according to the formula

$$x(t_{j+1}) \;=\; x(t_j) + h\,F(x(t_j)). \tag{7.6}$$

- **Backward Euler approximations.** The values of the approximate solution at the times t_j are determined as

$$x(t_{j+1}) \;=\; x(t_j) + h\,F(x(t_{j+1})). \tag{7.7}$$

In both cases, after the values $x(t_j)$ have been computed on the discrete set of times t_j, one can extend the approximate solution to all real values of $t \geq 0$, letting $t \mapsto x(t)$ be an affine function on each interval $[t_{j-1},\, t_j]$.

Forward Euler approximations are easy to construct: during the whole time interval $[t_j, t_{j+1}]$ we let the derivative $\dot{x}(t)$ be constant, equal to the value of F at the *initial point*:

$$\dot{x}(t) \;=\; F(x(t_j)), \qquad t \in [t_j,\, t_{j+1}].$$

On the other hand, in a backward Euler approximation, for $t \in [t_j, t_{j+1}]$ we let the derivative $\dot{x}(t)$ be constant, equal to the value of F at the *terminal point*:

$$\dot{x}(t) \;=\; F(x(t_{j+1})), \qquad t \in [t_j,\, t_{j+1}].$$

In this case, given $x(t_j)$, to find $x(t_{j+1})$ one needs to solve the implicit equation (7.7). Backward Euler approximations thus require more computational work. On the other hand, they often have much better stability and convergence properties.

7.1.1. Linear homogeneous ODEs. Let $A : \mathbb{R}^n \mapsto \mathbb{R}^n$ be a linear operator. Then the solution to the linear Cauchy problem

$$\dot{x} = Ax, \qquad x(0) = \bar{x} \tag{7.8}$$

is the map $t \mapsto e^{tA}\bar{x}$, where

$$e^{tA} \doteq \sum_{k=0}^{\infty} \frac{t^k A^k}{k!}. \tag{7.9}$$

This same formula remains valid for any bounded linear operator A on a Banach space X. We observe that the series in (7.9) is absolutely convergent for every $t \in \mathbb{R}$. Moreover, the exponential map has the following properties:

(i) $e^{0A} = I$, the identity map.

(ii) $e^{sA}e^{tA} = e^{(s+t)A}$ (semigroup property).

(iii) For every $\bar{x} \in X$, the map $t \mapsto e^{tA}\bar{x}$ is continuous.

According to (i)–(ii), the family $\{e^{tA}\,;\ t \geq 0\}$ is a group of linear operators.

More generally, the theory of linear semigroups studies the correspondence

$$A \longleftrightarrow \{e^{tA}\,;\ t \geq 0\}$$

between a linear operator and its exponential.

When A is a bounded linear operator, its exponential function is computed by the convergent series (7.9). Conversely, given the family of operators e^{tA}, one can recover A as the limit

$$A = \lim_{t \to 0+} \frac{e^{tA} - I}{t}.$$

There are important cases where the operators e^{tA} are bounded for every $t \geq 0$, while A is an unbounded operator. These are indeed the most interesting applications of semigroup theory, useful in the analysis of PDEs of parabolic or hyperbolic type.

Example 7.2. If $A : \mathbb{R}^n \mapsto \mathbb{R}^n$ is a diagonal matrix, then its exponential can be readily computed. Indeed

$$A = \begin{pmatrix} \lambda_1 & \cdots & 0 \\ \vdots & \ddots & \vdots \\ 0 & \cdots & \lambda_n \end{pmatrix}, \qquad e^{tA} = \begin{pmatrix} e^{t\lambda_1} & \cdots & 0 \\ \vdots & \ddots & \vdots \\ 0 & \cdots & e^{t\lambda_n} \end{pmatrix}.$$

Observe that the corresponding operator norms are

$$\|A\| = \max_k |\lambda_k|, \qquad \|e^{tA}\| = \max_k |e^{t\lambda_k}|.$$

Example 7.3. Let ℓ^1 be the Banach space of all sequences of complex numbers $x = (x_1, x_2, x_3, \ldots)$ with norm

$$\|x\|_{\ell^1} \doteq \sum_k |x_k| .$$

Given any sequence of complex numbers $(\lambda_k)_{k\geq 1}$, consider the linear operator

$$Ax \doteq (\lambda_1 x_1,\ \lambda_2 x_2,\ \lambda_3 x_3, \ldots). \tag{7.10}$$

The corresponding exponential operators are

$$e^{tA}x = (e^{t\lambda_1}x_1,\ e^{t\lambda_2}x_2,\ e^{t\lambda_3}x_3, \ldots) .$$

It is important to observe that the quantity

$$\|A\| = \sup_k |\lambda_k|$$

may well be infinite, but at the same time the norm

$$\|e^{tA}\| = \sup_k |e^{t\lambda_k}|$$

can be bounded, for every $t \geq 0$. Indeed, as shown in Figure 7.1.2, assume that the real part of all eigenvalues λ_k is uniformly bounded above, say

$$\lambda_k = \alpha_k + i\beta_k , \qquad \alpha_k \leq \omega \qquad \text{for all } k \geq 1 ,$$

for some constant $\omega \in \mathbb{R}$. Then

$$|e^{t\lambda_k}| = |e^{t\alpha_k}| \leq e^{t\omega} \qquad \text{for all } t \geq 0. \tag{7.11}$$

Therefore, for each $t \geq 0$ the operator e^{tA} is bounded. Namely $\|e^{tA}\| \leq e^{t\omega}$.

Two further cases are worth exploring.

CASE 1: Assume that all real parts of the eigenvalues λ_k are bounded above and below, say

$$-\omega \leq \alpha_k \leq \omega$$

for some constant $\omega > 0$ and all $k \geq 1$. Then, for all $t \in \mathbb{R}$ we have the estimate

$$|e^{t\lambda_k}| = |e^{t\alpha_k}| \leq e^{|t|\omega}. \tag{7.12}$$

Hence the operators e^{tA} are all bounded, also for $t < 0$. This shows that the differential equation (7.8) can be solved both forward and backward in time. The family of operators $\{e^{tA}\,;\ t \in \mathbb{R}\}$ forms a group of bounded linear operators.

CASE 2: Assume that all eigenvalues $\lambda_k = \alpha_k + i\beta_k$ are contained in a sector of the complex plane. More precisely, assume that there exists $\omega \in \mathbb{R}$

and an angle $0 < \bar{\theta} < \pi/2$ such that, for every k, we can write

$$\alpha_k = \omega - r_k \cos\theta_k, \qquad \beta_k = -r_k \sin\theta_k \tag{7.13}$$

for some $r_k \geq 0$ and $\theta_k \in [-\bar{\theta}, \bar{\theta}]$.

In this case, since $\alpha_k \leq \omega$, it is clear that (7.11) holds and the operator $\|e^{tA}\|$ is bounded, for every $t \geq 0$. However, much more is true: for every $t > 0$, the composed operator Ae^{tA} is bounded. Indeed

$$\begin{aligned}\|Ae^{tA}\| &= \sup_k |\lambda_k e^{t\lambda_k}| \leq \sup_k (\omega + r_k)e^{(\omega - r_k\cos\theta_k)t} \\ &\leq \sup_{r\geq 0,\ |\theta|\leq\bar{\theta}} (\omega + r)e^{(\omega - r\cos\theta)t} \leq \sup_{r\geq 0} (\omega + r)e^{(\omega - r\cos\bar{\theta})t} < +\infty,\end{aligned}$$

because $\cos\bar{\theta} > 0$. In this case, A is called a sectorial operator.

Even if the initial datum $\bar{x}$ does not lie in the domain of the operator A, for every $t > 0$ we have $x(t) = e^{tA}\bar{x} \in \mathrm{Dom}(A)$.

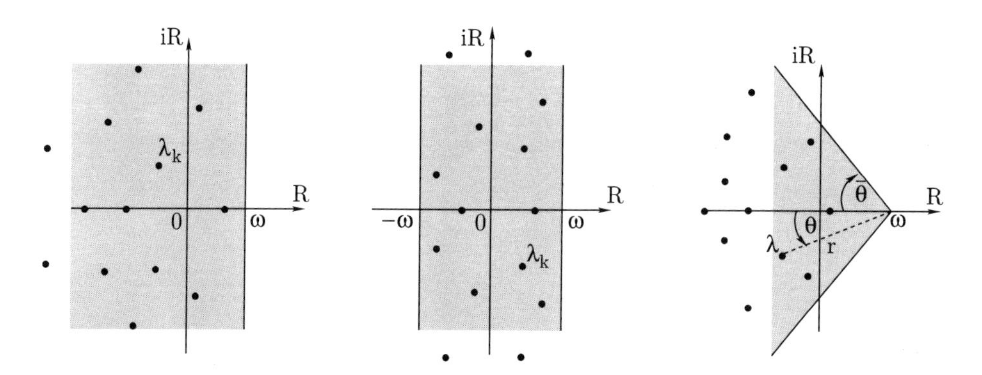

Figure 7.1.2. Left: when all the eigenvalues λ_k of the operator A in (7.10) have real part $\mathrm{Re}(\lambda_k) \leq \omega$, then for every $t \geq 0$ the exponential operator is bounded: $\|e^{tA}\| \leq e^{t\omega}$. Center: if all eigenvalues λ_k satisfy $\mathrm{Re}(\lambda_k) \in [-\omega, \omega]$, then for every $t \in \mathbb{R}$ the exponential operator is bounded: $\|e^{tA}\| \leq e^{|t|\omega}$. Right: if all eigenvalues λ_k lie in a sector of the complex plane with angle $\bar{\theta} < \pi/2$, then for each $t > 0$ the operator Ae^{tA} is bounded as well.

7.2. Semigroups of linear operators

Consider a linear evolution equation in a Banach space X, say

$$\frac{d}{dt}u(t) = Au(t), \qquad u(0) = \bar{u} \in X. \tag{7.14}$$

For $t \geq 0$, one would like to express the solution as $u(t) = e^{tA}\bar{u}$, for some family of linear operators $\{e^{tA};\ t \geq 0\}$.

Example 7.4. Consider the partial differential equation $u_t - u_x = 0$. This can be recast in the form (7.14), where $A = \frac{\partial}{\partial x}$ is a differential operator. As function space, we can take $X = \mathbf{L}^p(\mathbb{R})$, for some $p \in [1, \infty[$. Clearly, the differential operator A is unbounded. Its domain is the set of absolutely continuous functions $u \in \mathbf{L}^p(\mathbb{R})$ with derivative $u_x \in \mathbf{L}^p(\mathbb{R})$. On the other hand, for every initial data $\bar{u} \in \mathbf{L}^p$, the solution of the initial value problem

$$u_t = u_x, \qquad u(0,x) = \bar{u}(x)$$

can be explicitly computed:

$$u(t,x) = \bar{u}(x+t), \qquad t \in \mathbb{R}.$$

This indicates that, although the differential operator $A = \frac{\partial}{\partial x}$ is unbounded, the corresponding exponential operators e^{tA} are uniformly bounded:

$$(e^{tA}\bar{u})(x) = \bar{u}(x+t)$$

and hence $\|e^{tA}\| = 1$ for every $t \in \mathbb{R}$.

In the theory of linear semigroups one considers two basic problems:

(1) Given a semigroup of linear operators $\{S_t\,;\ t \geq 0\}$, find its generator, i.e., the operator A such that $S_t = e^{tA}$.

(2) Given a linear operator A, decide whether it generates a semigroup $\{e^{tA}\,;\ t \geq 0\}$ and establish the properties of this semigroup.

For applications, **(2)** is clearly more important. However, for the development of the theory it is convenient to begin with **(1)**.

7.2.1. Definition and basic properties of semigroups.

Definition 7.5. Let X be a Banach space. A **strongly continuous semigroup of linear operators** on X is a family of linear maps $\{S_t\,;\ t \geq 0\}$ with the following properties.

(i) Each $S_t : X \mapsto X$ is a bounded linear operator.

(ii) For every $s, t \geq 0$, the composition satisfies $S_t S_s = S_{t+s}$ (semigroup property). Moreover $S_0 = I$ (the identity operator).

(iii) For every $u \in X$, the map $t \mapsto S_t u$ is continuous from $[0, \infty[$ into X.

We say that $\{S_t\,;\ t \geq 0\}$ is a **semigroup of type** ω if, in addition, the linear operators S_t satisfy the bounds

$$\|S_t\| \leq e^{t\omega} \qquad \text{for all } t \geq 0. \tag{7.15}$$

A semigroup of type $\omega = 0$ is called a **contractive semigroup**. Indeed, in this case $\|S_t\| \le 1$ for every $t \ge 0$, hence

$$\|S_t u - S_t v\| \le \|u - v\| \qquad \text{for all } u, v \in X,\ t \ge 0.$$

The linear operator

$$Au \doteq \lim_{t\to 0+} \frac{S_t u - u}{t} \tag{7.16}$$

is called the **generator of the semigroup** $\{S_t\,;\ t \ge 0\}$. Its domain $\mathrm{Dom}(A)$ is the set of all $u \in X$ for which the limit in (7.16) exists.

For a given $\bar u \in X$, we regard the map $t \mapsto S_t \bar u$ as the solution to the linear ODE (7.14). Notice that, in a sense, here we are approaching the problem backwards: given the solution $u(t) = S_t \bar u$, we seek to reconstruct the evolution equation, finding the operator A. Some elementary properties of the semigroup S and its generator A are now derived.

Theorem 7.6 (Properties of semigroups). *Let $\{S_t\,;\ t \ge 0\}$ be a strongly continuous semigroup and let A be its generator. Assume $\bar u \in \mathrm{Dom}(A)$. Then*

(i) *For every $t \ge 0$ one has $S_t \bar u \in \mathrm{Dom}(A)$ and $AS_t \bar u = S_t A \bar u$.*

(ii) *The map $t \mapsto u(t) \doteq S_t \bar u$ is continuously differentiable and provides a solution to the Cauchy problem (7.14).*

Proof. 1. Let $\bar u \in \mathrm{Dom}(A)$, so that the limit in (7.16) exists. Then

$$\lim_{s\to 0+} \frac{S_s S_t \bar u - S_t \bar u}{s} = \lim_{s\to 0+} \frac{S_t S_s \bar u - S_t \bar u}{s} = S_t \lim_{s\to 0+} \frac{S_s \bar u - \bar u}{s} = S_t A \bar u.$$

Therefore $S_t \bar u \in \mathrm{Dom}(A)$ and $AS_t \bar u = S_t A \bar u$, proving (i).

2. Next, assume $\bar u \in \mathrm{Dom}(A)$, $t > 0$. Then, by the semigroup property,

$$\lim_{h\to 0+} \left\{ \frac{S_t \bar u - S_{t-h} \bar u}{h} - S_t A \bar u \right\} = \lim_{h\to 0+} \left\{ S_{t-h} \left(\frac{S_{t-h} \bar u - \bar u}{h} \right) - S_t A \bar u \right\}$$

$$= \lim_{h\to 0+} \left\{ S_{t-h} \left(\frac{S_h \bar u - \bar u}{h} - A \bar u \right) + (S_{t-h} A \bar u - S_t A \bar u) \right\} = 0.$$

Indeed, $\frac{S_h \bar u - \bar u}{u} \to A \bar u$, while $\|S_{t-h}\| \le e^{t\omega}$. Moreover, since $A\bar u \in X$ is well defined, the map $s \mapsto S_s A \bar u$ is continuous. The above computation shows that the map $t \mapsto u(t) = S_t \bar u$ has a left derivative:

$$\lim_{h\to 0+} \frac{S_t \bar u - S_{t-h} \bar u}{h} = S_t A \bar u\,.$$

The right derivative is computed as

$$\lim_{h\to 0+} \frac{S_{t+h}\bar{u} - S_t\bar{u}}{h} = S_t \lim_{h\to 0+} \frac{S_h\bar{u} - \bar{u}}{h} = S_t A\bar{u}\,.$$

Therefore, for every $t > 0$, the map $t \mapsto S_t\bar{u}$ is differentiable, with derivative $\frac{d}{dt}S_t\bar{u} = S_tA\bar{u} = AS_t\bar{u}$. Since $A\bar{u} \in X$, by the definition of semigroup the map $t \mapsto S_t(A\bar{u})$ must be continuous. Hence the map $t \mapsto S_t\bar{u}$ is continuously differentiable. □

The following theorem collects some properties of the generator A. We recall that a linear operator $A : X \mapsto X$ is **closed** if its graph

$$\text{Graph}(A) = \Big\{(x,y) \in X \times X\,;\ x \in \text{Dom}(A),\ y = Ax\Big\}$$

is a closed subset of the product space $X \times X$.

Theorem 7.7 (Properties of generators). *Let $\{S_t\,;\ t \geq 0\}$ be a strongly continuous semigroup on the Banach space X, and let A be its generator. Then*

(i) *The domain of A is dense in X.*

(ii) *The operator A is closed.*

Proof. 1. Fix any $u \in X$ and consider the approximation $U_\varepsilon \doteq \varepsilon^{-1}\int_0^\varepsilon S_s u\,ds$. Since the map $t \mapsto S_t u$ is continuous, we have $U_\varepsilon \to u$ as $\varepsilon \to 0+$. To prove (i), we now show that $U_\varepsilon \in \text{Dom}(A)$ for every $\varepsilon > 0$. Since $\text{Dom}(A)$ is a vector subspace, it suffices to show that

$$u_\varepsilon \doteq \varepsilon U_\varepsilon = \int_0^\varepsilon S_s u\,ds \in \text{Dom}(A).$$

For $0 < h < \varepsilon$ we have

$$\begin{aligned}
\frac{S_h u_\varepsilon - u_\varepsilon}{h} &= \frac{1}{h}\left\{S_h\Big(\int_0^\varepsilon S_s u\,ds\Big) - \Big(\int_0^\varepsilon S_s u\,ds\Big)\right\}\\
&= \frac{1}{h}\int_0^\varepsilon (S_{s+h}u - S_s u)ds\\
&= \frac{1}{h}\int_\varepsilon^{\varepsilon+h} S_s u\,ds - \frac{1}{h}\int_0^h S_s u\,ds \ \to\ S_\varepsilon u - u \qquad \text{as } h \to 0+\,.
\end{aligned}$$

This shows that $u_\varepsilon \in \text{Dom}(A)$ for every $\varepsilon > 0$, proving (i).

2. Next, we prove that the graph of A is closed. Let (u_k, v_k) be a sequence of points on the graph of A. More precisely, let $u_k \in \text{Dom}A$, $v_k = Au_k$, and assume the convergence $u_k \to u$, $v_k \to v$, for some $u, v \in X$. We need to show that $u \in \mathrm{Dom}(A)$ and $Au = v$. For each $k \geq 1$, since $u_k \in \mathrm{Dom}(A)$, the previous theorem implies

$$S_h u_k - u_k = \int_0^h \left(\frac{d}{dt} S_t u_k\right) dt = \int_0^h S_t A u_k \, dt.$$

Letting $k \to \infty$, we obtain

$$S_h u - u = \int_0^h S_t v \, dt.$$

Therefore,

$$\lim_{h \to 0+} \frac{S_h u - u}{h} = \lim_{h \to 0+} \frac{1}{h} \int_0^h S_t v \, dt = v.$$

By definition, this means that $u \in \mathrm{Dom}(A)$ and $Au = v$. □

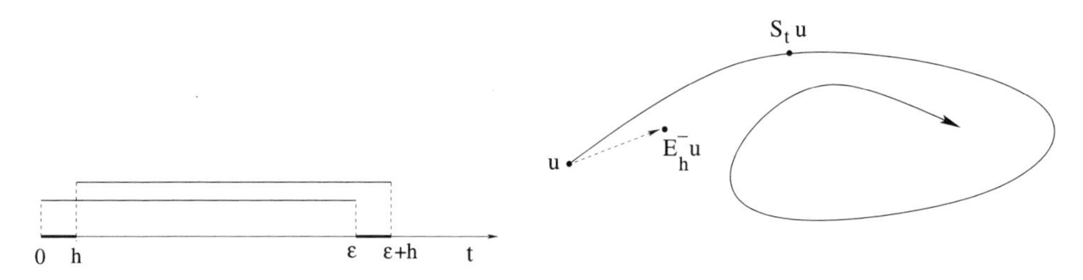

Figure 7.2.1. Left: the intervals of integration, in the proof of Theorem 7.7 (i). Right: according to (7.20) or (7.28), the backward Euler approximation with step $h > 0$ is the weighted average of points on the trajectory $t \mapsto S_t u$, with exponentially decreasing weight $w(t) = e^{-t/h}/h$.

7.3. Resolvents

The crucial link between a semigroup $\{S_t\,;\ t \geq 0\}$ and its generator A is provided by the so-called "resolvent". This can be best understood in terms of backward Euler approximations.

Assume that we want to solve (7.14) approximately, by backward Euler approximations. We thus fix a time step $h > 0$ and iteratively solve

$$u(t+h) = u(t) + h\, Au(t+h).$$

At each step, given a value $u(t) \in X$, we thus need to compute

$$u(t+h) = (I - hA)^{-1} u(t).$$

The **backward Euler operator**

$$E_h^- \doteq (I - hA)^{-1}, \tag{7.17}$$

with $h > 0$ small, plays a crucial role in semigroup theory. Given A, there are two main ways to recover the solution of the Cauchy problem (7.14), in terms of this operator.

(I): As a limit of backward Euler approximations.

For a fixed $\tau > 0$, we consider the time step $h = \tau/n$. After n steps, the backward Euler approximation scheme yields

$$u(\tau) \approx E^-_{\tau/n} \circ \cdots \circ E^-_{\tau/n} = \left(I - \frac{\tau}{n}A\right)^{-n} \bar{u}\,.$$

Keeping τ fixed and letting $n \to \infty$, we expect to recover the exact solution of (7.14) in the limit:

$$u(\tau) = S_\tau \bar{u} \doteq \lim_{n\to\infty} \left(I - \frac{\tau}{n}A\right)^{-n} \bar{u}. \tag{7.18}$$

(II): As a limit of solutions to approximate evolution equations.

Fix again a time step $h > 0$ and set $\lambda = 1/h$. We then define the operator $A_\lambda : X \mapsto X$ as

$$A_\lambda u \doteq A\, E^-_h u = A(I - hA)^{-1} u.$$

In other words, $A_\lambda u$ is the value of A computed not at u but at the nearby point $E^-_h u$, i.e., at the first step in a backward Euler approximation, with $h = 1/\lambda$. It turns out that, for $h > 0$ small enough, $A_\lambda = A_{1/h}$ is a well defined, bounded linear operator. We can thus consider the exponential operators $e^{tA_\lambda} = \sum_{k\geq 0}(tA_\lambda)^k/k!$ and define

$$u(t) = S_t \bar{u} \doteq \lim_{\lambda\to\infty} e^{tA_\lambda}\bar{u}. \tag{7.19}$$

Example 7.8. Consider the scalar ODE

$$\dot{x} = ax, \qquad\qquad x(0) = \bar{x}\,.$$

Its solution is $x(t) = e^{ta}\bar{x}$. In this case we trivially have

$$a_\lambda = a_{1/h} = \frac{a}{1 - ha}, \qquad e^{at}\bar{x} = \lim_{h\to 0} e^{ta_{1/h}}\bar{x} = \lim_{h\to 0+} e^{ta/(1-ha)}\bar{x}\,.$$

This corresponds to the limit (7.19). On the other hand, the limit (7.18) is related to the identity

$$e^{\tau a} = \lim_{n\to\infty} \left(1 - \frac{\tau}{n}a\right)^{-n}.$$

Notice that, for $0 < h < a^{-1}$, one also has the identities

$$\int_0^\infty \frac{e^{-t/h}}{h}\, dt = 1, \qquad (1 - ha)^{-1}\bar{x} = \int_0^\infty \left(\frac{e^{-t/h}}{h}\right) e^{ta}\bar{x}\, dt\,. \tag{7.20}$$

The second identity shows that the backward Euler operator $E_h^-\bar{x} = (1-ha)^{-1}\bar{x}$ can be obtained by taking an average of points along the trajectory $t \mapsto e^{ta}\bar{x}$, with the exponentially decreasing weight $w(t) = h^{-1}e^{-t/h}$.

Motivated by the previous analysis, we introduce some definitions.

Definition 7.9. Let A be a linear operator on a Banach space X. The **resolvent set** of A is the set $\rho(A)$ of all real numbers λ such that the operator

$$\lambda I - A \ : \mathrm{Dom}(A) \mapsto X$$

is one-to-one and onto. If $\lambda \in \rho(A)$, the **resolvent operator** $R_\lambda : X \mapsto X$ is defined by

$$R_\lambda u \doteq (\lambda I - A)^{-1} u.$$

We remark that, if A is a closed operator, then by the closed graph theorem the operator $R_\lambda : X \mapsto \mathrm{Dom}(A) \subseteq X$ is a bounded linear operator. Moreover

$$AR_\lambda u = R_\lambda A u \qquad \text{if } u \in \mathrm{Dom}(A).$$

There is a close connection between resolvents and backward Euler operators. Namely

$$\lambda R_\lambda = E^-_{1/\lambda}\,.$$

Theorem 7.10 (Resolvent identities). *Let A be a closed linear operator. If $\lambda, \mu \in \rho(A)$, then*

$$R_\lambda - R_\mu = (\mu - \lambda) R_\lambda R_\mu\,, \tag{7.21}$$

$$R_\lambda R_\mu = R_\mu R_\lambda\,. \tag{7.22}$$

Proof. For any $u \in X$ one has

$$v \doteq R_\lambda u - R_\mu u = (\lambda I - A)^{-1}u - (\mu I - A)^{-1}u \ \in \mathrm{Dom}(A),$$

$$(\lambda I - A)v = u - (\lambda I - \mu I + \mu I - A)(\mu I - A)^{-1}u = (\mu - \lambda)(\mu I - A)^{-1}u\,.$$

Applying the operator $(\lambda I - A)^{-1}$ to both sides of the above identity, one obtains (7.21).

Next, using (7.21), for any $\lambda \neq \mu$ in the resolvent set $\rho(A)$ we obtain

$$R_\lambda R_\mu = \frac{R_\lambda - R_\mu}{\lambda - \mu} = \frac{R_\mu - R_\lambda}{\mu - \lambda} = R_\mu R_\lambda\,. \qquad \square$$

Theorem 7.11 (Integral formula for the resolvent). *Let $\{S_t\,;\ t \geq 0\}$ be a semigroup of type ω, and let A be its generator. Then for every $\lambda > \omega$ one has $\lambda \in \rho(A)$ and*

$$R_\lambda u \;=\; \int_0^\infty e^{-t\lambda} S_t u\, dt\,. \tag{7.23}$$

Moreover,

$$\|R_\lambda\| \;\leq\; \frac{1}{\lambda - \omega}\,. \tag{7.24}$$

Proof. 1. By assumption, the semigroup satisfies the bounds $\|S_t\| \leq e^{t\omega}$. Therefore the integral in (7.23) is absolutely convergent:

(7.25)

$$\left\|\int_0^\infty e^{-t\lambda} S_t u\, dt\right\| \;\leq\; \int_0^\infty e^{t\lambda}\|S_t\|\,\|u\|\, dt \;\leq\; \int_0^\infty e^{(\omega-\lambda)t}\|u\|\, dt \;\leq\; \frac{1}{\lambda-\omega}\,\|u\|.$$

Define

$$\widetilde{R}_\lambda u \;\doteq\; \int_0^\infty e^{-t\lambda} S_t u\, dt\,.$$

The above estimate shows that $\widetilde{R}_\lambda$ is a bounded linear operator, with norm

$$\|\widetilde{R}_\lambda\| \;\leq\; \frac{1}{\lambda-\omega}\,.$$

2. We now show that

$$(\lambda I - A)\widetilde{R}_\lambda u \;=\; u \qquad \text{for all} \ \ u \in X\,. \tag{7.26}$$

Indeed, for any $h > 0$ we have

$$\begin{aligned}
\frac{S_h \widetilde{R}_\lambda u - \widetilde{R}_\lambda u}{h} &= \frac{1}{h}\int_0^\infty e^{-\lambda t}\,(S_{t+h}u - S_t u)\, dt \\
&= \frac{1}{h}\int_0^\infty \left(e^{-\lambda(t-h)} - e^{-\lambda t}\right) S_t u\, dt - \frac{1}{h}\int_0^h e^{-\lambda(t-h)}\, S_t u\, dt \\
&= \left(\frac{e^{\lambda h}-1}{h}\right)\int_0^\infty e^{-\lambda t}\, S_t u\, dt - \frac{e^{\lambda h}}{h}\int_0^h e^{-\lambda t}\, S_t u\, dt\,.
\end{aligned}$$

Taking the limit as $h \to 0+$ we obtain

$$\lim_{h\to 0+} \frac{S_h \widetilde{R}_\lambda u - \widetilde{R}_\lambda u}{h} \;=\; \lambda \widetilde{R}_\lambda u - u\,.$$

By the definition of generator, this means that

$$\widetilde{R}_\lambda u \in \mathrm{Dom}(A), \qquad A\widetilde{R}_\lambda u \;=\; \lambda\widetilde{R}_\lambda u - u\,,$$

proving (7.26).

3. The identity (7.26) already shows that the map $v \mapsto (\lambda I - A)v$ from $\mathrm{Dom}(A)$ into X is surjective. We now prove that this map is one-to-one. If $u \in \mathrm{Dom}(A)$, then

$$\begin{aligned} A\widetilde{R}_\lambda u &= A\int_0^\infty e^{-\lambda t} S_t u\,dt = \int_0^\infty e^{-\lambda t} A S_t u\,dt \\ &= \int_0^\infty e^{-\lambda t} S_t A u\,dt = \widetilde{R}_\lambda A u\,. \end{aligned}$$

This proves the commutativity relation

$$\widetilde{R}_\lambda(\lambda I - A)u = (\lambda I - A)\widetilde{R}_\lambda u \qquad \text{for all } u \in \mathrm{Dom}(A). \tag{7.27}$$

If now $(\lambda I - A)u = (\lambda I - A)v$, using (7.27) and (7.26) we obtain

$$u = \widetilde{R}_\lambda(\lambda I - A)u = \widetilde{R}_\lambda(\lambda I - A)v = v,$$

proving that the map $(\lambda I - A)$ is one-to-one.

We thus conclude that $\lambda \in \rho(A)$ and $\widetilde{R}_\lambda = (\lambda I - A)^{-1} = R_\lambda$. □

Remark 7.12. According to the integral representation formula (7.23), the resolvent operators R_λ provide the Laplace transform of the semigroup S. Taking $0 < h = \lambda^{-1}$, this same formula shows that the backward Euler approximations can be obtained as

$$E_h^- u \doteq (I - hA)^{-1} u = \int_0^\infty \frac{e^{-t/h}}{h} S_t u\,dt\,. \tag{7.28}$$

Here the integral is convergent, provided that h is sufficiently small: if S is a semigroup of type ω, we need $h < \omega^{-1}$.

Generalizing the identity (7.20), we see that the first step in an Euler backward approximation coincides with an averaged value of the entire trajectory $t \mapsto S_t\bar{u}$, with the exponentially decreasing weight $w(t) = h^{-1}e^{-t/h}$.

7.4. Generation of a semigroup

In this section we tackle the most important question. Namely, given an operator A from $\mathrm{Dom}(A) \subset X$ into X, under which conditions does there exist a semigroup $\{S_t\,;\ t \geq 0\}$ generated by A?

Theorem 7.13 (Existence of the semigroup generated by a linear operator). *Let A be a linear operator on a Banach space X. Then the following are equivalent.*

(i) *A is the generator of a semigroup of linear operators $\{S_t\,;\ t \geq 0\}$, of type ω.*

(ii) *A is a closed, densely defined operator. Moreover, every real number $\lambda > \omega$ is in the resolvent set of A, and*

$$\left\|(\lambda I - A)^{-1}\right\| \leq \frac{1}{\lambda - \omega} \qquad \text{for all } \lambda > \omega\,. \tag{7.29}$$

Proof. The fact that (i) implies (ii) has already been proved. We shall thus assume that (ii) holds and prove that A generates a semigroup of type ω. The proof will be achieved in several steps.

1. By assumption, for each $\lambda > \omega$ the resolvent operator $R_\lambda = (\lambda I - A)^{-1}$ is well defined. We can thus consider the bounded linear operator

$$A_\lambda \doteq -\lambda I + \lambda^2 R_\lambda = \lambda A R_\lambda. \tag{7.30}$$

Notice that, setting $h = 1/\lambda$, we can write

$$A_\lambda u = A(I - hA)^{-1}u = A(E_h^- u)\,.$$

In other words, $A_\lambda u$ is the value of A computed not at the point u but at the point $E_h^- u$ obtained by a backward Euler step of size $h = 1/\lambda$, starting at u. It is thus expected that A_λ should be a good approximation for the operator A, at least when λ is large (so that $h = 1/\lambda$ is small).

2. Since each A_λ is bounded, we can construct its exponential as

$$e^{tA_\lambda} \doteq \sum_{k=0}^{\infty} \frac{(tA_\lambda)^k}{k!} = e^{-\lambda t} e^{\lambda^2 t R_\lambda} = e^{-\lambda t} \sum_{k=0}^{\infty} \frac{(\lambda^2 t)^k R_\lambda^k}{k!}\,.$$

We observe that, if A is unbounded, then $\|A_\lambda\| \to \infty$ as $\lambda \to \infty$. However, for $t \geq 0$ the norms of the exponential operators remain uniformly bounded as $\lambda \to \infty$. Indeed, the assumption (7.29) together with the definition (7.30) yields

$$\|e^{tA_\lambda}\| \leq e^{-\lambda t} \sum_{k=0}^{\infty} \frac{(\lambda^2 t)^k \|R_\lambda\|^k}{k!} \leq e^{-\lambda t} e^{\lambda^2 t/(\lambda - \omega)} = e^{\lambda \omega t/(\lambda-\omega)}\,.$$

In particular, as soon as $\lambda \geq 2\omega$, we have

$$\|e^{tA_\lambda}\| \leq e^{2\omega t} \qquad \text{for all } t \geq 0\,. \tag{7.31}$$

3. In this step we establish the limit

$$\lim_{\lambda \to \infty} A_\lambda v = Av \qquad \text{for all } v \in \mathrm{Dom}(A). \tag{7.32}$$

To prove (7.32), we start with the identity

$$\lambda R_\lambda u - u = A R_\lambda u = R_\lambda A u\,,$$

which is valid for all $u \in \mathrm{Dom}(A)$. This yields

$$\|\lambda R_\lambda u - u\| \;=\; \|R_\lambda A u\| \;\le\; \|R_\lambda\|\,\|Au\| \;\le\; \frac{1}{\lambda-\omega}\|Au\| \;\to\; 0 \quad \text{as } \lambda\to\infty.$$

Hence

$$\lim_{\lambda\to\infty} \lambda R_\lambda u \;=\; u \qquad \text{for all } u \in \mathrm{Dom}(A).$$

Using the fact that $\mathrm{Dom}(A)$ is dense in X, we now prove that the above limit holds more generally for all $u \in X$. Indeed, for every $u \in X$ and $\varepsilon > 0$, there exists $v \in \mathrm{Dom}(A)$ such that $\|u - v\| < \varepsilon$. Using the triangle inequality, we obtain

$$\begin{aligned}
&\limsup_{\lambda\to\infty} \|\lambda R_\lambda u - u\| \\
&\qquad \le\; \limsup_{\lambda\to\infty} \|\lambda R_\lambda u - \lambda R_\lambda v\| + \limsup_{\lambda\to\infty} \|\lambda R_\lambda v - v\| + \|v - u\| \\
&\qquad \le\; \limsup_{\lambda\to\infty} \|\lambda R_\lambda\|\,\|u - v\| + 0 + \|v - u\| \\
&\qquad \le\; \limsup_{\lambda\to\infty} \frac{\lambda}{\lambda-\omega}\,\varepsilon + \varepsilon \;=\; 2\varepsilon\,.
\end{aligned}$$

Since $\varepsilon > 0$ was arbitrary, this proves that

$$\lim_{\lambda\to\infty} \lambda R_\lambda u \;=\; u \qquad \text{for all } u \in X. \tag{7.33}$$

For the backward Euler approximations with time step $h = 1/\lambda$, (7.33) yields

$$\lim_{h\to 0+} E_h^- u \;=\; u \qquad \text{for all } u \in X\,.$$

If $v \in \mathrm{Dom}(A)$, then in (7.33) we can take $u = Av$ and conclude that

$$\lim_{\lambda\to\infty} A_\lambda v \;=\; \lim_{\lambda\to\infty} \lambda A R_\lambda v \;=\; \lim_{\lambda\to\infty} \lambda R_\lambda A v \;=\; \lim_{\lambda\to\infty} \lambda R_\lambda u \;=\; u \;=\; Av\,.$$

This proves (7.32).

4. Fix any $t \ge 0$. We claim that, as $\lambda \to \infty$, the family of uniformly bounded operators e^{tA_λ} converges to some linear operator, which we call S_t.

To prove this, for $\lambda, \mu > 2\omega$ we shall estimate the difference $e^{tA_\lambda} - e^{tA_\mu}$.

The commutativity relation $R_\lambda R_\mu = R_\mu R_\lambda$ implies $A_\lambda A_\mu = A_\mu A_\lambda$. Therefore

$$A_\mu e^{tA_\lambda} = A_\mu \sum_{k=0}^{\infty} \frac{(tA_\lambda)^k}{k!} = e^{tA_\lambda} A_\mu .$$

For every $u \in X$ we thus have

$$e^{tA_\lambda}u - e^{tA_\mu}u = \int_0^t \frac{d}{ds}\Big[e^{(t-s)A_\mu}e^{sA_\lambda}u\Big]\,ds$$

$$= \int_0^t e^{(t-s)A_\mu}(A_\lambda - A_\mu)e^{sA_\lambda}u\,ds = \int_0^t e^{(t-s)A_\mu}e^{sA_\lambda}(A_\lambda u - A_\mu u)\,ds .$$

By (7.31), this implies

$$\|e^{tA_\lambda}u - e^{tA_\mu}u\| \le \int_0^t e^{2(t-s)\omega}e^{2s\omega}\|A_\lambda u - A_\mu u\|\,ds = t\,e^{2\omega t}\,\|A_\lambda u - A_\mu u\| .$$

If $u \in \mathrm{Dom}(A)$, then (7.32) yields $A_\lambda u \to Au$ and $A_\mu u \to Au$ as $\lambda, \mu \to \infty$. Therefore

$$\limsup_{\lambda,\mu\to\infty} \|e^{tA_\lambda}u - e^{tA_\mu}u\| \le t\,e^{2\omega t} \limsup_{\lambda,\mu\to\infty} \|A_\lambda u - A_\mu u\| = 0.$$

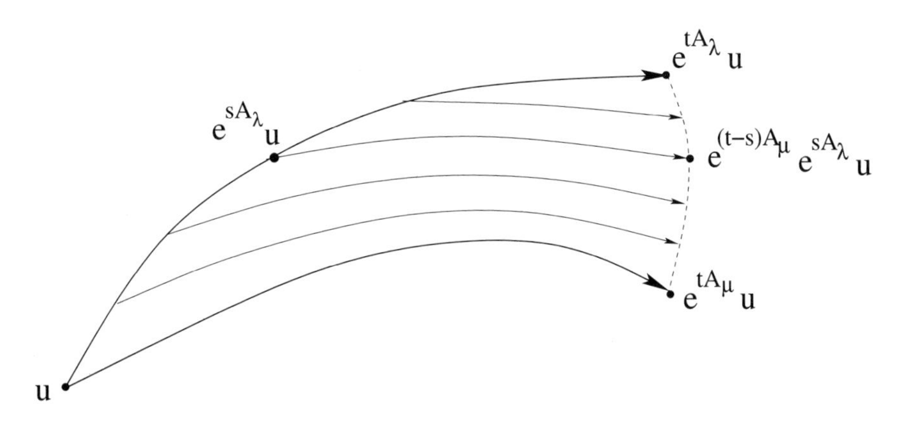

Figure 7.4.1. Proving the convergence of the approximations e^{tA_λ} as $\lambda \to \infty$. The difference $\|e^{tA_\lambda}u - e^{tA_\mu}u\|$ can be estimated by the length of the curve $s \mapsto e^{(t-s)A_\mu}e^{sA_\lambda}u$, for $s \in [0,t]$.

Using the triangle inequality, we now show that the same limit is valid more generally for every $v \in X$, uniformly as t ranges in bounded intervals.

Indeed, given $\varepsilon > 0$, choose $u \in \mathrm{Dom}(A)$ with $\|u - v\| \le \varepsilon$. Then

$$\limsup_{\lambda,\mu\to\infty} \|e^{tA_\lambda}v - e^{tA_\mu}v\|$$

$$\le \limsup_{\lambda\to\infty} \|e^{tA_\lambda}v - e^{tA_\lambda}u\| + \limsup_{\lambda,\mu\to\infty} \|e^{tA_\lambda}u - e^{tA_\mu}u\|$$

$$+ \limsup_{\mu\to\infty} \|e^{tA_\mu}u - e^{tA_\mu}v\|$$

$$\le 2\limsup_{\lambda\to\infty} \|e^{tA_\lambda}\|\,\|v - u\| \le 2e^{2\omega t}\,\varepsilon\,.$$

Since $\varepsilon > 0$ was arbitrary, our claim is proved.

5. By the previous step, for every $t \ge 0$ and $u \in X$ the following limit is well defined:

$$S_t u \doteq \lim_{\lambda\to\infty} e^{tA_\lambda}u\,.$$

We claim that $\{S_t\,;\ t \ge 0\}$ is a strongly continuous semigroup of type ω.

Indeed, the semigroup property follows from

$$S_t\,S_s\,u = \lim_{\lambda\to\infty} e^{tA_\lambda}e^{sA_\lambda}u = \lim_{\lambda\to\infty} e^{(t+s)A_\lambda}u = S_{t+s}u\,.$$

For a fixed $u \in X$, the map $t \mapsto S_t u$ is continuous, being the limit of the continuous maps $t \mapsto e^{tA_\lambda}u$, uniformly for t in bounded intervals.

Finally, for every $t \ge 0$ and $u \in X$ with $\|u\| \le 1$ we have the estimate

$$\|S_t u\| = \lim_{\lambda\to\infty} \|e^{tA_\lambda}u\| \le \limsup_{\lambda\to\infty} \|e^{tA_\lambda}\|\,\|u\|$$
$$\le \lim_{\lambda\to\infty} e^{t\lambda\omega/(\lambda-\omega)}\|u\| = e^{t\omega}\|u\|.$$

This proves that $\|S_t\| \le e^{t\omega}$, i.e., that the semigroup is of type ω.

6. It remains to prove that the linear operator A is indeed the generator of the semigroup. Toward this goal, call B the generator of the semigroup $\{S_t\,;\ t \ge 0\}$. By our earlier analysis, we know that B is a linear, closed operator, with domain $\mathrm{Dom}(B)$ dense in X. We need to show that $B = A$.

Since A_λ is the generator of the semigroup $\{e^{tA_\lambda}\,;\ t \ge 0\}$, for every $\lambda > \omega$ we have

$$e^{tA_\lambda}u - u = \int_0^t e^{sA_\lambda}A_\lambda u\,ds\,. \tag{7.34}$$

Moreover, for $u \in \mathrm{Dom}(A)$, the triangle inequality yields

$$\|e^{sA_\lambda}A_\lambda u - S_s Au\| \le \|e^{sA_\lambda}\|\,\|A_\lambda u - Au\| + \|e^{sA_\lambda}Au - S_s Au\|. \tag{7.35}$$

As $\lambda \to \infty$, the right-hand side of (7.35) goes to zero, uniformly for s in bounded intervals. Taking the limit as $\lambda \to \infty$ in (7.34) we thus obtain

$$S_t u - u = \int_0^t S_s A u \, ds \qquad \text{for all } t \geq 0, \ u \in \mathrm{Dom}(A). \tag{7.36}$$

As a consequence, $\mathrm{Dom}(B) \supseteq \mathrm{Dom}(A)$ and

$$Bu = \lim_{h \to 0+} \frac{S_t u - u}{t} = \lim_{t \to 0+} \frac{1}{t} \int_0^t S_s A u \, ds = Au \quad \text{for all } u \in \mathrm{Dom}(A).$$

To prove that $A = B$, it remains to show that $\mathrm{Dom}(B) \subseteq \mathrm{Dom}(A)$. For this purpose, choose any $\lambda > \omega$. We know that the operators

$$\lambda I - A : \mathrm{Dom}(A) \mapsto X \qquad \text{and} \qquad \lambda I - B : \mathrm{Dom}(B) \mapsto X$$

are both one-to-one and surjective. In particular, the restriction of $\lambda I - B$ to $\mathrm{Dom}(A)$ coincides with $\lambda I - A$ and is thus surjective. Hence this operator $\lambda I - B$ cannot be extended to any domain strictly larger than $\mathrm{Dom}(A)$, preserving the one-to-one property. This shows that $\mathrm{Dom}(B) = \mathrm{Dom}(A)$, completing the proof. □

We now investigate uniqueness. Given a semigroup of linear operators $\{S_t\,;\ t \geq 0\}$, its generator is uniquely defined by the limit in (7.16). Conversely, given a linear operator A satisfying conditions (ii) in Theorem 7.13, we now show that the semigroup generated by A is uniquely determined.

Theorem 7.14 (Uniqueness of the semigroup). *Let $\{S_t\}$, $\{\widetilde{S}_t\}$ be two strongly continuous semigroups of linear operators, having the same generator A. Then $S_t = \widetilde{S}_t$ for every $t \geq 0$.*

Proof. Let $u \in \mathrm{Dom}(A)$. Then $\widetilde{S}_s u \in \mathrm{Dom}(A)$ and $S_{t-s}\widetilde{S}_s u \in \mathrm{Dom}(A)$ for every $0 \leq s \leq t$. We can thus estimate

$$\widetilde{S}_t u - S_t u = \int_0^t \frac{d}{ds}\Big[S_{t-s}\widetilde{S}_s u\Big]\, ds = \int_0^t \Big[S_{t-s} A \widetilde{S}_s u - A S_{t-s}\widetilde{S}_s u\Big]\, ds = 0.$$

Indeed,

$$\begin{aligned}
\frac{d}{ds}\Big[S_{t-s}\widetilde{S}_s u\Big] &= \lim_{h \to 0} \frac{S_{t-s-h}(\widetilde{S}_{s+h} u) - S_{t-s}\widetilde{S}_s u}{h} \\
&= \lim_{h \to 0} \frac{S_{t-s-h}(\widetilde{S}_{s+h} u - \widetilde{S}_s u)}{h} + \lim_{h \to 0} \frac{S_{t-s-h}(\widetilde{S}_s u) - S_{t-s}(\widetilde{S}_s u)}{h} \\
&= S_{t-s}(A\widetilde{S}_s u) - A S_{t-s}(\widetilde{S}_s u) = 0.
\end{aligned}$$

This shows that, for every fixed $t \geq 0$, the bounded linear operators S_t and $\widetilde{S}_t$ coincide on the dense subset $\mathrm{Dom}(A)$. Hence $S_t = \widetilde{S}_t$. □

Applications of the theory of semigroups to the solution of partial differential equations of evolutionary type will be illustrated in Chapter 9.

7.5. Problems

1. Assume that the operator A in (7.10) is sectorial, so that (7.13) holds, for some $\omega \in \mathbb{R}$ and some angle $\bar{\theta} < \pi/2$.

(i) For $t > 0$, give an a priori estimate on the norm of the operator $\|Ae^{tA}\|$, depending only on $\omega, \bar{\theta}$.

(ii) More generally, for $t > 0$ and $k \geq 1$, prove that the operator $A^k e^{tA}$ is also bounded. Estimate the norm $\|A^k e^{tA}\|$.

2. Let A be the generator of a contractive semigroup. Show that, for every $\lambda > 0$, the bounded linear operator A_λ in (7.30) also generates a contractive semigroup.

3. Fix $\lambda > 0$ and consider the weight function

$$w(t) = \begin{cases} \lambda e^{-\lambda t} & \text{if } t \geq 0, \\ 0 & \text{if } t < 0. \end{cases}$$

By induction on $n \geq 1$, check that the n-fold convolution $w_n \doteq w * w * \cdots * w$ is given by

$$w_n(t) = \begin{cases} \dfrac{\lambda^n}{(n-1)!} t^{n-1} e^{-\lambda t} & \text{if } t \geq 0, \\ 0 & \text{if } t < 0. \end{cases} \tag{7.37}$$

Show that $w(\cdot)$ is the density of a probability measure with

$$\text{mean value} = \int_0^\infty t\, w(t)\, dt = \frac{1}{\lambda},$$

$$\text{variance} = \int_0^\infty \left(t - \frac{1}{\lambda}\right)^2 w(t)\, dt = \int_0^\infty \left(t^2 - \frac{1}{\lambda^2}\right) w(t)\, dt = \frac{1}{\lambda^2}.$$

Therefore, $w_n(\cdot)$ is the density of a probability measure with mean n/λ and variance n/λ^2.

Now fix $T > 0$ and let $\lambda_n \doteq n/T$. For every $n \geq 1$, set

$$W_n(t) = \begin{cases} \dfrac{(n/T)^n}{(n-1)!} t^{n-1} e^{-(n/T)t} & \text{if } t \geq 0, \\ 0 & \text{if } t < 0. \end{cases} \tag{7.38}$$

Show that $W_n(\cdot)$, plotted in Figure 7.5.1, is the density of a probability measure with mean value $n/\lambda_n = T$ and variance $n/(n/T)^2 = T^2/n$. In particular, for every $\varepsilon > 0$, one has

(7.39)

$$\lim_{n\to\infty} \left[\int_0^{T-\varepsilon} W_n(t)\,dt + \int_{T+\varepsilon}^{\infty} W_n(t)\,dt \right] = 0, \qquad \lim_{n\to\infty} \int_{T-\varepsilon}^{T+\varepsilon} W_n(t)\,dt = 1.$$

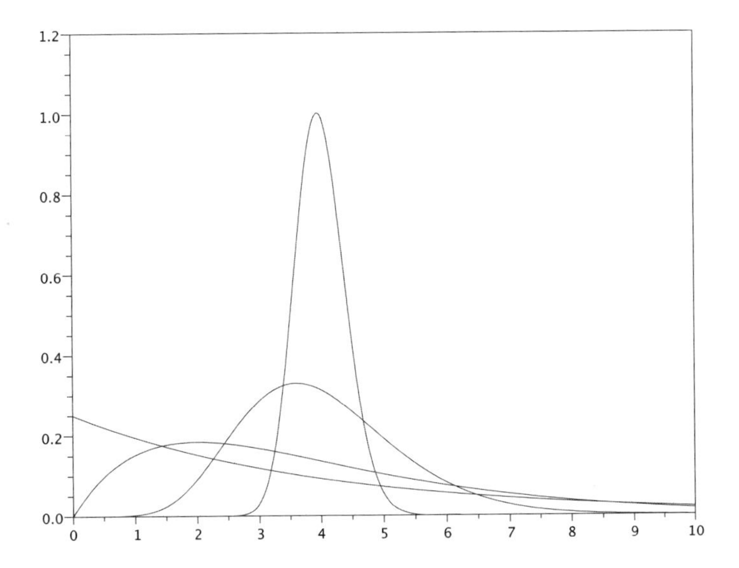

Figure 7.5.1. A plot of the weight functions W_n defined in (7.38). Here $T = 4$, while $n = 1, 2, 10, 100$. All these probability distributions have average value 4, while their variance $(= 16/n)$ decreases to zero as $n \to \infty$.

4. (Convergence of backward Euler approximations) Let $\{S_t\,;\ t \geq 0\}$ be a contractive semigroup on a Banach space X, generated by the linear operator A. Fix $h > 0$ and set $\lambda = 1/h$. By induction on $n \geq 1$, extend formula (7.28) for backward Euler approximations. Namely, prove that

$$(I - hA)^{-n} u = \int_0^{\infty} w_n(t) S_t u\,dt \qquad \text{for all } u \in X,$$

where w_n is the weight function defined in (7.37).

Set $h = T/n$, so that $\lambda_n = n/T$. Using (7.39) prove that, for every $u \in X$, one has the convergence

$$\left(I - \frac{T}{n} A\right)^{-n} u = \int_0^{\infty} W_n(t) S_t u\,dt \to S_T u \qquad \text{as } n \to \infty.$$

5. (Positively invariant sets) Let A be the generator of a contractive semigroup $\{S_t\,;\ t \geq 0\}$ on a Banach space X. Let $\Omega \subset X$ be a closed, convex subset. Prove that the following are equivalent.

(i) Ω is positively invariant: if $u \in \Omega$ then $S_t u \in \Omega$ for all $t \geq 0$.

(ii) Ω is invariant with respect to the backward Euler operator: if $u \in \Omega$, then for every $h > 0$ one has $(I - hA)^{-1} u \in \Omega$.

6. Let $\{S_t\,;\ t \geq 0\}$ be a semigroup of type ω, and let A be its generator. Prove that, for every $\gamma \in \mathbb{R}$, the family of bounded linear operators $\{e^{\gamma t} S_t\,;\ t \geq 0\}$ is a semigroup of type $\omega + \gamma$, having $A + \gamma I$ as its generator.

7. On the space $\mathbf{L}^p(\mathbb{R})$, consider the linear operators S_t defined by

$$(S_t f)(x) \;=\; e^{-2t} f(x+t)\,.$$

(i) Prove that the family of linear operators $\{S_t\,;\ t \geq 0\}$ is a strongly continuous, contractive semigroup on $\mathbf{L}^p(\mathbb{R})$, for all $1 \leq p < \infty$. Find the generator A of this semigroup. What is $\mathrm{Dom}(A)$?

(ii) Show that the family of operators $\{S_t\,;\ t \geq 0\}$ is NOT a strongly continuous semigroup on $\mathbf{L}^\infty(\mathbb{R})$.

8. Let $\{S_t\,;\ t \geq 0\}$ be a strongly continuous semigroup of linear operators on $\mathbb{R}^n$. Prove that there exists an $n \times n$ matrix A such that $S_t = e^{tA}$ for every $t \geq 0$.

9. (Semilinear equations) Let A be a linear operator on a Banach space X, generating the contractive semigroup $\{S_t\,;\ t \geq 0\}$. We say that a map $u : [0,T] \mapsto X$ is a **mild solution** of the semilinear Cauchy problem

$$\frac{d}{dt}u(t) \;=\; Au(t) + f(t, u(t))\,, \qquad u(0) = \bar{u} \tag{7.40}$$

if

$$u(t) \;=\; S_t \bar{u} + \int_0^t S_{t-s} f(s, u(s))\, ds \qquad \text{for all } \ t \in [0,T]\,. \tag{7.41}$$

(i) Assuming that $f : [0,T] \times X \mapsto X$ is continuous and satisfies the bounds

$$\|f(t,x)\| \;\leq\; M\,, \qquad \|f(t,x) - f(t,y)\| \;\leq\; L\,\|x - y\| \qquad \text{for all } \ t, x, y,$$

prove that the Cauchy problem (7.40) admits a unique mild solution.

(ii) In the special case where A is a bounded operator, so that $S_t = e^{tA}$, prove that $u(\cdot)$ satisfies (7.41) if and only if u is a continuously differentiable solution to the Cauchy problem (7.40).

10. Fix a time $T > 0$. On the space $X = \mathbf{L}^1([0,1])$, construct a strongly continuous, contractive semigroup of linear operators $\{S_t\,;\ t \geq 0\}$ such that $\|S_t\| = 1$ for $0 \leq t < T$ but $\|S_t\| = 0$ for $t \geq T$.

11. Let $\{S_t\,;\ t \geq 0\}$ be a strongly continuous, contractive semigroup of linear operators on a Banach space X. Assume that, for a given time $\tau > 0$, the operator S_τ is compact. Prove that, for every $t > \tau$, the operator S_t compact as well. Construct an example showing that the operators S_t may not be compact for $0 \leq t < \tau$.

12. On the space $\mathbf{L}^1(\mathbb{R})$, consider the operator $Au = \frac{\partial}{\partial x} u$ with domain

$$\mathrm{Dom}(A) \;=\; \Big\{ u \in \mathbf{L}^1(\mathbb{R})\,;\ u \text{ is absolutely continuous, } u_x \in \mathbf{L}^1(\mathbb{R}) \Big\}.$$

(i) Describe the semigroup $\{S_t\,;\ t \geq 0\}$ generated by A.

(ii) For any $u \in \mathbf{L}^1(\mathbb{R})$ and $h > 0$, construct the backward Euler approximation $E_h^- u$.

(iii) Let $u \in \mathcal{C}_c^\infty(\mathbb{R})$ be a smooth function with compact support, say with $u(x) = 0$ for $x \notin [a, b]$. Given any time step $h > 0$, show that the forward Euler approximations $(E_h^+)^n u \;=\; (I + hA)^n u$ are well defined and have support contained in $[a, b]$.

(iv) Using (iii) show that, for every time $\tau > 0$ sufficiently large, the functions $\big(I + \frac{\tau}{n} A\big)^n u$ cannot converge to $S_\tau u$ as $n \to \infty$.

Chapter 8

Sobolev Spaces

The present chapter covers the basic theory of Sobolev spaces, which provide a very useful abstract framework for the analysis of both linear and nonlinear PDEs. We begin with an introduction to the theory of distributions and some motivating examples.

8.1. Distributions and weak derivatives

In the following, $\mathbf{L}^1_{\text{loc}}(\mathbb{R})$ denotes the space of **locally summable functions** $f : \mathbb{R} \mapsto \mathbb{R}$. These are the Lebesgue measurable functions which are summable over every bounded interval. The **support** of a function ϕ, denoted by $\text{Supp}(\phi)$, is the closure of the set $\{x\,;\ \phi(x) \neq 0\}$ where ϕ does not vanish. By $\mathcal{C}^\infty_c(\mathbb{R})$ we denote the space of continuous functions with compact support, having continuous derivatives of every order.

Every locally summable function $f \in \mathbf{L}^1_{\text{loc}}(\mathbb{R})$ determines a linear functional $\Lambda_f : \mathcal{C}^\infty_c(\mathbb{R}) \mapsto \mathbb{R}$, namely

$$\Lambda_f(\phi) \doteq \int_{\mathbb{R}} f(x)\phi(x)\,dx\,. \tag{8.1}$$

Notice that this integral is well defined for every $\phi \in \mathcal{C}^\infty_c(\mathbb{R})$. Moreover, if $\text{Supp}(\phi) \subseteq [a,b]$, we have the estimate

$$|\Lambda_f(\phi)| \leq \left(\int_a^b |f(x)|\,dx\right) \|\phi\|_{\mathcal{C}^0}\,. \tag{8.2}$$

Next, assume that f is continuously differentiable. Then its derivative

$$f'(x) = \lim_{h\to 0} \frac{f(x+h) - f(x)}{h}$$

is continuous, hence locally summable. In turn, f' also determines a linear functional on $\mathcal{C}_c^\infty(\mathbb{R})$, namely

$$\Lambda_{f'}(\phi) \doteq \int_{\mathbb{R}} f'(x)\,\phi(x)\,dx = -\int_{\mathbb{R}} f(x)\,\phi'(x)\,dx\,. \tag{8.3}$$

At this stage, a key observation is that the first integral in (8.3) is defined only if $f'(x)$ exists for a.e. x and is locally summable. However, the second integral is well defined for every locally summable function f, even if f does not have a pointwise derivative at any point. Moreover, if $\mathrm{Supp}(\phi) \subseteq [a,b]$, we have the estimate

$$|\Lambda_{f'}(\phi)| \leq \left(\int_a^b |f(x)|\,dx\right) \|\phi\|_{\mathcal{C}^1}\,.$$

This construction can also be performed for higher-order derivatives.

Definition 8.1. Given an integer $k \geq 1$, the **distributional derivative** of order k of $f \in \mathbf{L}^1_{\mathrm{loc}}(\mathbb{R})$ is the linear functional

$$\Lambda_{D^k f}(\phi) \doteq (-1)^k \int_{\mathbb{R}} f(x) D^k\phi(x)\,dx\,.$$

If there exists a locally summable function g such that $\Lambda_g = \Lambda_{D^k f}$, namely

$$\int_{\mathbb{R}} g(x)\phi(x)\,dx = (-1)^k \int_{\mathbb{R}} f(x) D^k\phi(x)\,dx \qquad \text{for all } \phi \in \mathcal{C}_c^\infty(\mathbb{R}),$$

then we say that g is the **weak derivative** of order k of f.

Remark 8.2. Classical derivatives are defined pointwise, as limits of difference quotients. On the other hand, weak derivatives are defined only in an integral sense, up to a set of measure zero. By arbitrarily changing the function f on a set of measure zero we do not affect its weak derivatives in any way.

Example 8.3. Consider the function

$$f(x) \doteq \begin{cases} 0 & \text{if } x \leq 0, \\ x & \text{if } x > 0\,. \end{cases}$$

Its distributional derivative is the map

$$\Lambda(\phi) = -\int_0^\infty x \cdot \phi'(x)\,dx = \int_{\mathbb{R}} H(x)\,\phi(x)\,dx\,,$$

where

$$H(x) = \begin{cases} 0 & \text{if } x \leq 0, \\ 1 & \text{if } x > 0\,. \end{cases} \tag{8.4}$$

In this case, the Heaviside function H in (8.4) is the weak derivative of f.

Example 8.4. The function H in (8.4) is locally summable. Its distributional derivative is the linear functional

$$\Lambda(\phi) \doteq -\int_{\mathbb{R}} H(x)\,\phi'(x)\,dx = -\int_0^\infty \phi'(x)\,dx = \phi(0).$$

This corresponds to the Dirac measure, concentrating a unit mass at the origin. We claim that the function H does not have any weak derivative, i.e., there cannot exist any locally summable function g such that

$$\int g(x)\,\phi(x)\,dx = \phi(0) \qquad \text{for all } \phi \in \mathcal{C}_c^\infty. \tag{8.5}$$

Indeed, if (8.5) holds, then by the Lebesgue dominated convergence theorem

$$\lim_{h\to 0}\int_{-h}^{h} |g(x)|\,dx = 0.$$

Hence we can choose $\delta > 0$ so that $\int_{-\delta}^{\delta} |g(x)|\,dx \leq 1/2$. Let $\phi : \mathbb{R} \mapsto [0,1]$ be a smooth function, with $\phi(0) = 1$ and with support contained in the interval $[-\delta, \delta]$. We now reach a contradiction by writing

$$1 = \phi(0) = \Lambda(\phi) = \int_{\mathbb{R}} g(x)\phi(x)\,dx = \int_{-\delta}^{\delta} g(x)\phi(x)\,dx$$
$$\leq \max_x|\phi(x)|\cdot\int_{-\delta}^{\delta}|g(x)|\,dx \leq \frac{1}{2}.$$

Example 8.5. Consider the function

$$f(x) \doteq \begin{cases} 0 & \text{if } x \text{ is rational,} \\ 2+\sin x & \text{if } x \text{ is irrational.} \end{cases}$$

Being discontinuous at every point, the function f is nowhere differentiable. On the other hand, the function $g(x) = \cos x$ provides a weak derivative for f. Indeed, the behavior of f on the set of rational points (having measure zero) is irrelevant. We thus have

$$-\int f(x)\phi'(x)\,dx = -\int (2+\sin x)\,\phi'(x)\,dx = \int (\cos x)\,\phi(x)\,dx.$$

Example 8.6. Consider the Cantor function $f : \mathbb{R} \mapsto [0,1]$, defined by

$$f(x) = \begin{cases} 0 & \text{if } x \leq 0, \\ 1 & \text{if } x \geq 1, \\ 1/2 & \text{if } x \in [1/3,\ 2/3], \\ 1/4 & \text{if } x \in [1/9,\ 2/9], \\ 3/4 & \text{if } x \in [7/9,\ 8/9], \\ & \dots \end{cases} \tag{8.6}$$

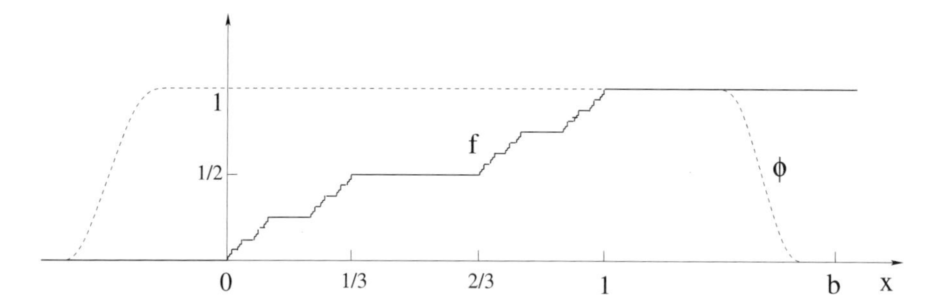

Figure 8.1.1. The Cantor function f and a test function ϕ showing that $g(x) \equiv 0$ cannot be the weak derivative of f.

(see Figure 8.1.1). This provides a standard example of a continuous function which is not absolutely continuous. We claim that f does not have a weak derivative. Indeed, assume on the contrary that $g \in \mathbf{L}^1_{\text{loc}}$ is a weak derivative of f. Since f is constant on each of the open sets

$$]-\infty, 0[\,, \qquad]1, +\infty[\,, \qquad \left]\frac{1}{3}, \frac{2}{3}\right[\,, \qquad \left]\frac{1}{9}, \frac{2}{9}\right[\,, \qquad \left]\frac{7}{9}, \frac{8}{9}\right[\,, \quad \ldots,$$

we must have $g(x) = f'(x) = 0$ on the union of these open intervals. Hence $g(x) = 0$ for a.e. $x \in \mathbb{R}$. To obtain a contradiction, it remains to show that the function $g \equiv 0$ is NOT the weak derivative of f. As shown in Figure 8.1.1, let $\phi \in \mathcal{C}^\infty_c$ be a test function such that $\phi(x) = 1$ for $x \in [0, 1]$ while $\phi(x) = 0$ for $x \geq b$. Then

$$\int g(x)\phi(x)\,dx \;=\; 0 \;\neq\; 1 \;=\; -\int f(x)\phi'(x)\,dx\,.$$

8.1.1. Distributions. The construction described in the previous section can be extended to any open domain in a multi-dimensional space. Let $\Omega \subseteq \mathbb{R}^n$ be an open set. By $\mathbf{L}^1_{\text{loc}}(\Omega)$ we denote the space of locally summable functions on Ω. These are the measurable functions $f : \Omega \mapsto \mathbb{R}$ which are summable restricted to every compact subset $K \subset \Omega$.

Example 8.7. The functions e^x and $\ln|x|$ are in $\mathbf{L}^1_{\text{loc}}(\mathbb{R})$, while $x^{-1} \notin \mathbf{L}^1_{\text{loc}}(\mathbb{R})$. On the other hand, the function $f(x) = x^\gamma$ is in $\mathbf{L}^1_{\text{loc}}(]0, \infty[)$ for every (positive or negative) exponent $\gamma \in \mathbb{R}$. In several space dimensions, the function $f(x) = |x|^{-\gamma}$ is in $\mathbf{L}^1_{\text{loc}}(\mathbb{R}^n)$ provided that $\gamma < n$. One should keep in mind that the pointwise values of a function $f \in \mathbf{L}^1_{\text{loc}}$ on a set of measure zero are irrelevant.

By $\mathcal{C}^\infty_c(\Omega)$ we denote the space of continuous functions $\phi : \Omega \mapsto \mathbb{R}$ having continuous partial derivatives of all orders and whose support is a compact subset of Ω. Functions $\phi \in \mathcal{C}^\infty_c(\Omega)$ are usually called *"test functions"*. We recall that the **support** of a function ϕ is the closure of the set where ϕ does not vanish:

$$\mathrm{Supp}\,(\phi) \;\doteq\; \overline{\{x \in \Omega;\ \phi(x) \neq 0\}}.$$

We shall need an efficient way to denote higher-order derivatives of a function f. A **multi-index** $\alpha = (\alpha_1, \alpha_2, \ldots, \alpha_n)$ is an n-tuple of nonnegative integer numbers. Its length is defined as

$$|\alpha| \;\doteq\; \alpha_1 + \alpha_2 + \cdots + \alpha_n\,.$$

Each multi-index α determines a partial differential operator of order $|\alpha|$, namely

$$D^\alpha f \;=\; \left(\frac{\partial}{\partial x_1}\right)^{\alpha_1} \left(\frac{\partial}{\partial x_2}\right)^{\alpha_2} \cdots \left(\frac{\partial}{\partial x_n}\right)^{\alpha_n} f\,.$$

Definition 8.8. A **distribution** on the open set $\Omega \subseteq \mathbb{R}^n$ is a linear functional $\Lambda : \mathcal{C}_c^\infty(\Omega) \mapsto \mathbb{R}$ such that the following boundedness property holds: For every compact $K \subset \Omega$ there exists an integer $N \geq 0$ and a constant C such that

$$(8.7)\qquad |\Lambda(\phi)| \;\leq\; C\,\|\phi\|_{\mathcal{C}^N} \qquad \text{for every } \phi \in \mathcal{C}^\infty \text{ with } \mathrm{Supp}(\phi) \subseteq K\,.$$

In other words, for all test functions ϕ which vanish outside a given compact set K, the value $\Lambda(\phi)$ should be bounded in terms of the maximum value of derivatives of ϕ, up to a certain order N.

Notice that here both N and C depend on the compact subset K. If there exists an integer $N \geq 0$ independent of K such that (8.7) holds (with $C = C_K$ possibly still depending on K), we say that the distribution has finite order. The smallest such integer N is called the **order of the distribution**.

Example 8.9. Let Ω be an open subset of $\mathbb{R}^n$ and consider a function $f \in \mathbf{L}^1_{\mathrm{loc}}(\Omega)$. Then the linear map $\Lambda_f : \mathcal{C}_c^\infty(\Omega) \mapsto \mathbb{R}$ defined by

$$(8.8)\qquad \Lambda_f(\phi) \;\doteq\; \int_\Omega f\,\phi\,dx$$

is a distribution. Indeed, it is clear that Λ_f is well defined and linear. Given a compact subset $K \subset \Omega$, for every test function ϕ with $\mathrm{Supp}(\phi) \subseteq K$ we have the estimate

$$|\Lambda_f(\phi)| \;=\; \left|\int_K f\,\phi\,dx\right| \;\leq\; \int_K |f(x)|\,dx \cdot \max_{x\in K}|\phi(x)| \;\leq\; C\,\|\phi\|_{\mathcal{C}^0}\,.$$

Hence the bound (8.7) holds with $C = \int_K |f|\,dx$ and $N = 0$. This provides an example of a distribution of order zero.

The family of all distributions on Ω is clearly a vector space. A remarkable fact is that, while a function f may not admit any partial derivative (in the classical sense), for a distribution Λ an appropriate notion of derivative can always be defined.

Definition 8.10. Given a distribution Λ and a multi-index α, we define the distribution $D^\alpha \Lambda$ by setting

$$D^\alpha \Lambda(\phi) \doteq (-1)^{|\alpha|} \Lambda(D^\alpha \phi). \tag{8.9}$$

It is easy to check that $D^\alpha \Lambda$ is itself a distribution. Indeed, the linearity of the map $\phi \mapsto D^\alpha \Lambda(\phi)$ is clear. Next, let K be a compact subset of Ω and let ϕ be a test function with support contained in K. By assumption, there exists a constant C and an integer $N \geq 0$ such that (8.7) holds. In turn, this implies

$$|D^\alpha \Lambda(\phi)| = |\Lambda(D^\alpha \phi)| \leq C \|D^\alpha \phi\|_{\mathcal{C}^N} \leq C \|\phi\|_{\mathcal{C}^{N+|\alpha|}}.$$

Hence $D^\alpha \Lambda$ also satisfies (8.7), with N replaced by $N + |\alpha|$.

Notice that, if Λ_f is the distribution in (8.8) corresponding to a function f which is $|\alpha|$-times continuously differentiable, then we can integrate by parts and obtain

$$\begin{aligned} D^\alpha \Lambda_f(\phi) &= (-1)^{|\alpha|} \Lambda_f(D^\alpha \phi) = (-1)^{|\alpha|} \int f(x) D^\alpha \phi(x)\, dx \\ &= \int D^\alpha f(x) \phi(x)\, dx = \Lambda_{D^\alpha f}(\phi). \end{aligned}$$

This justifies the formula (8.9).

8.1.2. Weak derivatives. For every locally summable function f and every multi-index $\alpha = (\alpha_1, \ldots, \alpha_n)$, the distribution Λ_f always admits a distributional derivative $D^\alpha \Lambda_f$, defined according to (8.9). In some cases, one can find a locally summable function g such that the distribution $D^\alpha \Lambda_f$ coincides with the distribution Λ_g. This leads to the concept of weak derivative.

Definition 8.11. Let $f \in \mathbf{L}^1_{\mathrm{loc}}(\Omega)$ be a locally summable function on the open set $\Omega \subseteq \mathbb{R}^n$ and let Λ_f be the corresponding distribution, as in (8.8). Given a multi-index $\alpha = (\alpha_1, \ldots, \alpha_n)$, if there exists a locally summable function $g \in \mathbf{L}^1_{\mathrm{loc}}(\Omega)$ such that $D^\alpha \Lambda_f = \Lambda_g$, i.e.,

$$\int f\, D^\alpha \phi\, dx = (-1)^{|\alpha|} \int g\, \phi\, dx \qquad \text{for all test functions } \phi \in \mathcal{C}^\infty_c(\Omega), \tag{8.10}$$

then we say that g is the **weak α-th derivative of** f and write $g = D^\alpha f$.

In general, a weak derivative may not exist. In particular, Example 8.4 shows that the Heaviside function does not admit a weak derivative. Indeed, its distributional derivative is a Dirac measure (concentrating a unit mass at the origin), not a locally summable function. On the other hand, if a weak derivative does exist, then it is unique (up to a set of measure zero).

Lemma 8.12 (Uniqueness of weak derivatives). *Assume $f \in \mathbf{L}^1_{\rm loc}(\Omega)$ and let $g, \tilde{g} \in \mathbf{L}^1_{\rm loc}(\Omega)$ be the weak α-th derivatives of f, so that*

$$\int f\, D^\alpha \phi\, dx \;=\; (-1)^{|\alpha|} \int g\, \phi\, dx \;=\; (-1)^{|\alpha|} \int \tilde{g}\, \phi\, dx$$

for all test functions $\phi \in \mathcal{C}^\infty_c(\Omega)$. Then $g(x) = \tilde{g}(x)$ for a.e. $x \in \Omega$.

Proof. By the assumptions, the function $(g - \tilde{g}) \in \mathbf{L}^1_{\rm loc}(\Omega)$ satisfies

$$\int (g - \tilde{g})\phi\, dx \;=\; 0 \qquad \text{for all test functions } \phi \in \mathcal{C}^\infty_c(\Omega).$$

By Corollary A.17 in the Appendix, we thus have $g(x) - \tilde{g}(x) = 0$ for a.e. $x \in \Omega$. □

If a function f is twice continuously differentiable, a basic theorem of calculus states that partial derivatives commute: $f_{x_j x_k} = f_{x_k x_j}$. This property remains valid for weak derivatives. To state this result in full generality, we recall that the sum of two multi-indices $\alpha = (\alpha_1, \ldots, \alpha_n)$ and $\beta = (\beta_1, \ldots, \beta_n)$ is defined as $\alpha + \beta = (\alpha_1 + \beta_1, \ldots, \alpha_n + \beta_n)$.

Lemma 8.13 (Weak derivatives commute). *Assume that $f \in \mathbf{L}^1_{\rm loc}(\Omega)$ has weak derivatives $D^\alpha f$ for every $|\alpha| \le k$. Then, for every pair of multi-indices α, β with $|\alpha| + |\beta| \le k$ one has*

$$D^\alpha(D^\beta f) \;=\; D^\beta(D^\alpha f) \;=\; D^{\alpha+\beta} f\,. \tag{8.11}$$

Proof. Consider any test function $\phi \in \mathcal{C}^\infty_c(\Omega)$. Using the fact that $D^\beta \phi \in \mathcal{C}^\infty_c(\Omega)$ is a test function as well, we obtain

$$\begin{aligned}
\int_\Omega D^\alpha f\, D^\beta \phi\, dx \;&=\; (-1)^{|\alpha|} \int_\Omega f\,(D^{\alpha+\beta}\phi)\, dx \\
&=\; (-1)^{|\alpha|}(-1)^{|\alpha+\beta|} \int_\Omega (D^{\alpha+\beta} f)\, \phi\, dx \\
&=\; (-1)^{|\beta|} \int_\Omega (D^{\alpha+\beta} f)\, \phi\, dx\,.
\end{aligned}$$

By definition, this means that $D^{\alpha+\beta}f = D^\beta(D^\alpha f)$. Exchanging the roles of the multi-indices α and β in the previous computation, one obtains $D^{\alpha+\beta}f = D^\alpha(D^\beta f)$, completing the proof. □

The next lemma extends another familiar result, stating that the weak derivative of a limit coincides with the limit of the weak derivatives.

Lemma 8.14 (Convergence of weak derivatives). *Consider a sequence of functions $f_n \in \mathbf{L}^1_{\mathrm{loc}}(\Omega)$. For a fixed multi-index α, assume that each f_n admits the weak derivative $g_n = D^\alpha f_n$. If $f_n \to f$ and $g_n \to g$ in $\mathbf{L}^1_{\mathrm{loc}}(\Omega)$, then $g = D^\alpha f$.*

Proof. For every test function $\phi \in \mathcal{C}_c^\infty(\Omega)$, a direct computation yields

$$\begin{aligned}\int_\Omega g\,\phi\,dx &= \lim_{n\to\infty}\int_\Omega g_n\,\phi\,dx = \lim_{n\to\infty}(-1)^{|\alpha|}\int_\Omega f_n\,D^\alpha\phi\,dx\\ &= (-1)^{|\alpha|}\int_\Omega f\,D^\alpha\phi\,dx\,.\end{aligned}$$

By definition, this means that g is the α-th weak derivative of f. □

8.2. Mollifications

As usual, let $\Omega \subseteq \mathbb{R}^n$ be an open set. For a given $\varepsilon > 0$, define the open subset

$$\Omega_\varepsilon \doteq \{x \in \mathbb{R}^n\,;\ \overline{B}(x,\varepsilon) \subset \Omega\}\,. \tag{8.12}$$

For every $u \in \mathbf{L}^1_{\mathrm{loc}}(\Omega)$ the mollification

$$u_\varepsilon(x) \doteq (J_\varepsilon * u)(x) = \int_{B(x,\varepsilon)} J_\varepsilon(x-y)u(y)\,dy$$

is well defined for every $x \in \Omega_\varepsilon$. Moreover, $u_\varepsilon \in \mathcal{C}^\infty(\Omega_\varepsilon)$. A very useful property of the mollification operator is that it commutes with weak differentiation.

Lemma 8.15 (Mollifications). *Let $\Omega_\varepsilon \subset \Omega$ be as in (8.12). Assume that a function $u \in \mathbf{L}^1_{\mathrm{loc}}(\Omega)$ admits a weak derivative $D^\alpha u$, for some multi-index α. Then the derivative of the mollification (which exists in the classical sense) coincides with the mollification of the weak derivative:*

$$D^\alpha(J_\varepsilon * u) = J_\varepsilon * D^\alpha u \qquad \textit{for all } x \in \Omega_\varepsilon\,. \tag{8.13}$$

Proof. Observe that, for each fixed $x \in \Omega_\varepsilon$, the function $\phi(y) \doteq J_\varepsilon(x-y)$ is in $\mathcal{C}_c^\infty(\Omega)$. Hence we can apply the definition of weak derivative $D^\alpha u$ using ϕ as a test function. Writing D_x^α and D_y^α to distinguish differentiation with respect to the variables x or y, we thus obtain

$$\begin{aligned} D^\alpha u_\varepsilon(x) &= D_x^\alpha\left(\int_\Omega J_\varepsilon(x-y)\,u(y)\,dy\right) \\ &= \int_\Omega D_x^\alpha J_\varepsilon(x-y)\,u(y)\,dy \\ &= (-1)^{|\alpha|}\int_\Omega D_y^\alpha J_\varepsilon(x-y)\,u(y)\,dy \\ &= (-1)^{|\alpha|+|\alpha|}\int_\Omega J_\varepsilon(x-y)\,D_y^\alpha u(y)\,dy \\ &= \Big(J_\varepsilon * D^\alpha u\Big)(x). \end{aligned}$$

□

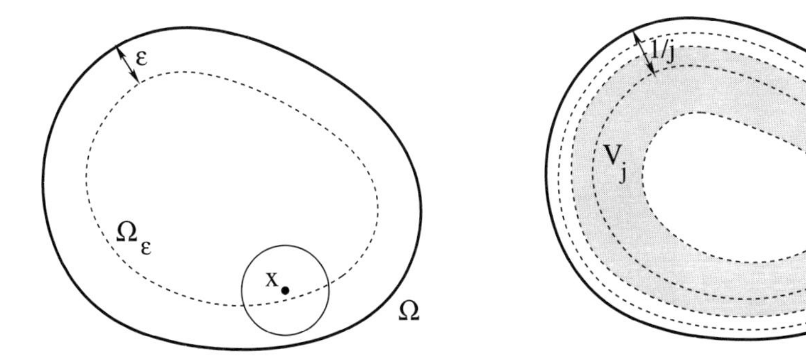

Figure 8.2.1. Left: the open subset $\Omega_\varepsilon \subset \Omega$ of points having distance $> \varepsilon$ from the boundary. Right: the domain Ω can be covered by countably many open subdomains $V_j = \Omega_{1/(j-1)} \setminus \overline{\Omega}_{1/(j+1)}$.

This property of mollifications stated in Lemma 8.15 provides the key tool to relate weak derivatives with partial derivatives in the classical sense. As a first application, we prove

Corollary 8.16 (Constant functions). *Let $\Omega \subseteq \mathbb{R}^n$ be an open, connected set, and assume $u \in L^1_{\mathrm{loc}}(\Omega)$. If the first-order weak derivatives of u satisfy*

$$D_{x_i}u(x) \;=\; 0 \qquad \text{for } i = 1,2,\ldots,n \text{ and a.e. } x \in \Omega\,,$$

then u coincides a.e. with a constant function.

Proof. 1. For $\varepsilon > 0$, consider the mollified function $u_\varepsilon = J_\varepsilon * u$. By the previous analysis, $u_\varepsilon : \Omega_\varepsilon \mapsto \mathbb{R}$ is a smooth function whose derivatives $D_{x_i} u_\varepsilon$ vanish identically on Ω_ε. Therefore, u_ε must be constant on each connected component of Ω_ε.

2. Now consider any two points $x, y \in \Omega$. Since the open set Ω is connected, there exists a polygonal path Γ joining x with y and remaining inside Ω. Let $\delta \doteq \min_{z \in \Gamma} d(z, \partial\Omega) > 0$ be the minimum distance of points in Γ to the boundary of Ω. Then for every $\varepsilon < \delta$ the whole polygonal curve Γ is in Ω_ε. Hence x, y lie in the same connected component of Ω_ε. In particular, $u_\varepsilon(x) = u_\varepsilon(y)$.

3. Let $\tilde{u}(x) \doteq \lim_{\varepsilon \to 0} u_\varepsilon(x)$. By the previous step, $\tilde{u}$ is a constant function on Ω. Moreover, $\tilde{u}(x) = u(x)$ at every Lebesgue point of u, hence almost everywhere on Ω. This concludes the proof. □

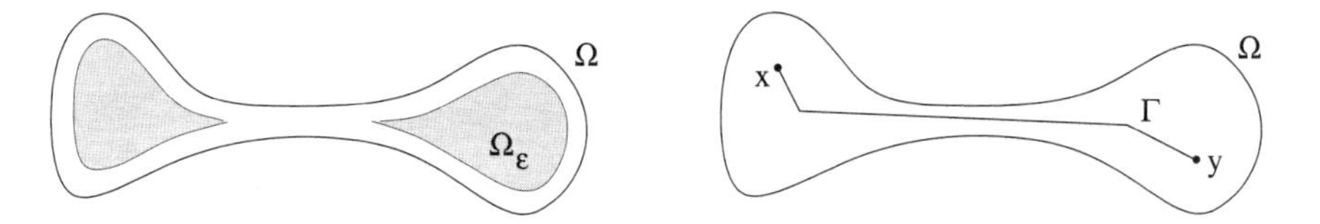

Figure 8.2.2. Left: even if Ω is connected, the subdomain $\Omega_\varepsilon \doteq \{x \in \Omega;\ \overline{B}(x, \varepsilon) \subseteq \Omega\}$ may not be connected. Right: any two points $x, y \in \Omega$ can be connected by a polygonal path Γ remaining inside Ω. Hence, if $\varepsilon > 0$ is sufficiently small, x and y belong to the same connected component of Ω_ε.

In the one-dimensional case, relying again on Lemma 8.15, we now characterize the set of functions having a weak derivative in $\mathbf{L}^1$.

Corollary 8.17 (Absolutely continuous functions). *Consider an open interval* $]a, b[$ *and assume that* $u \in \mathbf{L}^1_{\text{loc}}(]a, b[)$ *has a weak derivative* $v \in \mathbf{L}^1(]a, b[)$. *Then there exists an absolutely continuous function* $\tilde{u}$ *such that*

$$\tilde{u}(x) = u(x) \qquad \text{for a.e. } x \in]a, b[, \tag{8.14}$$

$$v(x) = \lim_{h \to 0} \frac{\tilde{u}(x+h) - \tilde{u}(x)}{h} \qquad \text{for a.e. } x \in]a, b[. \tag{8.15}$$

Proof. Let $x_0 \in]a,b[$ be a Lebesgue point of u, and define

$$\tilde{u}(x) \doteq u(x_0) + \int_{x_0}^{x} v(y)\, dy\,.$$

Clearly $\tilde{u}$ is absolutely continuous and satisfies (8.15).

In order to prove (8.14), let J_ε be the standard mollifier and let $u_\varepsilon \doteq J_\varepsilon * u$, $v_\varepsilon \doteq J_\varepsilon * v$. Then $u_\varepsilon, v_\varepsilon \in \mathcal{C}^\infty(]a+\varepsilon\,,\ b-\varepsilon[)$, while Lemma 8.15 yields

$$u_\varepsilon(x) = u_\varepsilon(x_0) + \int_{x_0}^{x} v_\varepsilon(y)\, dy \qquad \text{for all } x \in]a+\varepsilon\,,\ b-\varepsilon[\,. \tag{8.16}$$

Letting $\varepsilon \to 0$, we obtain $u_\varepsilon(x_0) \to u(x_0)$ because x_0 is a Lebesgue point. Moreover, the right-hand side of (8.16) converges to $\tilde{u}(x)$ for every $x \in]a,b[$, while the left-hand side converges to $u(x)$ at every Lebesgue point of u (and hence almost everywhere). Therefore (8.14) holds. □

If $f, g \in \mathbf{L}^1_{\text{loc}}(\Omega)$ are weakly differentiable functions, for any constants $a, b \in \mathbb{R}$ it is clear that the linear combination $af + bg$ is also weakly differentiable. Indeed,

$$D_{x_i}(af + bg) = a\, D_{x_i} f + b\, D_{x_i} g\,. \tag{8.17}$$

We now consider products and compositions of weakly differentiable functions. One should be aware that, in general, the product of two functions $f, g \in \mathbf{L}^1_{\text{loc}}$ may not be locally summable. Similarly, the product of two weakly differentiable functions on $\mathbb{R}^n$ may not be weakly differentiable (see problem 20). For this reason, in the next lemma we shall assume that one of the two functions is continuously differentiable with uniformly bounded derivatives.

Given two multi-indices $\alpha = (\alpha_1, \ldots, \alpha_n)$ and $\beta = (\beta_1, \ldots, \beta_n)$, we recall that the notation $\beta \leq \alpha$ means $\beta_i \leq \alpha_i$ for every $i = 1, \ldots, n$. Moreover,

$$\binom{\alpha}{\beta} \doteq \frac{\alpha!}{\beta!\,(\alpha-\beta)!} \doteq \frac{\alpha_1!}{\beta_1!\,(\alpha_1-\beta_1)!}\cdot\frac{\alpha_2!}{\beta_2!\,(\alpha_2-\beta_2)!}\cdots\frac{\alpha_2!}{\beta_2!\,(\alpha_2-\beta_2)!}\,.$$

Lemma 8.18 (Products and compositions of weakly differentiable functions). *Let $\Omega \subseteq \mathbb{R}^n$ be any open set and consider a function $u \in \mathbf{L}^1_{\text{loc}}(\Omega)$ having weak derivatives $D^\alpha u$ of every order $|\alpha| \leq k$.*

(i) *If $\eta \in \mathcal{C}^k(\Omega)$, then the product ηu admits weak derivatives up to order k. These are given by the Leibniz formula*

$$D^\alpha(\eta u) = \sum_{\beta \leq \alpha} \binom{\alpha}{\beta} D^\beta \eta\, D^{\alpha-\beta} u\,. \tag{8.18}$$

(ii) *Let $\Omega' \subseteq \mathbb{R}^n$ be an open set and let $\varphi : \Omega' \mapsto \Omega$ be a $\mathcal{C}^k$ bijection whose Jacobian matrix has a uniformly bounded inverse. Then the composition $u \circ \varphi$ is a function in $\mathbf{L}^1_{\mathrm{loc}}(\Omega')$ which admits weak derivatives up to order k.*

Proof. 1. To prove (i), let J_ε be the standard mollifier and set $u_\varepsilon \doteq J_\varepsilon * u$. Since the Leibniz formula holds for the product of smooth functions, for every $\varepsilon > 0$ we obtain

$$D^\alpha(\eta u_\varepsilon) = \sum_{\beta \le \alpha} \binom{\alpha}{\beta} D^\beta \eta \, D^{\alpha-\beta} u_\varepsilon \,. \tag{8.19}$$

For every test function $\phi \in \mathcal{C}_c^\infty(\Omega)$, we thus have

$$\begin{aligned} (-1)^{|\alpha|} \int_\Omega (\eta u_\varepsilon) D^\alpha \phi \, dx &= \int_\Omega D^\alpha(\eta u_\varepsilon) \phi \, dx \\ &= \sum_{\beta \le \alpha} \binom{\alpha}{\beta} \int_\Omega \left(D^\beta \eta \, D^{\alpha-\beta} u_\varepsilon \right) \phi \, dx \,. \end{aligned}$$

Notice that, if $\varepsilon > 0$ is small enough so that $\mathrm{Supp}(\phi) \subset \Omega_\varepsilon$, then the above integrals are well defined. Letting $\varepsilon \to 0$, we obtain

$$(-1)^{|\alpha|} \int_\Omega (\eta u) D^\alpha \phi \, dx = \int_\Omega \left(\sum_{\beta \le \alpha} \binom{\alpha}{\beta} D^\beta \eta \, D^{\alpha-\beta} u \right) \phi \, dx \,.$$

By the definition of weak derivative, (8.18) holds.

2. We prove (ii) by induction on k. Call y the variable in Ω' and $x = \varphi(y)$ the variable in Ω, as shown in Figure 8.2.3. By assumption, the $n \times n$ Jacobian matrix $\left(\frac{\partial \varphi_i}{\partial y_j}\right)_{i,j=1,\ldots,n}$ has a uniformly bounded inverse. Hence the composition $u \circ \varphi$ lies in $\mathbf{L}^1_{\mathrm{loc}}(\Omega')$, proving the theorem in the case $k = 0$.

Next, assume that the result is true for all weak derivatives of order $|\alpha| \le k - 1$. Consider any test function $\phi \in \mathcal{C}_c^\infty(\Omega')$ and define the mollification $u_\varepsilon \doteq J_\varepsilon * u$. For any $\varepsilon > 0$ small enough so that $\varphi\big(\mathrm{Supp}(\phi)\big) \subset \Omega_\varepsilon$, we have

$$\begin{aligned} -\int_{\Omega'} (u_\varepsilon \circ \varphi) \cdot D_{y_i} \phi \, dy &= \int_{\Omega'} D_{y_i}(u_\varepsilon \circ \varphi) \cdot \phi \, dy \\ &= \int_{\Omega'} \left(\sum_{j=1}^n D_{x_j} u_\varepsilon(\varphi(y)) \cdot D_{y_i} \varphi_j(y) \right) \cdot \phi(y) \, dy \,. \end{aligned}$$

Letting $\varepsilon \to 0$, we conclude that the composition $u \circ \varphi$ admits a first-order weak derivative, given by

$$D_{y_i}(u \circ \varphi)(y) = \sum_{j=1}^{n} D_{x_j} u(\varphi(y)) \cdot D_{y_i} \varphi_j(y). \tag{8.20}$$

By the inductive assumption, each function $D_{x_j}(u \circ \varphi)$ admits weak derivatives up to order $k-1$, while $D_{y_i}\varphi_j \in \mathcal{C}^{k-1}(\Omega')$. By part (i) of the lemma, all the products on the right-hand side of (8.20) have weak derivatives up to order $k-1$. Using Lemma 8.13, we conclude that the composition $u \circ \varphi$ admits weak derivatives up to order k. By induction, this concludes the proof. □

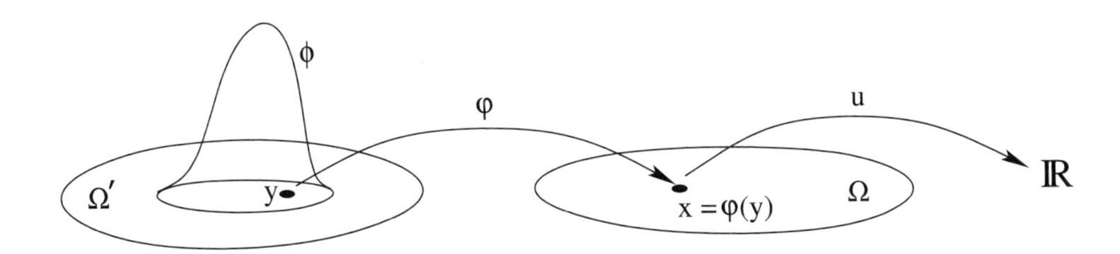

Figure 8.2.3. The mappings considered in part (ii) of Lemma 8.18.

8.3. Sobolev spaces

Consider an open set $\Omega \subseteq \mathbb{R}^n$, fix $p \in [1, \infty]$, and let k be a nonnegative integer.

Definition 8.19. (i) The **Sobolev space** $W^{k,p}(\Omega)$ is the space of all locally summable functions $u : \Omega \mapsto \mathbb{R}$ such that, for every multi-index α with $|\alpha| \leq k$, the weak derivative $D^\alpha u$ exists and belongs to $\mathbf{L}^p(\Omega)$.

On $W^{k,p}$ we shall use the norm

$$\|u\|_{W^{k,p}} \doteq \left(\sum_{|\alpha| \leq k} \int_\Omega |D^\alpha u|^p \, dx \right)^{1/p} \qquad \text{if } 1 \leq p < \infty, \tag{8.21}$$

$$\|u\|_{W^{k,\infty}} \doteq \sum_{|\alpha| \leq k} \operatorname*{ess\,sup}_{x \in \Omega} |D^\alpha u| \qquad \text{if } p = \infty. \tag{8.22}$$

(ii) The subspace $W_0^{k,p}(\Omega) \subseteq W^{k,p}(\Omega)$ is defined as the closure of $\mathcal{C}_c^\infty(\Omega)$ in $W^{k,p}(\Omega)$. More precisely, $u \in W_0^{k,p}(\Omega)$ if and only if there exists a sequence of functions $u_n \in \mathcal{C}_c^\infty(\Omega)$ such that

$$\|u - u_n\|_{W^{k,p}} \to 0.$$

(iii) In addition, $W_{\mathrm{loc}}^{k,p}(\Omega)$ denotes the space of functions which are locally in $W^{k,p}$. These are the functions $u : \Omega \mapsto \mathbb{R}$ satisfying the following property: If Ω' is an open set compactly contained[1] in Ω, then the restriction of u to Ω' is in $W^{k,p}(\Omega')$.

Intuitively, one can think of the closed subspace $W_0^{1,p}(\Omega)$ as the space of all functions $u \in W^{1,p}(\Omega)$ which vanish along the boundary of Ω. More generally, $W_0^{k,p}(\Omega)$ is a space of functions whose derivatives $D^\alpha u$ vanish along $\partial\Omega$, for $|\alpha| \leq k-1$.

Definition 8.20. In the special case where $p = 2$, we define the **Hilbert-Sobolev** space $H^k(\Omega) \doteq W^{k,2}(\Omega)$. The space $H^k(\Omega)$ is endowed with the inner product

$$\langle u, v\rangle_{H^k} \doteq \sum_{|\alpha|\leq k} \int_\Omega D^\alpha u \, D^\alpha v \, dx \,. \tag{8.23}$$

Similarly, we define $H_0^k(\Omega) \doteq W_0^{k,2}(\Omega)$.

Theorem 8.21 (Basic properties of Sobolev spaces).

(i) *Each Sobolev space $W^{k,p}(\Omega)$ is a Banach space.*

(ii) *The space $W_0^{k,p}(\Omega)$ is a closed subspace of $W^{k,p}(\Omega)$. Hence it is a Banach space, with the same norm.*

(iii) *The spaces $H^k(\Omega)$ and $H_0^k(\Omega)$ are Hilbert spaces.*

Proof. 1. Let $u, v \in W^{k,p}(\Omega)$. For $|\alpha| \leq k$, call $D^\alpha u$, $D^\alpha v$ their weak derivatives. Then, for any $\lambda, \mu \in \mathbb{R}$, the linear combination $\lambda u + \mu v$ is a locally summable function. One easily checks that its weak derivatives are

$$D^\alpha(\lambda u + \mu v) = \lambda D^\alpha u + \mu D^\alpha v \,. \tag{8.24}$$

Therefore, $D^\alpha(\lambda u+\mu v) \in \mathbf{L}^p(\Omega)$ for every $|\alpha| \leq k$. This proves that $W^{k,p}(\Omega)$ is a vector space.

[1]By definition, a set Ω' is *compactly contained in* Ω if the closure $\overline{\Omega'}$ is a compact subset of Ω.

2. Next, we check that (8.21) and (8.22) satisfy the conditions (N1)–(N3) defining a norm. For $\lambda \in \mathbb{R}$ and $u \in W^{k,p}$ one has

$$\|\lambda u\|_{W^{k,p}} = |\lambda| \, \|u\|_{W^{k,p}} ,$$

$$\|u\|_{W^{k,p}} \geq \|u\|_{\mathbf{L}^p} \geq 0 ,$$

with equality holding if and only if $u = 0$.

Moreover, if $u, v \in W^{k,p}(\Omega)$, then for $1 \leq p < \infty$ Minkowski's inequality yields

$$\begin{aligned} \|u+v\|_{W^{k,p}} &= \left(\sum_{|\alpha| \leq k} \|D^\alpha u + D^\alpha v\|_{\mathbf{L}^p}^p \right)^{1/p} \\ &\leq \left(\sum_{|\alpha| \leq k} \Big(\|D^\alpha u\|_{\mathbf{L}^p} + \|D^\alpha v\|_{\mathbf{L}^p} \Big)^p \right)^{1/p} \\ &\leq \left(\sum_{|\alpha| \leq k} \|D^\alpha u\|_{\mathbf{L}^p}^p \right)^{1/p} + \left(\sum_{|\alpha| \leq k} \|D^\alpha v\|_{\mathbf{L}^p}^p \right)^{1/p} \\ &= \|u\|_{W^{k,p}} + \|v\|_{W^{k,p}}. \end{aligned}$$

In the case $p = \infty$, the above computation is replaced by

$$\begin{aligned} \|u+v\|_{W^{k,\infty}} &= \sum_{|\alpha| \leq k} \|D^\alpha u + D^\alpha v\|_{\mathbf{L}^\infty} \leq \sum_{|\alpha| \leq k} \Big(\|D^\alpha u\|_{\mathbf{L}^\infty} + \|D^\alpha v\|_{\mathbf{L}^\infty} \Big) \\ &= \|u\|_{W^{k,\infty}} + \|v\|_{W^{k,\infty}}. \end{aligned}$$

3. To conclude the proof of (i), we need to show that the space $W^{k,p}(\Omega)$ is complete; hence it is a Banach space.

Let $(u_n)_{n \geq 1}$ be a Cauchy sequence in $W^{k,p}(\Omega)$. For any multi-index α with $|\alpha| \leq k$, the sequence of weak derivatives $D^\alpha u_n$ is Cauchy in $\mathbf{L}^p(\Omega)$. Since the space $\mathbf{L}^p(\Omega)$ is complete, there exist functions u and u_α, such that

$$(8.25) \qquad \|u_n - u\|_{\mathbf{L}^p} \to 0, \qquad \|D^\alpha u_n - u_\alpha\|_{\mathbf{L}^p} \to 0 \qquad \text{for all } |\alpha| \leq k .$$

By Lemma 8.14, the limit function u_α is precisely the weak derivative $D^\alpha u$. Since this holds for every multi-index α with $|\alpha| \leq k$, the convergence $u_n \to u$ holds in $W^{k,p}(\Omega)$. This completes the proof of (i).

4. The fact that $W_0^{k,p}(\Omega)$ is a closed subspace of $W^{k,p}(\Omega)$ follows immediately from the definition. It is also straightforward to check that (8.23) is an inner product, yielding the norm $\|\cdot\|_{W^{k,2}}$. □

Example 8.22. Let $\Omega =]a,b[$ be an open interval. By Corollary 8.17, each element of the space $W^{1,p}(]a,b[)$ coincides a.e. with an absolutely continuous function $f:]a,b[\mapsto \mathbb{R}$ having derivative $f' \in \mathbf{L}^p(]a,b[)$.

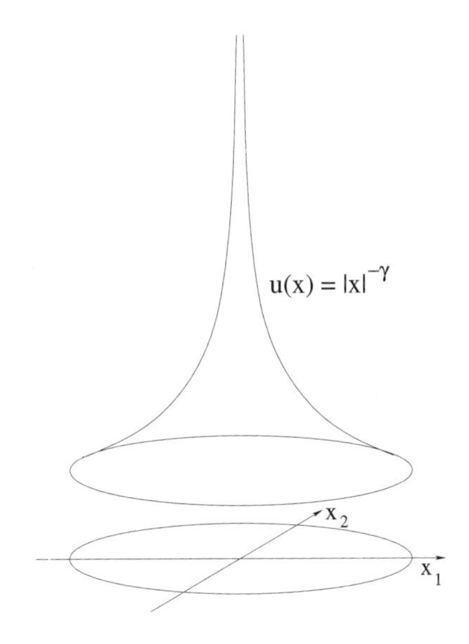

Figure 8.3.1. For certain values of p, n a function $u \in W^{1,p}(\mathbb{R}^n)$ may not be continuous, or bounded.

Example 8.23. Let $\Omega = B(0,1) \subset \mathbb{R}^n$ be the open ball centered at the origin with radius one. Fix $\gamma > 0$ and consider the radially symmetric function

$$u(x) \doteq |x|^{-\gamma} = \left(\sum_{i=1}^n x_i^2\right)^{-\gamma/2}, \qquad 0 < |x| < 1.$$

Observe that $u \in \mathcal{C}^1(\Omega \setminus \{0\})$. We claim that, for $1 \le p < \infty$,

$$u \in W^{1,p}(\Omega) \qquad \text{if and only if} \qquad \gamma < \frac{n-p}{p}. \tag{8.26}$$

Outside the origin, the partial derivatives are computed as

$$u_{x_i} = -\frac{\gamma}{2}\left(\sum_{i=1}^n x_i^2\right)^{-(\gamma/2)-1} 2x_i = \frac{-\gamma x_i}{|x|^{\gamma+2}}. \tag{8.27}$$

Hence, the gradient $\nabla u = (u_{x_1}, \ldots, u_{x_n})$ has Euclidean norm

$$|\nabla u(x)| = \left(\sum_{i=1}^{n} |u_{x_i}(x)|^2 \right)^{1/2} = \frac{\gamma}{|x|^{\gamma+1}}.$$

Calling σ_n the $(n-1)$-dimensional measure of the unit sphere $\{x \in \mathbb{R}^n\,;\ |x| = 1\} \subset \mathbb{R}^n$, we compute

$$\begin{aligned} \int_\Omega \left(\frac{1}{|x|^{\gamma+1}} \right)^p dx &= \int_{x\in\mathbb{R}^n,\ |x|<1} |x|^{-p(\gamma+1)}\, dx \\ &= \sigma_n \int_0^1 r^{n-1} r^{-p(\gamma+1)}\, dr\,. \end{aligned} \tag{8.28}$$

The right-hand side of (8.28) is finite if and only if $n - 1 - p(\gamma + 1) > -1$, i.e., $\gamma < \frac{n-p}{p}$. After a similar computation for $|u|^p$, we conclude that

$$\int_\Omega \Big(|u|^p + |\nabla u|^p \Big)\, dx < \infty \qquad \text{if and only if} \qquad \gamma < \frac{n-p}{p}.$$

To complete the proof of (8.26), we need to show that, if $0 < \gamma < \frac{n-p}{p}$ (and hence $n \geq 2$), then the functions in (8.27) are indeed the weak derivatives of u on the entire ball Ω (and not only on the set $\Omega \setminus \{0\}$). For this purpose, consider any test function $\phi \in \mathcal{C}_c^\infty(\Omega)$. Fix $i \in \{1, \ldots, n\}$. For convenience, we extend ϕ to the entire space $\mathbb{R}^n$ by setting $\phi(x) = 0$ for $x \notin \Omega$. Since $n \geq 2$, the x_i-axis has n-dimensional Lebesgue measure zero. An integration by part yields

$$\begin{aligned} \int_\Omega \frac{-\gamma x_i}{|x|^{\gamma+2}}\, \phi\, dx &= \int_{\mathbb{R}^{n-1}\setminus\{0\}} \left(\int_{\mathbb{R}} \frac{-\gamma x_i}{|x|^{\gamma+2}}\, \phi\, dx_i \right) dx_1 \cdots dx_{i-1} dx_{i+1} \cdots dx_n \\ &= - \int_{\mathbb{R}^{n-1}\setminus\{0\}} \left(\int_{\mathbb{R}} |x|^{-\gamma}\, \phi_{x_i}\, dx_i \right) dx_1 \cdots dx_{i-1} dx_{i+1} \cdots dx_n \\ &= - \int_\Omega |x|^{-\gamma}\, \phi_{x_i}\, dx\,, \end{aligned}$$

completing the proof.

Observe that the previous computation relied on the fact that u is absolutely continuous (in fact, smooth) on a.e. line parallel to one of the coordinate axes. However, there is no way to change u on a set of measure zero, so that it becomes continuous on the whole domain Ω.

An important question in the theory of Sobolev spaces is whether one can estimate the norm of a function in terms of the norm of its first derivatives.

The following result provides an elementary estimate in this direction. It is valid for domains Ω which are contained in a slab, say

$$\Omega \subseteq \{x = (x_1, x_2, \ldots, x_n)\,;\ a < x_1 < b\}\,. \tag{8.29}$$

Theorem 8.24 (Poincaré's inequality. I). *Let $\Omega \subset \mathbb{R}^n$ be an open set which satisfies (8.29) for some $a, b \in \mathbb{R}$. Then every $u \in H_0^1(\Omega)$ satisfies*

$$\|u\|_{\mathbf{L}^2(\Omega)} \leq 2(b-a)\ \|D_{x_1}u\|_{\mathbf{L}^2(\Omega)}\,. \tag{8.30}$$

Proof. 1. Assume first that $u \in \mathcal{C}_c^\infty(\Omega)$. We extend u to the whole space $\mathbb{R}^n$ by setting $u(x) = 0$ for $x \notin \Omega$. Using the variables $x = (x_1, x')$ with $x' = (x_2, \ldots, x_n)$, we compute

$$u^2(x_1, x') = \int_a^{x_1} 2uu_{x_1}(t, x')\, dt\,.$$

An integration by parts yields

$$\|u\|_{\mathbf{L}^2}^2 = \int_{\mathbb{R}^n} u^2(x)\, dx = \int_{\mathbb{R}^{n-1}} \int_a^b 1 \cdot \left(\int_a^{x_1} 2uu_{x_1}(t, x')\, dt\right) dx_1\, dx'$$

$$= \int_{\mathbb{R}^{n-1}} \int_a^b (b - x_1)\, 2uu_{x_1}(x_1, x')\, dx_1\, dx' \leq 2(b-a) \int_{\mathbb{R}^n} |u|\, |u_{x_1}|\, dx$$

$$\leq 2(b-a)\|u\|_{\mathbf{L}^2}\|u_{x_1}\|_{\mathbf{L}^2}\,.$$

Dividing both sides by $\|u\|_{\mathbf{L}^2}$, we obtain (8.30) for every $u \in \mathcal{C}_c^\infty(\Omega)$.

2. Now consider any $u \in H_0^1(\Omega)$. By assumption there exists a sequence of functions $u_n \in \mathcal{C}_c^\infty(\Omega)$ such that $\|u_n - u\|_{H^1} \to 0$. By the previous step, this implies

$$\|u\|_{\mathbf{L}^2} = \lim_{n\to\infty} \|u_n\|_{\mathbf{L}^2} \leq \lim_{n\to\infty} 2(b-a)\|D_{x_1}u_n\|_{\mathbf{L}^2} = 2(b-a)\, \|D_{x_1}u\|_{\mathbf{L}^2}\,.$$

□

To proceed in the analysis of Sobolev spaces, we need to establish some additional facts about weak derivatives.

Theorem 8.25 (Properties of weak derivatives). *Let $\Omega \subseteq \mathbb{R}^n$ be an open set, let $p \in [1, \infty]$, and let $|\alpha| \leq k$. If $u, v \in W^{k,p}(\Omega)$, then:*

(i) *The restriction of u to any open set $\widetilde{\Omega} \subset \Omega$ is in the space $W^{k,p}(\widetilde{\Omega})$.*
(ii) $D^\alpha u \in W^{k-|\alpha|,p}(\Omega)$.

(iii) *If* $\eta \in \mathcal{C}^k(\Omega)$, *then the product satisfies* $\eta\, u \in W^{k,p}(\Omega)$. *Moreover there exists a constant* C, *depending on* Ω *and on* $\|\eta\|_{\mathcal{C}^k}$ *but not on* u, *such that*

$$\|\eta u\|_{W^{k,p}(\Omega)} \;\leq\; C\,\|u\|_{W^{k,p}(\Omega)}\,. \tag{8.31}$$

(iv) *Let* $\Omega' \subseteq \mathbb{R}^n$ *be an open set and let* $\varphi : \Omega' \mapsto \Omega$ *be a* $\mathcal{C}^k$ *diffeomorphism whose Jacobian matrix has a uniformly bounded inverse. Then the composition satisfies* $u \circ \varphi \in W^{k,p}(\Omega')$. *Moreover there exists a constant* C, *depending on* Ω' *and on* $\|\varphi\|_{\mathcal{C}^k}$ *but not on* u, *such that*

$$\|u \circ \varphi\|_{W^{k,p}(\Omega')} \;\leq\; C\,\|u\|_{W^{k,p}(\Omega)}\,. \tag{8.32}$$

Proof. Statement (i) is an obvious consequence of the definitions, while (ii) follows from Lemma 8.13.

To prove (iii), we observe that by assumption all derivatives of η are bounded, namely

$$\|D^\beta \eta\|_{\mathbf{L}^\infty} \;\leq\; \|\eta\|_{\mathcal{C}^k} \qquad \text{for all } |\beta| \leq k\,.$$

Hence the bound (8.31) follows from the representation formula (8.18).

Recalling part (ii) of Lemma 8.18, we prove (iv) by induction on k. By assumption, the $n \times n$ Jacobian matrix $(D_{x_i}\varphi_j)_{i,j=1,\ldots,n}$ has a uniformly bounded inverse. Hence the case $k = 0$ is clear.

Next, assume that the result is true for $k = 0, 1, \ldots, m-1$. If $u \in W^{m,p}(\Omega)$, we have

$$\|D_{x_i}(u \circ \varphi)\|_{W^{m-1,p}(\Omega')} \;\leq\; C'\,\|\nabla u\|_{W^{m-1,p}(\Omega)}\,\|\varphi\|_{\mathcal{C}^m(\Omega')} \;\leq\; C\,\|u\|_{W^{m,p}(\Omega)}\,,$$

showing that the result is also true for $k = m$. By induction, this achieves the proof. □

8.4. Approximations of Sobolev functions

If $u \in W^{k,p}(\mathbb{R}^n)$ is a function defined on the entire space $\mathbb{R}^n$, it can be approximated by smooth functions simply by taking mollifications: $u_\varepsilon = J_\varepsilon * u$. However, if u is only defined on some open subset $\Omega \subset \mathbb{R}^n$, a more careful construction is needed.

Theorem 8.26 (Approximation with smooth functions). *Let* $\Omega \subseteq \mathbb{R}^n$ *be an open set. Let* $u \in W^{k,p}(\Omega)$ *with* $1 \leq p < \infty$. *Then for any* $\varepsilon > 0$ *there exists a function* $U \in \mathcal{C}^\infty(\Omega)$ *such that* $\|U - u\|_{W^{k,p}(\Omega)} < \varepsilon$.

Proof. 1. Let $\varepsilon > 0$ be given. Consider the following locally finite open covering of the set Ω, shown in Figure 8.2.1:

$$V_1 \doteq \left\{x \in \Omega;\ d(x, \partial\Omega) > \frac{1}{2}\right\},$$

$$V_j \doteq \left\{x \in \Omega;\ \frac{1}{j+1} < d(x, \partial\Omega) < \frac{1}{j-1}\right\}, \qquad j = 2, 3, \ldots .$$

Using Theorem A.18 in the Appendix, let $\eta_1, \eta_2, \ldots$ be a smooth partition of unity subordinate to the above covering. By Theorem 8.25, for every $j \geq 0$ the product $\eta_j u$ is in $W^{k,p}(\Omega)$. By construction, it has support contained in V_j.

2. Consider the mollifications $J_\varepsilon * (\eta_j u)$. By Lemma 8.15 and by Theorem A.16 in the Appendix, for every $|\alpha| \leq k$ we have

$$D^\alpha(J_\varepsilon * (\eta_j u)) = J_\epsilon * (D^\alpha(\eta_j u)) \to D^\alpha(\eta_j u)$$

as $\varepsilon \to 0$. Since each η_j has compact support, here the convergence takes place in $\mathbf{L}^p(\Omega)$. Therefore, for each $j \geq 0$ we can find $0 < \varepsilon_j < 2^{-j}$ small enough so that

$$\|\eta_j u - J_{\varepsilon_j} * (\eta_j u)\|_{W^{k,p}(\Omega)} \leq \varepsilon\, 2^{-j}.$$

3. Consider the function

$$U \doteq \sum_{j=1}^{\infty} J_{\varepsilon_j} * (\eta_j u).$$

Notice that the above series may not converge in $W^{k,p}$. However, it is certainly pointwise convergent because every compact set $K \subset \Omega$ intersects finitely many of the sets V_j. Restricted to K, the above sum contains only finitely many nonzero terms. Since each term is smooth, this implies $U \in \mathcal{C}^\infty(\Omega)$.

4. Consider the subdomains

$$\Omega_{1/n} \doteq \left\{x \in \Omega;\ d(x, \partial\Omega) > \frac{1}{n}\right\}.$$

Recalling that $\sum_j \eta_j(x) \equiv 1$, for every $n \geq 1$ we obtain

$$\|U - u\|_{W^{k,p}(\Omega_{1/n})} \leq \sum_{j=1}^{n+2} \|\eta_j u - J_{\varepsilon_j} * (\eta_j u)\|_{W^{k,p}(\Omega_{1/n})} \leq \sum_{j=1}^{n+2} \varepsilon\, 2^{-j} \leq \varepsilon.$$

This yields

$$\|U - u\|_{W^{k,p}(\Omega)} = \sup_{n \geq 1} \|U - u\|_{W^{k,p}(\Omega_{1/n})} \leq \varepsilon\,.$$

Since $\varepsilon > 0$ was arbitrary, this proves that the set of $\mathcal{C}^\infty$ function is dense on $W^{k,p}(\Omega)$. □

Using the above approximation result, we obtain a first regularity theorem for Sobolev functions (see Figure 8.4.1).

Theorem 8.27 (Relation between weak and strong derivatives). *Let $u \in W^{1,1}(\Omega)$, where $\Omega \subseteq \mathbb{R}^n$ is an open set having the form*

$$(8.33)\quad \Omega = \Big\{x = (x, x')\,;\ x' \doteq (x_2, \ldots, x_n) \in \Omega'\,,\ \alpha(x') < x_1 < \beta(x')\Big\}$$

(possibly with $\alpha \equiv -\infty$ or $\beta \equiv +\infty$). Then there exists a function $\tilde{u}$ with $\tilde{u}(x) = u(x)$ for a.e. $x \in \Omega$, such that the following holds. For a.e. $x' = (x_2, \ldots, x_n) \in \Omega' \subset \mathbb{R}^{n-1}$ (with respect to the $(n-1)$-dimensional Lebesgue measure), the map

$$x_1 \mapsto \tilde{u}(x_1, x')$$

is absolutely continuous. Its derivative coincides a.e. with the weak derivative $D_{x_1}u$.

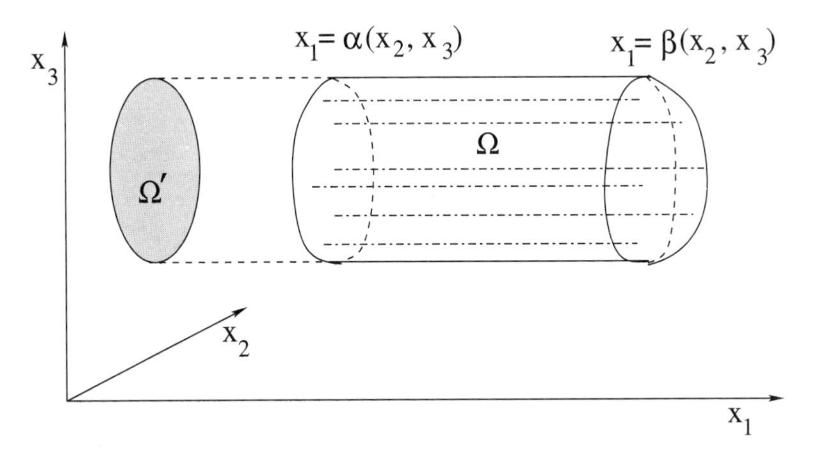

Figure 8.4.1. The domain Ω at (8.33). If u has a weak derivative $D_{x_1}u \in \mathbf{L}^1(\Omega)$, then (by possibly changing its values on a set of measure zero) the function u is absolutely continuous on almost every segment parallel to the x_1-axis and its partial derivative $\partial u/\partial x_1$ coincides a.e. with the weak derivative.

Proof. 1. By the previous theorem, there exists a sequence of functions $u_k \in \mathcal{C}^\infty(\Omega)$ such that

$$(8.34)\qquad \|u_k - u\|_{W^{1,1}} < 2^{-k}.$$

We claim that this implies the pointwise convergence

$$u_k(x) \to u(x), \qquad D_{x_1} u_k(x) \to D_{x_1} u(x) \qquad \text{for a.e. } x \in \Omega .$$

Indeed, consider the functions

$$(8.35) \qquad \begin{aligned} f(x) &\doteq |u_1(x)| + \sum_{k=1}^{\infty} |u_{k+1}(x) - u_k(x)|, \\ g(x) &\doteq |D_{x_1} u_1(x)| + \sum_{k=1}^{\infty} \Big| D_{x_1} u_{k+1}(x) - D_{x_1} u_k(x) \Big| . \end{aligned}$$

By (8.34),

$$\|u_k - u_{k+1}\|_{W^{1,1}} \leq 2^{1-k};$$

hence $f, g \in \mathbf{L}^1(\Omega)$ and the series in (8.35) are absolutely convergent for a.e. $x \in \Omega$. Therefore, they converge pointwise almost everywhere. Moreover, we have the bounds

$$(8.36) \quad |u_k(x)| \leq f(x), \qquad |D_{x_1} u_k(x)| \leq g(x) \qquad \text{for all } n \geq 1,\ x \in \Omega .$$

2. Since $f, g \in \mathbf{L}^1(\Omega)$, by Fubini's theorem there exists a null set $\mathcal{N} \subset \Omega'$ (with respect to the $(n-1)$-dimensional Lebesgue measure) such that, for every $x' \in \Omega' \setminus \mathcal{N}$, one has

$$(8.37) \qquad \int_{\alpha(x')}^{\beta(x')} f(x_1, x')\, dx_1 < \infty, \qquad \int_{\alpha(x')}^{\beta(x')} g(x_1, x')\, dx_1 < \infty.$$

Fix such a point $x' \in \Omega' \setminus \mathcal{N}$. Choose a point $y_1 \in]\alpha(x'), \beta(x')[$ where the pointwise convergence $u_k(y_1, x') \to u(y_1, x')$ holds. For every $\alpha(x') < x_1 < \beta(x')$, since u_k is smooth, we have

$$(8.38) \qquad u_k(x_1, x') = u_k(y_1, x') + \int_{y_1}^{x_1} D_{x_1} u_k(s, x')\, ds .$$

We now let $n \to \infty$ in (8.38). By (8.36) and (8.37), the functions $D_{x_1} u_k(\cdot, x')$ are all bounded by the integrable function $g(\cdot, x') \in \mathbf{L}^1$. By the dominated convergence theorem, the right-hand side of (8.38) thus converges to

$$(8.39) \qquad \tilde{u}(x_1, x') \doteq u(y_1, x') + \int_{y_1}^{x_1} D_{x_1} u(s, x')\, ds .$$

Clearly, the right-hand side of (8.39) is an absolutely continuous function of the scalar variable x_1. On the other hand, the left-hand side satisfies

$$\tilde{u}(x_1, x') \doteq \lim_{n\to\infty} u_k(x_1, x') = u(x_1, x') \qquad \text{for a.e. } x_1 \in [\alpha(x'), \beta(x')].$$

This achieves the proof. □

Remark 8.28. (i) It is clear that a similar result holds for any other derivative $D_{x_i}u$, with $i = 1, 2, \ldots, n$.

(ii) If $u \in W^{k,p}(\widetilde{\Omega})$ and $\Omega \subset \widetilde{\Omega}$, then the restriction of u to Ω lies in the space $W^{k,p}(\Omega)$. Even if the open set $\widetilde{\Omega}$ has a complicated topology, the result of Theorem 8.27 can be applied to any cylindrical subdomain $\Omega \subset \widetilde{\Omega}$ admitting the representation (8.33).

(iii) If $\Omega \subset \mathbb{R}^n$ is a bounded open set and $u \in W^{k,p}(\Omega)$, then $u \in W^{k,q}(\Omega)$ for every $q \in [1, p]$.

8.5. Extension operators

Let $\Omega \subset \mathbb{R}^n$ be a bounded open domain with $\mathcal{C}^1$ boundary.[2] Given a function u defined on Ω, the next theorem provides a way to extend it to the entire space $\mathbb{R}^n$, retaining some control on the $W^{1,p}$-norm.

Theorem 8.29 (Extension operators). *Let $\Omega \subset\subset \widetilde{\Omega} \subset \mathbb{R}^n$ be open sets, such that the closure of Ω is a compact subset of $\widetilde{\Omega}$. Moreover, assume that Ω has $\mathcal{C}^1$ boundary. Then there exists a bounded linear operator $E : W^{1,p}(\Omega) \mapsto W^{1,p}(\mathbb{R}^n)$ and a constant C such that*

(i) *$Eu(x) = u(x)$ for a.e. $x \in \Omega$,*

(ii) *$Eu(x) = 0$ for $x \notin \widetilde{\Omega}$,*

(iii) *one has the bound $\|Eu\|_{W^{1,p}(\mathbb{R}^n)} \leq C\|u\|_{W^{1,p}(\Omega)}$.*

Proof. 1. We first prove that the same conclusion holds in the case where the domain is a half-space: $\Omega = \{x = (x_1, x_2, \ldots, x_n)\,;\; x_1 > 0\}$, and $\widetilde{\Omega} = \mathbb{R}^n$. In this case, any function $u \in W^{1,p}(\Omega)$ can be extended to the whole space $\mathbb{R}^n$ by reflection, i.e., by setting

$$(E^\sharp u)(x_1, x_2, \ldots, x_n) \doteq u\left(|x_1|,\, x_2,\, x_3, \ldots,\, x_n\right). \tag{8.40}$$

By Theorem 8.27, for every $i \in \{1, \ldots, n\}$ the function u is absolutely continuous along a.e. line parallel to the x_i-axis. Hence the same is true of the extension $E^\sharp u$. A straightforward computation involving integration by parts shows that the first-order weak derivatives of $E^\sharp u$ exist on the entire space $\mathbb{R}^n$ and satisfy

$$\begin{cases} D_{x_1}E^\sharp u(-x_1, x_2, \ldots, x_n) &= -D_{x_1}u(x_1, x_2, \ldots, x_n), \\ D_{x_j}E^\sharp u(-x_1, x_2, \ldots, x_n) &= D_{x_j}u(x_1, x_2, \ldots, x_n) \qquad (j = 2, \ldots, n), \end{cases}$$

[2]Saying that Ω has $\mathcal{C}^1$ boundary means the following. For every boundary point $x \in \partial\Omega$ there exists a neighborhood V^x of x and a $\mathcal{C}^1$ map $\varphi : \mathbb{R}^n \mapsto \mathbb{R}$ such that $\nabla\varphi(x) \neq 0$ and $\Omega \cap V^x = \{y \in V^x\,;\; \varphi(y) > 0\}$.

for all $x_1 > 0$, $x_2, \ldots, x_n \in \mathbb{R}$. The extension operator $E^\sharp : W^{1,p}(\Omega) \mapsto W^{1,p}(\mathbb{R}^n)$ defined at (8.40) is clearly linear and bounded because

$$\|E^\sharp u\|_{W^{1,p}(\mathbb{R}^n)} \leq 2\|u\|_{W^{1,p}(\Omega)}\,.$$

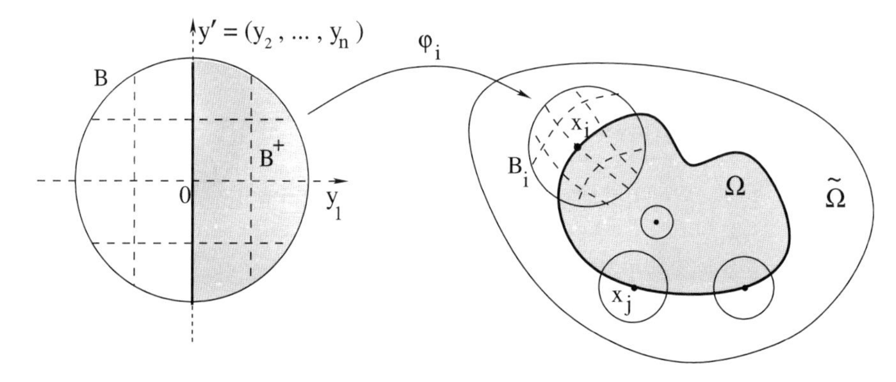

Figure 8.5.1. The open covering of the set Ω. For every ball $B_i = B(x_i, r_i)$ there is a $\mathcal{C}^1$ bijection φ_i mapping the open unit ball $B \subset \mathbb{R}^n$ onto B_i. For those balls B_i having center on the boundary Ω, the positive half-ball $B^+ = B \cap \{y_1 > 0\}$ is mapped onto the intersection $B_i \cap \Omega$.

2. To handle the general case, we use a partition of unity. For every $x \in \overline{\Omega}$ (the closure of Ω), choose a radius $r_x > 0$ such that the open ball $B(x, r_x)$ centered at x with radius $r_x 0$ satisfies the following conditions:

- If $x \in \Omega$, then $B(x, r_x) \subset \Omega$.

- If $x \in \partial\Omega$, then $B(x, r_x) \subset \widetilde{\Omega}$. Moreover, calling $B \doteq B(0,1)$ the open unit ball in $\mathbb{R}^n$, there exists a $\mathcal{C}^1$ bijection $\varphi^x : B \mapsto B(x, r_x)$, whose inverse is also $\mathcal{C}^1$, which maps the half-ball

$$B^+ \doteq \left\{ y = (y_1, y_2, \ldots, y_n)\,;\ \sum_{i=1}^n y_i^2 < 1\,,\ y_1 > 0 \right\}$$

onto the set $B(x, r^x) \cap \Omega$.

Choosing $r_x > 0$ sufficiently small, the existence of such a bijection follows from the assumption that Ω has a $\mathcal{C}^1$ boundary.

Since Ω is bounded, its closure $\overline{\Omega}$ is compact. Hence it can be covered with finitely many balls $B_i = B(x_i, r_i)$, $i = 1, \ldots, N$. Let $\varphi_i : B \mapsto B_i$ be the corresponding bijections. Recall that φ_i maps B^+ onto $B_i \cap \Omega$.

3. Let $\eta_1, \ldots, \eta_N$ be a smooth partition of unity subordinate to the above covering. For every $x \in \Omega$ we thus have

$$u(x) = \sum_{i=1}^{N} \eta_i(x)u(x). \tag{8.41}$$

We split the set of indices as

$$\{1, 2, \ldots, N\} = \mathcal{I} \cup \mathcal{J},$$

where $\mathcal{I}$ contains the indices with $x_i \in \Omega$ while $\mathcal{J}$ contains the indices with $x_i \in \partial\Omega$.

For every $i \in \mathcal{J}$, we have $\eta_i u \in W^{1,p}(B_i \cap \Omega)$. Hence by Theorem 8.25 (iv), one has $(\eta_i u) \circ \varphi_i \in W^{1,p}(B^+)$. Applying the extension operator $E^\sharp$ defined at (8.40), one obtains

$$E^\sharp\Big((\eta_i u) \circ \varphi_i\Big) \in W^{1,p}(B^+), \qquad E^\sharp\Big((\eta_i u) \circ \varphi_i\Big) \circ \varphi_i^{-1} \in W^{1,p}(B_i).$$

Summing all these extensions together, we define

$$Eu \doteq \sum_{i \in \mathcal{I}} \eta_i u + \sum_{i \in \mathcal{J}} E^\sharp\Big((\eta_i u) \circ \varphi_i\Big) \circ \varphi_i^{-1}.$$

It is now clear that the extension operator E satisfies all requirements. Indeed, (i) follows from (8.41). Property (ii) stems from the fact that, for every $u \in W^{1,p}(\Omega)$, the support of Eu is contained in $\bigcup_{i=1}^N B_i \subseteq \widetilde{\Omega}$. Finally, since E is defined as the sum of finitely many bounded linear operators, the bound (iii) holds, for some constant C. □

8.6. Embedding theorems

In one space dimension, a function $u : \mathbb{R} \mapsto \mathbb{R}$ which admits a weak derivative $Du \in \mathbf{L}^1(\mathbb{R})$ is absolutely continuous (after changing its values on a set of measure zero). On the other hand, if $\Omega \subseteq \mathbb{R}^n$ with $n \geq 2$, there exist functions $u \in W^{1,p}(\Omega)$ which are not continuous and not even bounded. This is indeed the case of the function $u(x) = |x|^{-\gamma}$, for $0 < \gamma < \frac{n-p}{p}$.

In several applications to PDEs or to the calculus of variations, it is important to understand the degree of regularity enjoyed by functions $u \in W^{k,p}(\mathbb{R}^n)$. We shall prove two basic results in this direction.

1. (Morrey) If $p > n$, then every function $u \in W^{1,p}(\mathbb{R}^n)$ is Hölder continuous (after a modification on a set of measure zero).

2. (Gagliardo-Nirenberg) If $p < n$, then every function $u \in W^{1,p}(\mathbb{R}^n)$ lies in the space $\mathbf{L}^{p^*}(\mathbb{R}^n)$, with the larger exponent $p^* = p + \frac{p^2}{n-p}$.

In both cases, the result can be stated as an embedding theorem: after a modification on a set of measure zero, each function $u \in W^{1,p}(\Omega)$ also lies in some other Banach space X. Here $X = \mathcal{C}^{0,\gamma}$ or $X = \mathbf{L}^q$ for some $q > p$. The basic approach is as follows:

(I): Prove an a priori bound valid for all smooth functions. Given any function $u \in \mathcal{C}^\infty(\Omega) \cap W^{k,p}(\Omega)$, one proves that u also lies in another Banach space X and that there exists a constant C depending on k, p, Ω but not on u, such that

$$\|u\|_X \leq C\,\|u\|_{W^{k,p}} \qquad \text{for all } u \in \mathcal{C}^\infty(\Omega) \cap W^{k,p}(\Omega)\,. \tag{8.42}$$

(II): Extend the embedding to the entire space, by continuity. Since $\mathcal{C}^\infty$ is dense in $W^{k,p}$, for every $u \in W^{k,p}(\Omega)$ we can find a sequence of functions $u_n \in \mathcal{C}^\infty(\Omega)$ such that $\|u - u_n\|_{W^{k,p}} \to 0$. By (8.42),

$$\limsup_{m,n\to\infty} \|u_m - u_n\|_X \leq \limsup_{m,n\to\infty} C\,\|u_m - u_n\|_{W^{k,p}} = 0\,.$$

Therefore the functions u_n also form a Cauchy sequence in the space X. By completeness, $u_n \to \tilde{u}$ for some $\tilde{u} \in X$. Observing that $\tilde{u}(x) = u(x)$ for a.e. $x \in \Omega$, we conclude that, up to a modification on a set $\mathcal{N} \subset \Omega$ of measure zero, each function $u \in W^{k,p}(\Omega)$ also lies in the space X.

8.6.1. Morrey's inequality. In this section we prove that, if $u \in W^{1,p}(\mathbb{R}^n)$, where the exponent p is bigger than the dimension n of the space, then u coincides a.e. with a Hölder continuous function.

Theorem 8.30 (Morrey's inequality). *Assume $n < p < \infty$ and set $\gamma \doteq 1 - \frac{n}{p} > 0$. Then there exists a constant C, depending only on p and n, such that*

$$\|u\|_{\mathcal{C}^{0,\gamma}(\mathbb{R}^n)} \leq C\,\|u\|_{W^{1,p}(\mathbb{R}^n)} \tag{8.43}$$

for every $u \in \mathcal{C}^1(\mathbb{R}^n) \cap W^{1,p}(\mathbb{R}^n)$.

Proof. Before giving the actual proof, we outline the underlying idea. From an integral estimate on the gradient of the function u, say

$$\int_{\mathbb{R}^n} |\nabla u(x)|^p\,dx \leq C_0, \tag{8.44}$$

we seek a pointwise estimate of the form

$$|u(y) - u(y')| \leq C_1|y - y'|^\gamma \qquad \text{for all } y, y' \in \mathbb{R}^n. \tag{8.45}$$

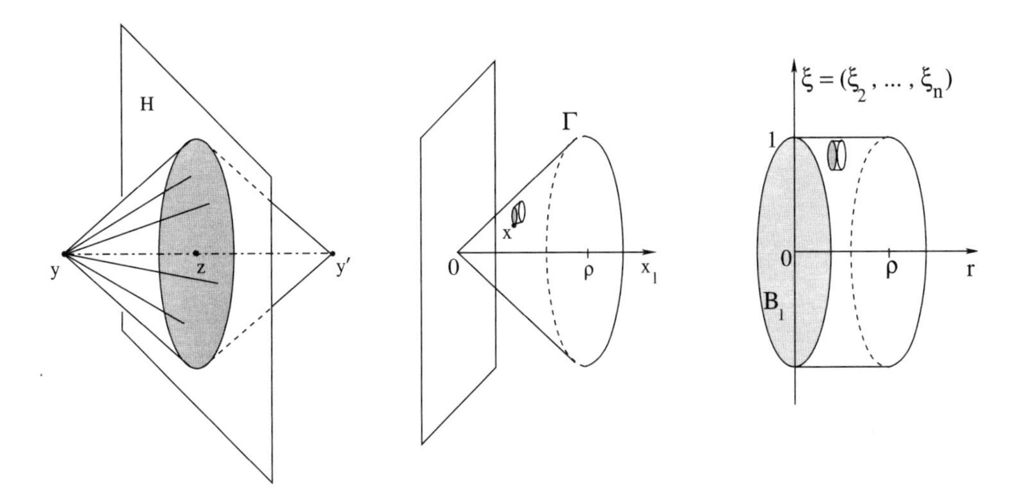

Figure 8.6.1. Proving Morrey's inequality. Left: the values $u(y)$ and $u(y')$ are compared with the average value of u on an $(n-1)$-dimensional ball (the shaded region) centered at the midpoint z and contained in the hyperplane H perpendicular to the vector $y - y'$. Center and right: a point x in the cone Γ is described in terms of the coordinates $(r, \xi) \in [0, \rho] \times B_1$.

To achieve (8.45), a natural attempt is to write

$$
\begin{aligned}
|u(y) - u(y')| &= \left| \int_0^1 \left[\frac{d}{d\theta} u\left(\theta y + (1-\theta) y'\right) \right] d\theta \right| \\
&\leq \int_0^1 \left| \nabla u(\theta y + (1-\theta) y') \right| |y - y'| \, d\theta \, .
\end{aligned}
\tag{8.46}
$$

However, the integral on the right-hand side of (8.46) only involves values of ∇u over the segment joining y with y'. If the dimension of the space is $n > 1$, this segment has zero measure. Hence the integral in (8.46) can be arbitrarily large, even if the integral in (8.44) is small. To address this difficulty, we shall compare both values $u(y)$, $u(y')$ with the average value u_A of the function u over an $(n-1)$-dimensional ball centered at the midpoint $z = \frac{y+y'}{2}$, as shown in Figure 8.6.1, left. Notice that the difference $|u(y) - u_A|$ can be estimated by an integral of $|\nabla u|$ ranging over a cone of dimension n. In this way the bound (8.44) can thus be brought into play.

1. We now begin the proof, with a preliminary computation. On $\mathbb{R}^n$, consider the cone

$$
\Gamma \doteq \left\{ x = (x_1, x_2, \ldots, x_n)\,; \quad \sum_{j=2}^{n} x_j^2 \leq x_1^2 \, , \quad 0 < x_1 < \rho \right\}
$$

and the function

$$\psi(x) = \frac{1}{x_1^{n-1}}. \tag{8.47}$$

Let $q = \frac{p}{p-1}$ be the conjugate exponent of p, so that $\frac{1}{p} + \frac{1}{q} = 1$. We compute

$$\begin{aligned} \|\psi\|^q_{\mathbf{L}^q(\Gamma)} &= \int_\Gamma \left(\frac{1}{x_1^{n-1}}\right)^q dx = \int_0^\rho c_{n-1} x_1^{n-1} \left(\frac{1}{x_1^{n-1}}\right)^q dx_1 \\ &= c_{n-1} \int_0^\rho s^{(n-1)(1-q)}\, ds, \end{aligned}$$

where the constant c_{n-1} gives the volume of the unit ball in $\mathbb{R}^{n-1}$. Therefore, $\psi \in \mathbf{L}^q(\Gamma)$ if and only if $n < p$. In this case,

$$\|\psi\|_{\mathbf{L}^q(\Gamma)} = \left(c_{n-1} \int_0^\rho s^{-\frac{n-1}{p-1}}\, ds\right)^{\frac{1}{q}} = c\left(\rho^{\frac{p-n}{p-1}}\right)^{\frac{p-1}{p}} = c\,\rho^{\frac{p-n}{p}}, \tag{8.48}$$

for some constant c depending only on n and p.

2. Consider any two distinct points $y, y' \in \mathbb{R}^n$. Let $\rho \doteq \frac{1}{2}|y - y'|$. The hyperplane passing through the midpoint $z \doteq \frac{y+y'}{2}$ and perpendicular to the vector $y - y'$ has equation

$$H = \Big\{x \in \mathbb{R}^n\,;\ \langle x - z\,,\ y - y'\rangle = 0\Big\}.$$

Inside H, consider the $(n-1)$-dimensional ball centered at z with radius ρ,

$$B_\rho \doteq \Big\{x \in H\,;\ |x - z| < \rho\Big\}.$$

Calling u_A the average value of u on the ball B_ρ, the difference $|u(y) - u(y')|$ will be estimated as

$$|u(y) - u(y')| \le |u(y) - u_A| + \big|u_A - u(y')\big|. \tag{8.49}$$

3. By a translation and a rotation of coordinates, we can assume

$$y = (0, \ldots, 0) \in \mathbb{R}^n, \qquad B_\rho = \Big\{x = (x_1, x_2, \ldots, x_n)\,;\ x_1 = \rho\,,\ \sum_{i=2}^n x_i^2 \le \rho^2\Big\}.$$

To compute the average value u_A, let B_1 be the unit ball in $\mathbb{R}^{n-1}$, and let c_{n-1} be its $(n-1)$-dimensional measure. Points in the cone Γ will be described using an alternative system of coordinates. To the point with coordinates $(r, \xi) = (r, \xi_2, \ldots, \xi_n) \in [0, \rho] \times B_1$ we associate the point $x(r, \xi) \in \Gamma$ with components

$$(x_1, x_2, \ldots, x_n) = (r,\ r\xi) = (r,\ r\xi_2, \ldots, r\xi_n). \tag{8.50}$$

Define $U(r, \xi) = u(r,\ r\xi)$, and observe that $U(0, \xi) = u(0)$ for every ξ.

Therefore

$$U(\rho,\xi) = U(0,\xi) + \int_0^\rho \left[\frac{\partial}{\partial r} U(r,\xi)\right] dr\,,$$

$$(8.51) \qquad u_A - u(0) = \frac{1}{c_{n-1}} \int_{B_1} \left(\int_0^\rho \left[\frac{\partial}{\partial r} U(r,\xi)\right] dr \right) d\xi\,.$$

We now change variables, transforming the integral (8.51) over $[0,\rho]\times B_1$ into an integral over the cone Γ. Computing the Jacobian matrix of the transformation (8.50), we find that its determinant is r^{n-1}; hence

$$dx_1\, dx_2 \;\cdots\; dx_n = r^{n-1} dr\, d\xi_2 \;\cdots\; d\xi_n\,.$$

Moreover, since $|\xi|\le 1$, the directional derivative of u in the direction of the vector $(1,\xi_2,\ldots,\xi_n)$ is estimated by

$$(8.52) \qquad \left|\frac{\partial}{\partial r} U(r,\xi)\right| = \left| u_{x_1} + \sum_{i=2}^n \xi_i u_{x_i}\right| \le 2|\nabla u(r,\xi)|\,.$$

Using (8.52) in (8.51) and using the estimate (8.48) on the $\mathbf{L}^q$ norm of the function $\psi(x) \doteq x_1^{1-n}$, we obtain

$$(8.53) \qquad \begin{aligned} \Big|u_A - u(0)\Big| &\le \frac{2}{c_{n-1}} \int_\Gamma \frac{1}{x_1^{n-1}} |\nabla u(x)|\, dx \\ &\le \frac{2}{c_{n-1}} \|\psi\|_{\mathbf{L}^q(\Gamma)} \|\nabla u\|_{\mathbf{L}^p(\Gamma)} \qquad \left(q = \frac{p}{p-1}\right) \\ &\le C\,\rho^{\frac{p-n}{p}} \|u\|_{W^{1,p}(\mathbb{R}^n)} \end{aligned}$$

for some constant C. Notice that the last two steps follow from Hölder's inequality and (8.48).

4. Using (8.53) to estimate each term on the right-hand side of (8.49) and recalling that $\rho = \frac{1}{2}|y-y'|$, we conclude that

$$(8.54) \qquad |u(y)-u(y')| \le 2C \left(\frac{|y-y'|}{2}\right)^{\frac{p-n}{p}} \|u\|_{W^{1,p}(\mathbb{R}^n)}.$$

This shows that u is Hölder continuous with exponent $\gamma = \frac{p-n}{p}$.

5. To estimate $\sup_y |u(y)|$, we observe that, by (8.54), for some constant C_1 one has

$$|u(y)| \le |u(x)| + C_1\|u\|_{W^{1,p}(\mathbb{R}^n)} \qquad \text{for all } x \in B(y,1)\,.$$

Taking the average of the right-hand side over the ball centered at y with radius one, we obtain
(8.55)
$$|u(y)| \leq \fint_{B(y,1)} |u(x)|\,dx + C_1\|u\|_{W^{1,p}(\mathbb{R}^n)} \leq C_2\|u\|_{\mathbf{L}^p(\mathbb{R}^n)} + C_1\|u\|_{W^{1,p}(\mathbb{R}^n)}.$$

6. Together, (8.54)–(8.55) yield

$$\|u\|_{\mathcal{C}^{0,\gamma}(\mathbb{R}^n)} \doteq \sup_y |u(y)| + \sup_{y\neq y'} \frac{|u(y)-u(y')|}{|y-y'|^\gamma} \leq C\|u\|_{W^{1,p}(\mathbb{R}^n)},$$

for some constant C depending only on p and n. □

Since $\mathcal{C}^\infty$ is dense in $W^{1,p}$, Morrey's inequality yields

Corollary 8.31 (Embedding). *Let $\Omega \subset \mathbb{R}^n$ be a bounded open set with $\mathcal{C}^1$ boundary. Assume $n < p < \infty$ and set $\gamma \doteq 1 - \frac{n}{p} > 0$. Then every function $f \in W^{1,p}(\Omega)$ coincides a.e. with a function $\tilde{f} \in \mathcal{C}^{0,\gamma}(\Omega)$. Moreover, there exists a constant C such that*

$$\|\tilde{f}\|_{\mathcal{C}^{0,\gamma}} \leq C\,\|f\|_{W^{1,p}} \qquad \textit{for all } f \in W^{1,p}(\Omega). \tag{8.56}$$

Proof. 1. Let $\widetilde{\Omega} \doteq \{x \in \mathbb{R}^n;\ d(x,\Omega) < 1\}$ be the open neighborhood of radius one around the set Ω. By Theorem 8.29 there exists a bounded extension operator E, which extends each function $f \in W^{1,p}(\Omega)$ to a function $Ef \in W^{1,p}(\mathbb{R}^n)$ with support contained inside $\widetilde{\Omega}$.

2. Since $\mathcal{C}^1(\mathbb{R}^n)$ is dense in $W^{1,p}(\mathbb{R}^n)$, we can find a sequence of functions $g_n \in \mathcal{C}^1(\mathbb{R}^n)$ converging to Ef in $W^{1,p}(\mathbb{R}^n)$. By Morrey's inequality

$$\limsup_{m,n\to\infty} \|g_m - g_n\|_{\mathcal{C}^{0,\gamma}(\mathbb{R}^n)} \leq C \limsup_{m,n\to\infty} \|g_m - g_n\|_{W^{1,p}(\mathbb{R}^n)} = 0.$$

This proves that the sequence $(g_n)_{n\geq 1}$ is also a Cauchy sequence in the space $\mathcal{C}^{0,\gamma}$. Therefore it converges to a limit function $g \in \mathcal{C}^{0,\gamma}(\mathbb{R}^n)$, uniformly for $x \in \mathbb{R}^n$.

3. Since $g_n \to Ef$ in $W^{1,p}(\mathbb{R}^n)$, we also have $g(x) = (Ef)(x)$ for a.e. $x \in \mathbb{R}^n$. In particular, $g(x) = f(x)$ for a.e. $x \in \Omega$. Since the extension operator E is bounded, from the bound (8.43) we deduce (8.56). □

Remark 8.32. The conclusion of Corollary 8.31 remains valid if $\Omega = \mathbb{R}^n$.

8.6.2. The Gagliardo-Nirenberg inequality. Next, we study the case $1 \leq p < n$. We define the **Sobolev conjugate** of p as

$$p^* \doteq \frac{np}{n-p} > p. \tag{8.57}$$

Notice that p^* depends not only on p but also on the dimension n of the space:

$$\frac{1}{p^*} = \frac{1}{p} - \frac{1}{n}. \tag{8.58}$$

As a preliminary, we describe a useful application of the generalized Hölder inequality; see (A.26) in the Appendix. Let $n-1$ nonnegative functions $g_1, g_2, \ldots, g_{n-1} \in \mathbf{L}^1(\Omega)$ be given. Since $g_i^{\frac{1}{n-1}} \in \mathbf{L}^{n-1}(\Omega)$ for each i, using the generalized Hölder inequality, one obtains

$$\int_\Omega g_1^{\frac{1}{n-1}} g_2^{\frac{1}{n-1}} \cdots g_{n-1}^{\frac{1}{n-1}}\, ds \leq \prod_{i=1}^{n-1} \|g_i^{\frac{1}{n-1}}\|_{\mathbf{L}^{n-1}} = \prod_{i=1}^{n-1} \|g_i\|_{\mathbf{L}^1}^{\frac{1}{n-1}}. \tag{8.59}$$

Theorem 8.33 (Gagliardo-Nirenberg inequality). *Assume $1 \leq p < n$. Then there exists a constant C, depending only on p and n, such that*

$$\|f\|_{\mathbf{L}^{p^*}(\mathbb{R}^n)} \leq C\,\|\nabla f\|_{\mathbf{L}^p(\mathbb{R}^n)} \qquad \textit{for all } f \in C_c^1(\mathbb{R}^n). \tag{8.60}$$

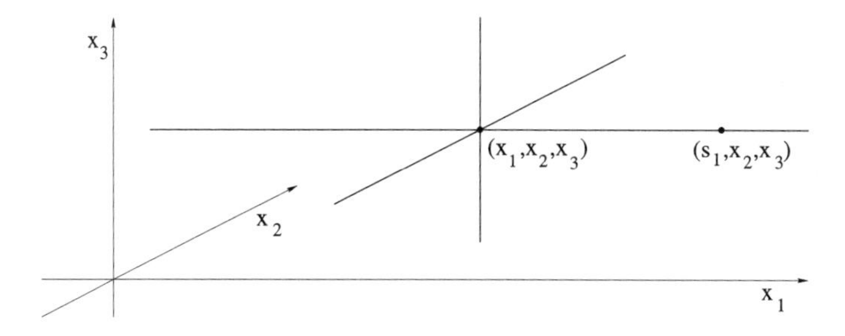

Figure 8.6.2. Proving the Gagliardo-Nirenberg inequality. The integral $\int_{-\infty}^{\infty} |D_{x_1} f(s_1, x_2, x_3)|\, ds_1$ depends on x_2, x_3 but not on x_1. Similarly, the integral $\int_{-\infty}^{\infty} |D_{x_2} f(x_1, s_2, x_3)|\, ds_2$ depends on x_1, x_3 but not on x_2.

Proof. 1. For each $i \in \{1, \ldots, n\}$ and every point $x = (x_1, \ldots, x_i, \ldots, x_n) \in \mathbb{R}^n$, since f has compact support, we can write

$$f(x_1, \ldots, x_i, \ldots, x_n) = \int_{-\infty}^{x_i} D_{x_i} f(x_1, \ldots, s_i, \ldots, x_n)\, ds_i.$$

In turn, this yields

$$|f(x_1,\ldots,x_n)| \le \int_{-\infty}^{\infty} \Big|D_{x_i} f(x_1,\ldots,s_i,\ldots,x_n)\Big|\, ds_i, \qquad 1\le i\le n\,,$$

$$(8.61) \qquad |f(x)|^{\frac{n}{n-1}} \le \prod_{i=1}^{n}\left(\int_{-\infty}^{\infty} |D_{x_i} f(x_1,\ldots,s_i,\ldots,x_n)|\, ds_i\right)^{\frac{1}{n-1}}.$$

We now integrate (8.61) with respect to x_1. Observe that the first factor on the right-hand side does not depend on x_1. This factor behaves like a constant and can be taken out of the integral. The product of the remaining $n-1$ factors is handled using (8.59). This yields

(8.62)

$$\int_{-\infty}^{\infty} |f|^{\frac{n}{n-1}}\, dx_1$$

$$\le \left(\int_{-\infty}^{\infty} |D_{x_1} f|\, ds_1\right)^{\frac{1}{n-1}} \int_{-\infty}^{\infty} \prod_{i=2}^{n}\left(\int_{-\infty}^{\infty} |D_{x_i} f|\, ds_i\right)^{\frac{1}{n-1}} dx_1$$

$$\le \left(\int_{-\infty}^{\infty} |D_{x_1} f|\, ds_1\right)^{\frac{1}{n-1}} \prod_{i=2}^{n}\left(\int_{-\infty}^{\infty}\int_{-\infty}^{\infty} |D_{x_i} f|\, ds_i\, dx_1\right)^{\frac{1}{n-1}}.$$

Notice that the second inequality was obtained by applying the generalized Hölder inequality to the $n-1$ functions $g_i = \int_{-\infty}^{\infty} |D_{x_i} f| ds_i$, $i = 2,\ldots,n$.

We now integrate both sides of (8.62) with respect to x_2. Observe that one of the factors appearing in the product on the right-hand side of (8.62) does not depend on the variable x_2 (namely, the one involving integration with respect to s_2). This factor behaves like a constant and can be taken out of the integral. The product of the remaining $n-2$ factors is again estimated using Hölder's inequality. This yields

(8.63)

$$\int_{-\infty}^{\infty}\int_{-\infty}^{\infty} |f|^{\frac{n}{n-1}}\, dx_1\, dx_2$$

$$\le \left(\int_{-\infty}^{\infty}\int_{-\infty}^{\infty} |D_{x_1} f|\, dx_1 dx_2\right)^{\frac{1}{n-1}} \left(\int_{-\infty}^{\infty}\int_{-\infty}^{\infty} |D_{x_2} f|\, dx_1 dx_2\right)^{\frac{1}{n-1}}$$

$$\times \prod_{i=3}^{n}\left(\int_{-\infty}^{\infty}\int_{-\infty}^{\infty}\int_{-\infty}^{\infty} |D_{x_i} f|\, ds_i\, dx_1 dx_2\right)^{\frac{1}{n-1}}.$$

Proceeding in the same way, after n integrations we obtain
(8.64)

$$\int_{-\infty}^{\infty} \cdots \int_{-\infty}^{\infty} |f|^{\frac{n}{n-1}}\, dx_1 \cdots dx_n$$

$$\leq \prod_{i=1}^{n} \left(\int_{-\infty}^{\infty} \cdots \int_{-\infty}^{\infty} |D_{x_i} f|\, dx_1 \cdots dx_n \right)^{\frac{1}{n-1}} \leq \left(\int_{\mathbb{R}^n} |\nabla f|\, dx \right)^{\frac{n}{n-1}}.$$

This already implies

$$\|f\|_{\mathbf{L}^{n/(n-1)}} = \left(\int_{\mathbb{R}^n} |f|^{\frac{n}{n-1}}\, dx \right)^{\frac{n-1}{n}} \leq \int_{\mathbb{R}^n} |\nabla f|\, dx, \tag{8.65}$$

proving the theorem in the case where $p = 1$ and $p^* = \frac{n}{n-1}$.

2. To cover the general case where $1 < p < n$, we apply (8.65) to the function

$$g \doteq |f|^{\beta} \qquad \text{with} \qquad \beta \doteq \frac{p(n-1)}{n-p}. \tag{8.66}$$

Using the standard Hölder inequality, one obtains
(8.67)

$$\left(\int_{\mathbb{R}^n} |f|^{\frac{\beta n}{n-1}}\, dx \right)^{\frac{n-1}{n}} \leq \int_{\mathbb{R}^n} \beta |f|^{\beta-1} |\nabla f|\, dx$$

$$\leq \beta \left(\int_{\mathbb{R}^n} |f|^{\frac{(\beta-1)p}{p-1}}\, dx \right)^{\frac{p-1}{p}} \left(\int_{\mathbb{R}^n} |\nabla f|^p\, dx \right)^{\frac{1}{p}}.$$

Our choice of β in (8.66) yields

$$\frac{(\beta-1)p}{p-1} = \frac{\beta n}{n-1} = \frac{np}{n-p} = p^*.$$

Therefore, from (8.67) it follows that

$$\left(\int_{\mathbb{R}^n} |f|^{p^*}\, dx \right)^{\frac{n-1}{n}} \leq \beta \left(\int_{\mathbb{R}^n} |f|^{p^*}\, dx \right)^{\frac{p-1}{p}} \left(\int_{\mathbb{R}^n} |\nabla f|^p\, dx \right)^{\frac{1}{p}}.$$

Observing that $\frac{n-1}{n} - \frac{p-1}{p} = \frac{n-p}{np} = \frac{1}{p^*}$, we conclude that

$$\left(\int_{\mathbb{R}^n} |f|^{p^*}\, dx \right)^{\frac{1}{p^*}} \leq C \left(\int_{\mathbb{R}^n} |\nabla f|^p\, dx \right)^{\frac{1}{p}}. \qquad \square$$

If the domain $\Omega \subset \mathbb{R}^n$ is bounded, then $\mathbf{L}^q(\Omega) \subseteq \mathbf{L}^{p^*}(\Omega)$ for every $q \in [1, p^*]$. Using the Gagliardo-Nirenberg inequality, we obtain

Corollary 8.34 (Embedding). *Let $\Omega \subset \mathbb{R}^n$ be a bounded open domain with $\mathcal{C}^1$ boundary, and assume $1 \leq p < n$. Then, for every $q \in [1, p^*]$ with $p^* \doteq \frac{np}{n-p}$, there exists a constant C such that*

$$\|f\|_{\mathbf{L}^q(\Omega)} \leq C\,\|f\|_{W^{1,p}(\Omega)} \qquad \text{for all } f \in W^{1,p}(\Omega)\,. \tag{8.68}$$

Proof. Let $\widetilde{\Omega} \doteq \{x \in \mathbb{R}^n;\ d(x,\Omega) < 1\}$ be the open neighborhood of radius one around the set Ω. By Theorem 8.29 there exists a bounded extension operator $E : W^{1,p}(\Omega) \mapsto W^{1,p}(\mathbb{R}^n)$, with the property that Ef is supported inside $\widetilde{\Omega}$, for every $f \in W^{1,p}(\Omega)$. Applying the Gagliardo-Nirenberg inequality to Ef, for suitable constants C_1, C_2, C_3 we obtain

$$\|f\|_{\mathbf{L}^q(\Omega)} \leq C_1\,\|f\|_{\mathbf{L}^{p^*}(\Omega)} \leq C_2\,\|Ef\|_{\mathbf{L}^{p^*}(\mathbb{R}^n)} \leq C_3\|f\|_{W^{1,p}(\Omega)}\,. \qquad \square$$

8.6.3. High-order Sobolev estimates. Let $\Omega \subset \mathbb{R}^n$ be a bounded open set with $\mathcal{C}^1$ boundary, and let $u \in W^{k,p}(\Omega)$. The number

$$k - \frac{n}{p}$$

will be called the **net smoothness** of u. As in Figure 8.6.3, let m be the integer part and let $0 \leq \gamma < 1$ be the fractional part of this number, so that

$$k - \frac{n}{p} = m + \gamma\,. \tag{8.69}$$

In the following, we say that a Banach space X is **continuously embedded** in a Banach space Y if $X \subseteq Y$ and there exists a constant C such that

$$\|u\|_Y \leq C\,\|u\|_X \qquad \text{for all } u \in X\,.$$

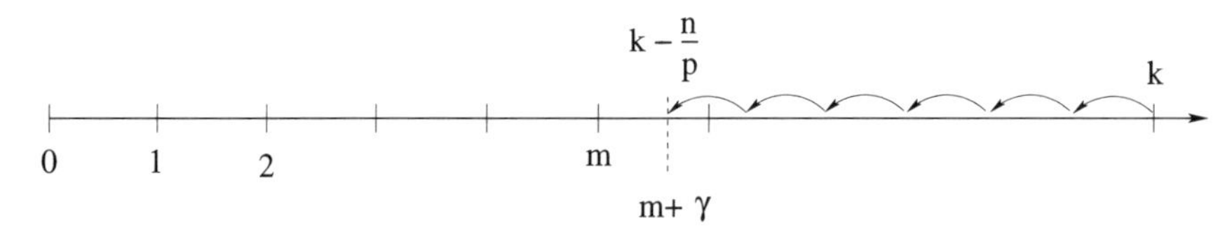

Figure 8.6.3. Computing the "net smoothness" of a function $f \in W^{k,p} \subset \mathcal{C}^{m,\gamma}$.

Theorem 8.35 (General Sobolev embeddings). *Let $\Omega \subset \mathbb{R}^n$ be a bounded open set with $\mathcal{C}^1$ boundary, and consider the space $W^{k,p}(\Omega)$. Let m, γ be as in (8.69). Then the following continuous embeddings hold.*

(i) *If* $k - \frac{n}{p} < 0$, *then* $W^{k,p}(\Omega) \subseteq \mathbf{L}^q(\Omega)$, *with* $\frac{1}{q} = \frac{1}{p} - \frac{k}{n} = \frac{1}{n}\left(\frac{n}{p} - k\right)$.

(ii) *If* $k - \frac{n}{p} = 0$, *then* $W^{k,p}(\Omega) \subseteq \mathbf{L}^q(\Omega)$ *for every* $1 \le q < \infty$.

(iii) *If* $m \ge 0$ *and* $\gamma > 0$, *then* $W^{k,p}(\Omega) \subseteq \mathcal{C}^{m,\gamma}(\Omega)$.

(iv) *If* $m \ge 1$ *and* $\gamma = 0$, *then for every* $0 \le \gamma' < 1$ *one has* $W^{k,p}(\Omega) \subseteq \mathcal{C}^{m-1,\gamma'}(\Omega)$.

Remark 8.36. Functions in a Sobolev space are only defined up to a set of measure zero. By saying that $W^{k,p}(\Omega) \subseteq \mathcal{C}^{m,\gamma}(\Omega)$ we mean the following. For every $u \in W^{k,p}(\Omega)$ there exists a function $\tilde{u} \in \mathcal{C}^{m,\gamma}(\Omega)$ such that $\tilde{u}(x) = u(x)$ for a.e. $x \in \Omega$. Moreover, there exists a constant C, depending on k, p, m, γ but not on u, such that

$$\|u\|_{\mathcal{C}^{m,\gamma}(\Omega)} \le C\,\|\tilde{u}\|_{W^{k,p}(\Omega)}\,.$$

Proof of the theorem. 1. We start by proving (i). Assume $k - \frac{n}{p} < 0$ and let $u \in W^{k,p}(\Omega)$. Since $D^\alpha u \in W^{1,p}(\Omega)$ for every $|\alpha| \le k-1$, the Gagliardo-Nirenberg inequality yields

$$\|D^\alpha u\|_{L^{p^*}(\Omega)} \le C\,\|u\|_{W^{k,p}(\Omega)}, \qquad |\alpha| \le k-1\,.$$

Therefore $u \in W^{k-1,p^*}(\Omega)$, where p^* is the Sobolev conjugate of p.

This argument can be iterated. Set $p_1 = p^*$, $p_2 \doteq p_1^*$, $\ldots$,$p_j \doteq p_{j-1}^*$. By (8.58) this means

$$\frac{1}{p_1} = \frac{1}{p} - \frac{1}{n}, \qquad \ldots, \qquad \frac{1}{p_j} = \frac{1}{p} - \frac{j}{n},$$

provided that $jp < n$. Using the Gagliardo-Nirenberg inequality several times, we obtain

$$W^{k,p}(\Omega) \subseteq W^{k-1,p_1}(\Omega) \subseteq W^{k-2,p_2}(\Omega) \subseteq \cdots \subseteq W^{k-j,p_j}(\Omega)\,. \tag{8.70}$$

After k steps we find that $u \in W^{0,p_k}(\Omega) = \mathbf{L}^{p_k}(\Omega)$, with $\frac{1}{p_k} = \frac{1}{p} - \frac{k}{n} = \frac{1}{q}$. Hence $p_k = q$ and (i) is proved.

2. In the special case $kp = n$, repeating the above argument, after $k-1$ steps we find

$$\frac{1}{p_{k-1}} = \frac{1}{p} - \frac{k-1}{n} = \frac{1}{n}\,.$$

Therefore $p_{k-1} = n$ and

$$W^{k,p}(\Omega) \subset W^{1,n}(\Omega) \subseteq W^{1,n-\varepsilon}(\Omega)$$

for every $\varepsilon > 0$. Using the Gagliardo-Nirenberg inequality once again, we obtain

$$u \in W^{1,n-\varepsilon}(\Omega) \subseteq \mathbf{L}^q(\Omega), \qquad q = \frac{n(n-\varepsilon)}{n-(n-\varepsilon)} = \frac{n^2-\varepsilon n}{\varepsilon}.$$

Since $\varepsilon > 0$ was arbitrary, this proves (ii).

3. To prove (iii), assume that $m \geq 0$ and $\gamma > 0$ and let $u \in W^{k,p}(\Omega)$. We use the embeddings (8.70), choosing j to be the smallest integer such that $p_j > n$. We thus have

$$\frac{1}{p} - \frac{j}{n} = \frac{1}{p_j} < \frac{1}{n} < \frac{1}{p} - \frac{j-1}{n}, \qquad u \in W^{k-j,p_j}(\Omega).$$

Hence, for every multi-index α with $|\alpha| \leq k-j-1$, Morrey's inequality yields

$$D^\alpha u \in W^{1,p_j}(\Omega) \subseteq \mathcal{C}^{0,\gamma}(\Omega) \qquad \text{with } \gamma = 1 - \frac{n}{p_j} = 1 - \frac{n}{p} + j.$$

Since α was any multi-index with length $\leq k-j-1$, the above implies

$$u \in \mathcal{C}^{k-j-1,\gamma}(\Omega).$$

To conclude the proof of (iii), it suffices to check that

$$k - \frac{n}{p} = (k-j-1) + \left(1 - \frac{n}{p} + j\right),$$

so that $m = k-j-1$ is the integer part of the number $k - \frac{n}{p}$, while γ is its fractional part.

4. To prove (iv), assume that $m \geq 1$ and that $j \doteq \frac{n}{p}$ is an integer. Let $u \in W^{k,p}(\Omega)$, and fix any multi-index with $|\alpha| \leq j-1$. Using the Gagliardo-Nirenberg inequality as in step **2**, we obtain

$$D^\alpha u \in W^{k-j,p}(\Omega) \subseteq W^{1,q}(\Omega)$$

for every $1 < q < \infty$. Hence, by Morrey's inequality

$$D^\alpha u \in W^{1,q}(\Omega) \subseteq \mathcal{C}^{0,1-\frac{n}{q}}(\Omega).$$

Since q can be chosen arbitrarily large, this proves (iv). □

Example 8.37. Let Ω be the open unit ball in $\mathbb{R}^5$, and assume $u \in W^{4,2}(\Omega)$. Applying the Gagliardo-Nirenberg inequality two times and then Morrey's inequality, we obtain

$$u \in W^{4,2}(\Omega) \subset W^{3,\frac{10}{3}}(\Omega) \subset W^{2,10}(\Omega) \subset C^{1,\frac{1}{2}}(\Omega).$$

Observe that the *net smoothness* of u is $k - \frac{n}{p} = 4 - \frac{5}{2} = 1 + \frac{1}{2}$.

8.7. Compact embeddings

Let $\Omega \subset \mathbb{R}^n$ be a bounded open set with $\mathcal{C}^1$ boundary. In this section we study the embedding $W^{1,p}(\Omega) \subset \mathbf{L}^q(\Omega)$ in greater detail and show that, when $\frac{1}{q} > \frac{1}{p} - \frac{1}{n}$, this embedding is compact. Namely, from any sequence $(u_m)_{m\geq 1}$ which is bounded in $W^{1,p}$ one can extract a subsequence which converges in $\mathbf{L}^q$.

As a preliminary we observe that, if $p > n$, then every function $u \in W^{1,p}(\Omega)$ is Hölder continuous. In particular, if $(u_m)_{m\geq 1}$ is a bounded sequence in $W^{1,p}(\Omega)$, then the functions u_m are equicontinuous and uniformly bounded. By Ascoli's compactness theorem we can extract a subsequence $(u_{m_j})_{j\geq 1}$ which converges to a continuous function u uniformly on Ω. Since Ω is bounded, this implies $\|u_{m_j} - u\|_{\mathbf{L}^q(\Omega)} \to 0$ for every $q \in [1, \infty]$. This already shows that the embedding $W^{1,p}(\Omega) \subset \mathbf{L}^q(\Omega)$ is compact whenever $p > n$ and $1 \leq q \leq \infty$.

In the remainder of this section we thus focus on the case $p < n$. By the Gagliardo-Nirenberg inequality, the space $W^{1,p}(\Omega)$ is continuously embedded in $\mathbf{L}^{p^*}(\Omega)$, where $p^* = \frac{np}{n-p}$. In turn, since Ω is bounded, for every $1 \leq q \leq p^*$ we have the continuous embedding $\mathbf{L}^{p^*}(\Omega) \subseteq \mathbf{L}^q(\Omega)$.

Theorem 8.38 (Rellich-Kondrachov compactness theorem). *Let $\Omega \subset \mathbb{R}^n$ be a bounded open set with $\mathcal{C}^1$ boundary. Assume $1 \leq p < n$. Then for each $1 \leq q < p^* \doteq \frac{np}{n-p}$ one has the compact embedding*

$$W^{1,p}(\Omega) \subset\subset \mathbf{L}^q(\Omega). \tag{8.71}$$

Proof. 1. Let $(u_m)_{m\geq 1}$ be a bounded sequence in $W^{1,p}(\Omega)$. Using Theorem 8.29 on the extension of Sobolev functions, we can assume that all functions u_m are defined on the entire space $\mathbb{R}^n$ and vanish outside a compact set K:

$$\mathrm{Supp}(u_m) \subseteq K \subset \widetilde{\Omega} \doteq B(\Omega, 1). \tag{8.72}$$

Here the right-hand side denotes the open neighborhood of radius one around the set Ω.

Since $q < p^*$ and $\widetilde{\Omega}$ is bounded, we have

$$\|u_m\|_{\mathbf{L}^q(\mathbb{R}^n)} = \|u_m\|_{\mathbf{L}^q(\widetilde{\Omega})} \leq C\, \|u_m\|_{\mathbf{L}^{p^*}(\widetilde{\Omega})} \leq C'\, \|u_m\|_{W^{1,p}(\widetilde{\Omega})}$$

for some constants C, C'. Hence the sequence u_m is uniformly bounded in $\mathbf{L}^q(\mathbb{R}^n)$.

2. Consider the mollified functions $u_m^\varepsilon \doteq J_\varepsilon * u_m$. By (8.72) we can assume that all these functions are supported inside $\widetilde{\Omega}$. We claim that

$$\text{(8.73)}\quad \|u_m^\varepsilon - u_m\|_{\mathbf{L}^q(\widetilde{\Omega})} \to 0 \quad as \quad \varepsilon \to 0, \qquad \text{uniformly with respect to } m.$$

Indeed, if u_m is smooth, then (performing the changes of variable $y' = \varepsilon y$ and $z = x - \varepsilon t y$)

$$\begin{aligned} u_m^\varepsilon(x) - u_m(x) &= \int_{|y'|<\varepsilon} J_\varepsilon(y')\,[u_m(x-y') - u_m(x)]\,dy' \\ &= \int_{|y|<1} J(y)[u_m(x-\varepsilon y) - u_m(x)]\,dy \\ &= \int_{|y|<1} J(y)\left(\int_0^1 \frac{d}{dt}\Big(u_m(x-\varepsilon t y)\Big)\,dt\right)dy \\ &= -\varepsilon\int_{|y|<1} J(y)\left(\int_0^1 \nabla u_m(x-\varepsilon t y)\cdot y\;dt\right)dy\,. \end{aligned}$$

In turn, this yields

$$\begin{aligned} \int_{\widetilde{\Omega}} |u_m^\varepsilon(x) - u_m(x)|\,dx &\le \varepsilon \int_{\widetilde{\Omega}}\int_{|y|\le 1} J(y)\left(\int_0^1 |\nabla u_m(x-\varepsilon t y)|\;dt\right)dy\,dx \\ &\le \varepsilon\int_{\widetilde{\Omega}} |\nabla u_m(z)|\,dz. \end{aligned}$$

By approximating u_m in $W^{1,p}$ with a sequence of smooth functions, we see that the same estimate remains valid for all functions $u_m \in W^{1,p}(\widetilde{\Omega})$. We have thus shown that

$$\text{(8.74)}\qquad \|u_m^\varepsilon - u_m\|_{\mathbf{L}^1(\widetilde{\Omega})} \le \varepsilon\|\nabla u_m\|_{\mathbf{L}^1(\widetilde{\Omega})} \le \varepsilon C\,\|u_m\|_{W^{1,p}(\widetilde{\Omega})},$$

for some constant C. Since the norms $\|u_m\|_{W^{1,p}}$ satisfy a uniform bound independent of m, this already proves our claim (8.73) in the case $q = 1$.

3. To prove (8.73) for $1 < q < p^*$ also, we now use the interpolation inequality for $\mathbf{L}^p$ norms (see (A.28) in the Appendix). Choose $0 < \theta < 1$ such that

$$\frac{1}{q} = \theta\cdot 1 + (1-\theta)\cdot\frac{1}{p^*}\,.$$

Then

$$\text{(8.75)}\quad \|u_m^\varepsilon - u_m\|_{\mathbf{L}^q(\widetilde{\Omega})} \le \|u_m^\varepsilon - u_m\|^{\theta}_{\mathbf{L}^1(\widetilde{\Omega})}\cdot\|u_m^\varepsilon - u_m\|^{1-\theta}_{\mathbf{L}^{p^*}(\widetilde{\Omega})} \le C_0\,\varepsilon^\theta$$

for some constant C_0 independent of m. Indeed, in the above expression, the $\mathbf{L}^1$ norm is bounded by (8.74), while the $\mathbf{L}^{p^*}$ norm is bounded by a constant, because of the Gagliardo-Nirenberg inequality.

4. Fix any $\delta > 0$, and choose $\varepsilon > 0$ small enough so that (8.75) yields

$$\|u_m^\varepsilon - u_m\|_{\mathbf{L}^q(\widetilde{\Omega})} \leq C_0\, \varepsilon^\theta \leq \frac{\delta}{2} \qquad \text{for all } m \geq 1\,.$$

Recalling that $u_m^\varepsilon = J_\varepsilon * u_m$, we have

$$\|u_m^\varepsilon\|_{\mathbf{L}^\infty} \leq \|J_\varepsilon\|_{\mathbf{L}^\infty} \|u_m\|_{\mathbf{L}^1} \leq C_1,$$

$$\|\nabla u_m^\varepsilon\|_{\mathbf{L}^\infty} \leq \|\nabla J_\varepsilon\|_{\mathbf{L}^\infty} \|u_m\|_{\mathbf{L}^1} \leq C_2,$$

where C_1, C_2 are constants depending on ε but not on m. The above inequalities show that, for each fixed $\varepsilon > 0$, the sequence $(u_m^\varepsilon)_{m\geq 1}$ is uniformly bounded and equicontinuous. By Ascoli's compactness theorem, there exists a subsequence $(u_{m_j}^\varepsilon)$ which converges uniformly on $\widetilde{\Omega}$ to some continuous function u^ε. We now have

$$\begin{aligned} &\limsup_{j,k\to\infty} \|u_{m_j} - u_{m_k}\|_{\mathbf{L}^q} \\ &\leq \limsup_{j,k\to\infty} \Big(\|u_{m_j} - u_{m_j}^\varepsilon\|_{\mathbf{L}^q} \|u_{m_j}^\varepsilon - u^\varepsilon\|_{\mathbf{L}^q} \\ &\qquad + \|u^\varepsilon - u_{m_k}^\varepsilon\|_{\mathbf{L}^q} + \|u_{m_k}^\varepsilon - u_{m_k}\|_{\mathbf{L}^q} \Big) \\ &\leq \frac{\delta}{2} + 0 + 0 + \frac{\delta}{2}\,. \end{aligned} \tag{8.76}$$

5. The proof is now concluded by a standard diagonalization argument. By the previous step we can find an infinite set of indices $I_1 \subset \mathbb{N}$ such that the subsequence $(u_m)_{m\in I_1}$ satifies

$$\limsup_{\ell,m\to\infty,\ \ell,m\in I_1} \|u_\ell - u_m\|_{\mathbf{L}^q} \leq 2^{-1}.$$

By induction on $j = 1, 2, \ldots$, after I_{j-1} has been constructed, we choose an infinite set of indices $I_j \subset I_{j-1} \subset \mathbb{N}$ such that the subsequence $(u_m)_{m\in I_j}$ satisfies

$$\limsup_{\ell,m\to\infty,\ \ell,m\in I_j} \|u_\ell - u_m\|_{\mathbf{L}^q} \leq 2^{-j}.$$

After the subsets I_j have been constructed for all $j \geq 1$, again by induction on j we choose a sequence of integers $m_1 < m_2 < m_3 < \cdots$ such that $m_j \in I_j$ for every j. The subsequence $(u_{m_j})_{j\geq 1}$ satisfies

$$\limsup_{j,k\to\infty} \|u_{m_j} - u_{m_k}\|_{\mathbf{L}^q} = 0\,.$$

Therefore this subsequence is Cauchy and converges to some limit $u \in \mathbf{L}^q$. This proves that the embedding (8.71) is compact when $p < n$. □

Corollary 8.39. *Let $\Omega \subset \mathbb{R}^n$ be a bounded open set with $\mathcal{C}^1$ boundary. Then one has the compact embedding*

$$H^1(\Omega) \subset\subset \mathbf{L}^2(\Omega).$$

Proof. For $n > 2$ the conclusion is a special case of Theorem 8.38 with $p = 2$. When $n = 1$, every function in $H^1(\Omega)$ is Hölder continuous and the result follows from Ascoli's theorem. When $n = 2$, we can apply the previous theorem with $p = 3/2$, $p^* = 6$, $q = 2$ and obtain

$$W^{1,2}(\Omega) \subset W^{1,3/2}(\Omega) \subset\subset \mathbf{L}^2(\Omega). \qquad \square$$

As an application of the compact embedding theorem, we now prove an estimate on the difference between a function u and its average value on a domain Ω.

Theorem 8.40 (Poincaré's inequality. II). *Let $\Omega \subset \mathbb{R}^n$ be a bounded, connected open set with $\mathcal{C}^1$ boundary, and let $p \in [1, \infty]$. Then there exists a constant C depending only on p and Ω such that*

$$\left\| u - \fint_\Omega u\,dx \right\|_{\mathbf{L}^p(\Omega)} \leq C \, \|\nabla u\|_{\mathbf{L}^p(\Omega)}, \tag{8.77}$$

for every $u \in W^{1,p}(\Omega)$.

Proof. If the conclusion were false, one could find a sequence of functions $u_k \in W^{1,p}(\Omega)$ with

$$\left\| u_k - \fint_\Omega u_k\,dx \right\|_{\mathbf{L}^p(\Omega)} > k \, \|\nabla u_k\|_{\mathbf{L}^p(\Omega)}$$

for every $k = 1, 2, \ldots$. Then the renormalized functions

$$v_k \doteq \frac{u_k - \fint_\Omega u_k\,dx}{\left\| u_k - \fint_\Omega u_k\,dx \right\|_{\mathbf{L}^p(\Omega)}}$$

satisfy

$$\fint_\Omega v_k\,dx = 0, \quad \|v_k\|_{\mathbf{L}^p(\Omega)} = 1, \quad \|Dv_k\|_{\mathbf{L}^p(\Omega)} < \frac{1}{k}, \qquad k = 1, 2, \ldots. \tag{8.78}$$

Since the sequence $(v_k)_{k\geq 1}$ is bounded in $W^{1,p}(\Omega)$, if $p < \infty$, we can use the Rellich-Kondrachov compactness theorem and find a subsequence that converges in $\mathbf{L}^p(\Omega)$ to some function v. If $p > n$, then by (8.56) the functions v_k are uniformly bounded and Hölder continuous. Using Ascoli's compactness theorem, we can thus find a subsequence that converges in $\mathbf{L}^\infty(\Omega)$ to some function v.

By (8.78), the sequence of weak gradients also converges, namely $\nabla v_k \to 0$ in $\mathbf{L}^p(\Omega)$. By Lemma 8.14, the zero function is the weak gradient of the limit function v.

We now have

$$\fint_\Omega v\,dx \;=\; \lim_{k\to\infty} \fint_\Omega v_k dx \;=\; 0.$$

Moreover, since $\nabla v = 0 \in \mathbf{L}^p(\Omega)$, by Corollary 8.16 the function v must be constant on the connected set Ω; hence $v(x) = 0$ for a.e. $x \in \Omega$. But this is in contradiction to

$$\|v\|_{\mathbf{L}^p(\Omega)} \;=\; \lim_{k\to\infty} \|v_k\|_{\mathbf{L}^p(\Omega)} \;=\; 1. \qquad \square$$

8.8. Differentiability properties

By Morrey's inequality, if $\Omega \subset \mathbb{R}^n$ and $w \in W^{1,p}(\Omega)$ with $p > n$, then w coincides a.e. with a Hölder continuous function. Indeed, after a modification on a set of measure zero, we have

$$|w(x) - w(y)| \;\leq\; C|x-y|^{1-\frac{n}{p}} \left(\int_{B(x,\,|y-x|)} |\nabla w(z)|^p\,dz \right)^{1/p}. \tag{8.79}$$

This by itself does not imply that u should be differentiable in a classical sense. Indeed, there exist Hölder continuous functions that are nowhere differentiable. However, for functions in a Sobolev space a better regularity result can be proved.

Theorem 8.41 (Almost everywhere differentiability). *Let $\Omega \subseteq \mathbb{R}^n$ and let $u \in W^{1,p}_{\text{loc}}(\Omega)$ for some $p > n$. Then u is differentiable at a.e. point $x \in \Omega$, and its gradient coincides with the weak gradient.*

Proof. Let $u \in W^{1,p}_{\text{loc}}(\Omega)$. Since the weak derivatives are in $\mathbf{L}^p_{\text{loc}}$, the same is true of the weak gradient $\nabla u \doteq (D_{x_1}u, \ldots, D_{x_n}u)$. By the Lebesgue differentiation theorem, for a.e. $x \in \Omega$ we have

$$\fint_{B(x,r)} |\nabla u(x) - \nabla u(z)|^p\,dz \;\to\; 0 \qquad \text{as } r \to 0\,. \tag{8.80}$$

Fix a point x for which (8.80) holds, and define

$$w(y) \;\doteq\; u(y) - u(x) - \nabla u(x)\cdot(y-x). \tag{8.81}$$

Observing that $w \in W^{1,p}_{\rm loc}(\Omega)$, we can apply the estimate (8.79) and obtain

$$\begin{aligned} |w(y) - w(x)| &= |w(y)| = |u(y) - u(x) - \nabla u(x) \cdot (y - x)| \\ &\leq C\,|y - x|^{1-\frac{n}{p}} \left(\int_{B(x,\,|y-x|)} |\nabla u(x) - \nabla u(z)|^p\, dz \right)^{1/p} \\ &\leq C'\,|y - x| \left(⨍_{B(x,\,|y-x|)} |\nabla u(x) - \nabla u(z)|^p\, dz \right)^{1/p} \end{aligned}$$

for suitable constants C, C'. Therefore

$$\frac{|w(y) - w(x)|}{|y - x|} \to 0 \qquad \text{as } |y - x| \to 0\,.$$

By the definition of w in (8.81), this means that u is differentiable at x in the classical sense and its gradient coincides with its weak gradient. □

8.9. Problems

1. Determine which of the following functionals define a distribution on $\Omega \subseteq \mathbb{R}$.

(i) $\Lambda(\phi) = \sum_{k=1}^{\infty} k!\, D^k\phi(k)$, with $\Omega = \mathbb{R}$.

(ii) $\Lambda(\phi) = \sum_{k=1}^{\infty} 2^{-k}\, D^k\phi(1/k)$, with $\Omega = \mathbb{R}$.

(iii) $\Lambda(\phi) = \sum_{k=1}^{\infty} \frac{\phi(1/k)}{k}$, with $\Omega = \mathbb{R}$.

(iv) $\Lambda(\phi) = \int_0^{\infty} \frac{\phi(x)}{x^2}\, dx$, with $\Omega =]0, \infty[$.

2. Give a direct proof that, if $f \in W^{1,p}(\,]a,b[\,)$ for some $a < b$ and $1 < p < \infty$, then, by possibly changing f on a set of measure zero, one has

$$|f(x) - f(y)| \leq C\,|x - y|^{1-\frac{1}{p}} \qquad \text{for all } x, y \in\,]a, b[\,.$$

Compute the best possible constant C.

3. Consider the open square

$$Q \doteq \{(x_1, x_2)\,;\ 0 < x_1 < 1,\ 0 < x_2 < 1\} \subset \mathbb{R}^2.$$

Let $f \in W^{1,1}(Q)$ be a function whose weak derivative satisfies $D_{x_1} f(x) = 0$ for a.e. $x \in Q$. Prove that there exists a function $g \in \mathbf{L}^1([0,1])$ such that

$$f(x_1, x_2) = g(x_2) \qquad \text{for a.e. } (x_1, x_2) \in Q\,.$$

4. Let $\Omega \subseteq \mathbb{R}^n$ be an open set and assume $f \in \mathbf{L}^1_{\rm loc}(\Omega)$. Let $g = D_{x_1} f$ be the weak derivative of f with respect to x_1. If f is $\mathcal{C}^1$ restricted to an open subset $\Omega' \subseteq \Omega$, prove that g coincides with the partial derivative $\partial f / \partial x_1$ at a.e. point $x \in \Omega'$.

5. (i) Prove that, if $u \in W^{1,\infty}(\Omega)$ for some open, convex set $\Omega \subseteq \mathbb{R}^n$, then u coincides a.e. with a Lipschitz continuous function.

(ii) Show that there exists a (nonconvex), connected open set $\Omega \subset \mathbb{R}^n$ and a function $u \in W^{1,\infty}(\Omega)$ that does not coincide a.e. with a Lipschitz continuous function.

6. **(Rademacher's theorem)** Let $\Omega \subset \mathbb{R}^n$ be an open set and let $u : \Omega \mapsto \mathbb{R}$ be a bounded, Lipschitz continuous function.

(i) Prove that $u \in W^{1,\infty}(\Omega)$.

(ii) Prove that u is differentiable at a.e. point $x \in \Omega$.

Hint for (i): Consider first the case where Ω is convex. To construct the weak derivative D_{x_i}, for any fixed $x_1, \ldots, x_{i-1}, x_{i+1}, \ldots, x_n$, consider the absolutely continuous function $s \mapsto u(x_1, \ldots, x_{i-1}, s, x_{i+1}, \ldots, x_n)$.

7. Let $\Omega = B(0,1)$ be the open unit ball in $\mathbb{R}^n$, with $n \geq 2$. Prove that the unbounded function $f(x) = \ln \ln \left(1 + \frac{1}{|x|}\right)$ is in $W^{1,n}(\Omega)$.

8. (i) Let $\Omega = \,]-1, 1[$. Consider the linear map $T : \mathcal{C}^1(\Omega) \mapsto \mathbb{R}$ defined by $Tf = f(0)$. Show that this map can be continuously extended, in a unique way, to a bounded linear functional $T : W^{1,1}(\Omega) \mapsto \mathbb{R}$.

(ii) Let $\Omega = B(0,1) \subset \mathbb{R}^2$ be the open unit disc. Consider again the linear map $T : \mathcal{C}^1(\Omega) \mapsto \mathbb{R}$ defined by $Tf = f(0)$. For which values of p can this map be continuously extended to a bounded linear functional $T : W^{1,p}(\Omega) \mapsto \mathbb{R}$?

9. Determine for which values of $p \geq 1$ a generic function $f \in W^{1,p}(\mathbb{R}^3)$ admits a trace along the x_1-axis. In other words, set $\Gamma \doteq \{(t, 0, 0)\,;\, t \in \mathbb{R}\} \subset \mathbb{R}^3$ and consider the map $T : \mathcal{C}^1_c(\mathbb{R}^3) \mapsto \mathbf{L}^p(\Gamma)$, where $Tf = f_{|\Gamma}$ is the restriction of f to Γ. Find values of p such that this map admits a continuous extension $T : W^{1,p}(\mathbb{R}^3) \mapsto \mathbf{L}^p(\Gamma)$.

10. Let $V \subset \mathbb{R}^n$ be a subspace of dimension m and let $V^\perp$ be the perpendicular subspace of dimension $n - m$. Let $u \in W^{1,p}(\mathbb{R}^n)$ with $m < p < n$. Show that, after a modification on a set of measure zero, the following hold.

(i) For a.e. $y \in V^\perp$ (with respect to the $(n-m)$-dimensional measure), the restriction of u to the affine subspace $y + V$ is Hölder continuous with exponent $\gamma = 1 - \frac{m}{p}$.

(ii) The pointwise value $u(y)$ is well defined for a.e. $y \in V^\perp$. Moreover

$$\|u(y)\|_{\mathbf{L}^p(V^\perp)} \leq C \cdot \|u\|_{W^{1,p}(\mathbb{R}^n)}$$

for some constant C depending on m, n, p but not on u.

11. Let $\Omega \subset \mathbb{R}^n$ be an arbitrary open set (without assuming any regularity of the boundary) and assume $p > n$. Show that every function $f \in W_0^{1,p}(\Omega)$ coincides a.e. with a continuous function $\tilde{f}$. Moreover, there exists a constant C, depending on p, n but not on Ω or f, such that

$$\|\tilde{f}\|_{\mathcal{C}^{0,\gamma}(\Omega)} \leq C \, \|f\|_{W^{1,p}(\Omega)} \qquad \text{with } \gamma = 1 - \frac{n}{p}.$$

12. When $k = 0$, by definition $W^{0,p}(\Omega) = \mathbf{L}^p(\Omega)$. If $1 \leq p < \infty$, prove that $W_0^{0,p}(\Omega) = \mathbf{L}^p(\Omega)$ as well. What is $W_0^{0,\infty}(\Omega)$?

13. Let $\varphi : \mathbb{R} \mapsto [0,1]$ be a smooth function such that

$$\varphi(r) = \begin{cases} 1 & \text{if } r \leq 0, \\ 0 & \text{if } r \geq 1. \end{cases}$$

Given any $f \in W^{k,p}(\mathbb{R}^n)$, prove that the functions $f_k(x) = f(x)\,\varphi(|x| - k)$ converge to f in $W^{k,p}(\mathbb{R}^n)$, for every $k \geq 0$ and $1 \leq p < \infty$. As a consequence, show that $W_0^{k,p}(\mathbb{R}^n) = W^{k,p}(\mathbb{R}^n)$.

14. Let $\mathbb{R}_+ \doteq \{x \in \mathbb{R}\,;\; x > 0\}$ and assume $u \in W^{2,p}(\mathbb{R}_+)$. Define the symmetric extension of u by setting $Eu(x) \doteq u(|x|)$. Prove that $Eu \in W^{1,p}(\mathbb{R})$ but $Eu \notin W^{2,p}(\mathbb{R})$, in general.

15. Let $u \in \mathcal{C}_c^1(\mathbb{R}^n)$ and fix $p, q \in [1, \infty[$. For a given $\lambda > 0$, consider the rescaled function $u_\lambda(x) \doteq u(\lambda x)$.

(i) Show that there exists an exponent α, depending on n, q, such that

$$\|u_\lambda\|_{\mathbf{L}^q(\mathbb{R}^n)} = \lambda^\alpha \|u\|_{\mathbf{L}^q(\mathbb{R}^n)}.$$

(ii) Show that there exists an exponent β, depending on n, p, such that

$$\|\nabla u_\lambda\|_{\mathbf{L}^p(\mathbb{R}^n)} = \lambda^\beta \|\nabla u\|_{\mathbf{L}^p(\mathbb{R}^n)}.$$

(iii) Determine for which values of n, p, q one has $\alpha = \beta$. Compare with (8.57).

16. Let $\Omega \subset \mathbb{R}^n$ be a bounded open set with $\mathcal{C}^1$ boundary. Let $(u_m)_{m \geq 1}$ be a sequence of functions which are uniformly bounded in $H^1(\Omega)$. Assuming that $\|u_m - u\|_{\mathbf{L}^2} \to 0$, prove that $u \in H^1(\Omega)$ and

$$\|u\|_{H^1} \leq \liminf_{m \to \infty} \|u_m\|_{H^1}.$$

17. Let $\Omega \doteq \{(x_1, x_2)\,;\; x_1^2 + x_2^2 < 1\}$ be the open unit disc in $\mathbb{R}^2$, and let $\Omega_0 \doteq \Omega \setminus \{(0,0)\}$ be the unit disc minus the origin. Consider the function $f(x) \doteq 1 - |x|$. Prove that (see Figure 8.9.1)

$$\begin{cases} f \in W_0^{1,p}(\Omega) & \text{for } 1 \leq p < \infty, \\ f \in W_0^{1,p}(\Omega_0) & \text{for } 1 \leq p \leq 2, \\ f \notin W_0^{1,p}(\Omega_0) & \text{for } 2 < p \leq \infty. \end{cases}$$

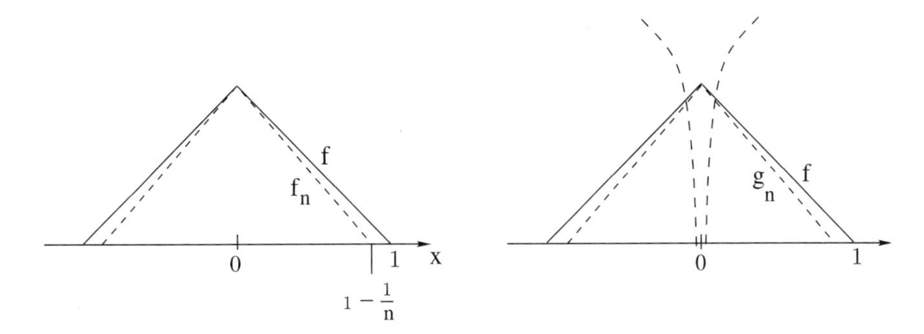

Figure 8.9.1. Left: the function f can be approximated in $W^{1,p}$ with functions f_n having compact support in Ω. Right: the function f can be approximated in $W^{1,2}$ with functions g_n having compact support in Ω_0.

18. Let $\Omega = \{(x_1, x_2)\,;\; 0 < x_1 < 1,\; 0 < x_2 < 1\}$ be the open unit square in $\mathbb{R}^2$.

(i) If $u \in H^1(\Omega)$ satisfies

$$\text{meas}\Big(\{x \in \Omega\,;\; u(x) = 0\}\Big) > 0\,, \qquad \nabla u(x) = 0 \text{ for a.e. } x \in \Omega\,,$$

prove that $u(x) = 0$ for a.e. $x \in \Omega$.

(ii) For every $\alpha > 0$, prove that there exists a constant C_α with the following property. If $u \in H^1(\Omega)$ is a function such that $\text{meas}(\{x \in \Omega\,;\; u(x) = 0\}) \geq \alpha$, then

$$\|u\|_{\mathbf{L}^2(\Omega)} \leq C_\alpha \, \|\nabla u\|_{\mathbf{L}^2(\Omega)}\,. \tag{8.82}$$

19. Let $\Omega \subseteq \mathbb{R}^n$ be an open set, and let $K \subset \Omega$ be a closed set which is "small", in the sense that its $(n-1)$-dimensional measure is $m_{n-1}(K) = 0$. More precisely, assume that the projection of K on every $(n-1)$-dimensional hyperplane has zero $(n-1)$-dimensional measure. Let u be continuously differentiable on the open set $\Omega \setminus K$, and assume $u \in \mathbf{L}^p(\Omega \setminus K)$, $\nabla u \in \mathbf{L}^p(\Omega \setminus K)$. Prove that $u \in W^{1,p}(\Omega)$.

20. (i) Find two functions $f, g \in \mathbf{L}^1_{\text{loc}}(\mathbb{R}^n)$ such that the product $f \cdot g$ is not locally summable.

(ii) Show that, if $f, g \in \mathbf{L}^1_{\text{loc}}(\mathbb{R})$ are both bounded and weakly differentiable, then the product $f \cdot g$ is also weakly differentiable and satisfies the usual product rule: $D_x(fg) = (D_x f) \cdot g + f \cdot (D_x g)$.

(iii) Find two functions $f, g \in \mathbf{L}^1_{\text{loc}}(\mathbb{R}^n)$ (with $n \geq 2$) with the following properties. For every $i = 1, \ldots, n$ the first-order weak derivatives $D_{x_i} f$, $D_{x_i} g$ are well defined. However, the product $f \cdot g$ does not have any weak derivative (on the entire space $\mathbb{R}^n$).

21. Let $\Omega \subset \mathbb{R}^n$ be a bounded, connected open set with $\mathcal{C}^1$ boundary, let $\Omega' \subset \Omega$ be a nonempty open subset, and let $p \in [1, \infty]$. Prove that there exists a constant

C such that

$$\|u\|_{\mathbf{L}^p(\Omega)} \leq C\Big(\|u\|_{\mathbf{L}^p(\Omega')} + \|\nabla u\|_{\mathbf{L}^p(\Omega)}\Big) \qquad \text{for every } u \in W^{1,p}(\Omega).$$

22. Consider the Banach space $X = \mathcal{C}^0(]0,1[)$ of all bounded continuous functions on the open interval $]0,1[$, with norm $\|f\|_{\mathcal{C}^0} = \sup_{0<x<1} |f(x)|$. For a fixed constant $M > 0$, let $S_M \subset X$ be the subset consisting of all functions $f \in X$ such that $\|f\|_{W^{2,\infty}} \leq M$. In other words, $f \in S_M$ provided that f has weak derivatives up to second-order and

$$\|f\|_{\mathbf{L}^\infty} \leq M, \qquad \|\partial_x f\|_{\mathbf{L}^\infty} \leq M, \qquad \|\partial_{xx} f\|_{\mathbf{L}^\infty} \leq M.$$

(i) Prove that S_M is a closed subset of X.

(ii) Prove that the differentiation operator $f \mapsto \partial_x f$ is continuous when restricted to the set S_M. In other words, if $\|f_n - f\|_{\mathcal{C}^0} \to 0$ and $f, f_n \in S_M$ for all $n \geq 1$, then $\|\partial_x f_n - \partial_x f\|_{\mathcal{C}^0} \to 0$.

23. Let $\Omega \subset \mathbb{R}^n$ be a bounded open set with $\mathcal{C}^1$ boundary. Given any $u \in W^{1,p}(\Omega)$ with $1 \leq p < \infty$, prove that there exists a sequence of smooth functions $u_k \in \mathcal{C}_c^\infty(\mathbb{R}^n)$ such that the restrictions of u_k to Ω satisfy

$$\lim_{k\to\infty} \|u_k - u\|_{W^{1,p}(\Omega)} = 0.$$

Moreover,

$$\|u_k\|_{W^{1,p}(\mathbb{R}^n)} \leq C\,\|u\|_{W^{1,p}(\Omega)},$$

for some constant C depending on p and Ω but not on u.

24. Let $f : \mathbb{R} \mapsto \mathbb{R}$ be a weakly differentiable function, with weak derivative $g \in \mathbf{L}^1_{\text{loc}}(\mathbb{R})$. Consider the sequence of divided differences $g_n(x) = n\left[f\left(x + \frac{1}{n}\right) - f(x)\right]$. Prove that $g_n(x) \to g(x)$ for a.e. $x \in \mathbb{R}$, and moreover $\|g_n - g\|_{\mathbf{L}^1([a,b])} \to 0$ for every bounded interval $[a, b]$.

25. Let $(u_n)_{n\geq 1}$ be a sequence of functions in the Hilbert space $H^2(\mathbb{R}^3) \doteq W^{2,2}(\mathbb{R}^3)$. Assume that

$$\lim_{n\to\infty} u_n(x) = u(x) \quad \text{for all } x \in \mathbb{R}^3, \qquad \|u_n\|_{H^2} \leq M \quad \text{for all } n.$$

Prove that the limit function u coincides a.e. with a continuous function.

Chapter 9

Linear Partial Differential Equations

The goal of this chapter is to illustrate how the abstract techniques of functional analysis can be applied to the solution of elliptic, parabolic, and hyperbolic PDEs.

A linear elliptic equation is defined by a second-order differential operator, which is linear but unbounded. As a first step, one must thus provide an alternative "weak" formulation of the boundary value problem, involving bounded linear operators.

In some cases, unique solutions can be obtained by applying the Lax-Milgram theorem to a suitable bilinear form on the Hilbert-Sobolev space H_0^1. More general situations can be studied using Fredholm's theory on the space $\mathbf{L}^2$. When the operator is selfadjoint, relying on the Hilbert-Schmidt theorem, we shall represent solutions by a series of mutually orthogonal eigenfunctions.

Evolution equations of parabolic and hyperbolic type will be studied by applying linear semigroup theory. When the defining operator is selfadjoint, solutions can again be obtained as sum of a series of eigenfunctions.

9.1. Elliptic equations

Let $\Omega \subset \mathbb{R}^n$ be a bounded open set. Given measurable functions $a^{ij}, b^i, c : \Omega \mapsto \mathbb{R}$, consider the linear, second-order differential operator

$$Lu \doteq -\sum_{i,j=1}^{n} (a^{ij}(x)u_{x_i})_{x_j} + \sum_{i=1}^{n} (b^i(x)u)_{x_i} + c(x)u\,. \tag{9.1}$$

We shall study solutions to the boundary value problem

$$\left\{\begin{array}{rcll} Lu & = & f, & x\in\Omega\,,\\ u & = & 0, & x\in\partial\Omega\,,\end{array}\right. \tag{9.2}$$

where $f\in\mathbf{L}^2(\Omega)$ is a given function. The requirement that u vanishes along $\partial\Omega$ is called *Dirichlet's boundary condition.*

For future reference, we collect the main hypotheses used throughout this chapter.

(H) *The domain $\Omega\subset\mathbb{R}^n$ is open and bounded. The coefficients of L in* (9.1) *satisfy*

$$a^{ij},b^i,c\ \in\ \mathbf{L}^\infty(\Omega)\,. \tag{9.3}$$

Moreover, the operator L is **uniformly elliptic**. *Namely, there exists a constant $\theta>0$ such that*

$$\sum_{i,j=1}^n a^{ij}(x)\xi_i\xi_j\ \geq\ \theta|\xi|^2\qquad\text{for all } x\in\Omega,\ \xi\in\mathbb{R}^n\,. \tag{9.4}$$

Remark 9.1. By definition, the uniform ellipticity of the operator L depends only on the coefficients a^{ij}. In the symmetric case where $a^{ij}=a^{ji}$, the above condition means that for every $x\in\Omega$ the $n\times n$ symmetric matrix $A(x)=(a^{ij}(x))$ is strictly positive definite and its smallest eigenvalue is $\geq\theta$.

9.1.1. Physical interpretation. As an example, consider a fluid moving with velocity $\mathbf{b}(x)=(b^1,b^2,b^3)(x)$ in a domain $\Omega\subset\mathbb{R}^3$. Let $u=u(t,x)$ describe the density of a chemical dispersed within the fluid.

Given any subdomain $V\subset\Omega$, assume that the total amount of chemical contained in V changes only due to the inward or outward flux through the boundary ∂V. Namely,

$$\frac{d}{dt}\int_V u\,dx\ =\ \int_{\partial V}\mathbf{n}\cdot(a\,\nabla u)\,dS-\int_{\partial V}\mathbf{n}\cdot(\mathbf{b}\,u)\,dS\,. \tag{9.5}$$

Here $\mathbf{n}(x)$ denotes the unit outer normal to the set V at a boundary point x, while $a>0$ is a constant diffusion coefficient. The first integral on the right-hand side of (9.5) describes how much chemical enters through the boundary by *diffusion.* Notice that this is positive at points where $\mathbf{n}\cdot\nabla u>0$. Roughly speaking, this is the case if the concentration of chemical outside the domain V is greater than inside. The second integral (with the minus sign in front) denotes the amount of chemical that moves out across the boundary of V by *advection*, being transported by the fluid in motion (Figure 9.1.1).

Using the divergence theorem, from (9.5) we obtain

$$\int_V u_t\,dx\ =\ \int_V a\,\Delta u\,dx-\int_V \mathrm{div}(\mathbf{b}\,u)\,dx\,. \tag{9.6}$$

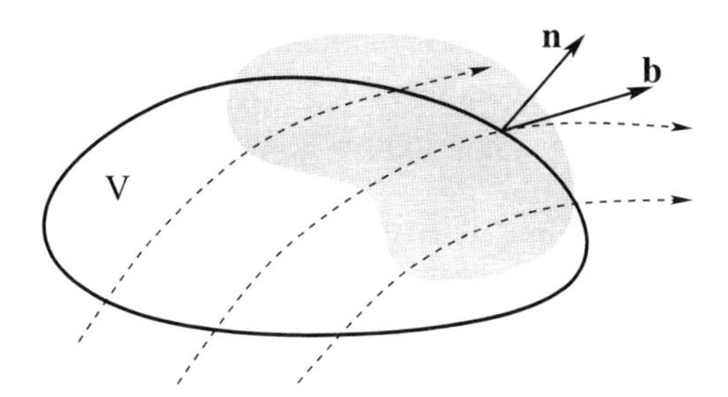

Figure 9.1.1. As the chemical is transported across the boundary of V, the total amount contained inside V changes in time.

Since the above identity holds on every subdomain $V \subset \Omega$, we conclude that $u = u(t,x)$ satisfies the parabolic PDE

$$\underset{}{u_t} \quad - \quad \underset{\text{[diffusion]}}{a\,\Delta u} \quad + \quad \underset{\text{[advection]}}{\operatorname{div}(\mathbf{b}\,u)} \quad = 0\,.$$

A more general model can describe the following situations:

- The diffusion is not uniform throughout the domain. In other words, the coefficient a is not a constant but depends on the location $x \in \Omega$. Moreover, the diffusion is not isotropic: in some directions it is faster than in others. All this can be modeled by replacing the constant diffusion matrix $A(x) \equiv aI$ with a more general symmetric matrix $A(x) = (a^{ij}(x))$.
- The total amount of u is not conserved. Additional terms are present, accounting for linear decay and for an external source.

In n space dimensions, this leads to a linear evolution equation of the form (9.7)

$$u_t \quad - \underset{\text{[diffusion]}}{\sum_{i,j=1}^{n} \left(a^{ij}(x)u_{x_i}\right)_{x_j}} \quad + \underset{\text{[advection]}}{\sum_{i=1}^{n} \left(b^i(x)u\right)_{x_i}} \quad = \underset{\text{[decay]}}{-c(x)\,u} \quad + \underset{\text{[source]}}{f(x)}\,.$$

Equation (9.7) can be used to model a variety of phenomena, such as mass transport, heat propagation, etc. In many situations, one is interested in *steady states*, i.e., in solutions which are independent of time. Setting $u_t = 0$ in (9.7), we obtain the linear elliptic equation

$$-\sum_{i,j=1}^{n} \left(a^{ij}(x)u_{x_i}\right)_{x_j} + \sum_{i=1}^{n} \left(b^i(x)u\right)_{x_i} + c(x)\,u \;=\; f(x)\,. \tag{9.8}$$

9.1.2. Classical and weak solutions. By a **classical solution** of the boundary value problem (9.2) we mean a function $u \in \mathcal{C}^2(\Omega)$ which satisfies the equation and the boundary conditions at every point. In general, due to the lack of regularity in the coefficients of the equation, problem (9.2) may not have classical solutions. A weaker concept of solution is thus needed.

Definition 9.2. A **weak solution** of (9.2) is a function $u \in H_0^1(\Omega)$ such that
(9.9)

$$\int_\Omega \Big(\sum_{i,j=1}^n a^{ij} u_{x_i} v_{x_j} - \sum_{i=1}^n b^i u\, v_{x_i} + cuv \Big)\, dx = \int_\Omega f v\, dx \qquad \text{for all } v \in H_0^1(\Omega)\,.$$

Remark 9.3 (On the concept of weak solution). The boundary condition $u = 0$ on $\partial\Omega$ is taken into account by requiring that $u \in H_0^1(\Omega)$. The equality (9.9) is formally obtained by writing

$$\int_\Omega (Lu)\, v\, dx = \int_\Omega f\, v\, dx \qquad \text{for all } v \in \mathcal{C}_c^\infty(\Omega) \tag{9.10}$$

and integrating by parts. Notice that, if (9.9) holds for every test function $v \in \mathcal{C}_c^\infty(\Omega)$, then by an approximation argument the same integral identity remains valid for every $v \in H_0^1(\Omega)$. It is important to observe that a function $u \in H_0^1$ may not have weak derivatives of second-order. However, the integral in (9.9) is always well defined, for all $u, v \in H_0^1$.

A convenient way to reformulate the concept of weak solution is the following. On the Hilbert space $H_0^1(\Omega)$, consider the bilinear form

$$B[u,v] \doteq \int_\Omega \left(\sum_{i,j=1}^n a^{ij}\, u_{x_i} v_{x_j} - \sum_{i=1}^n b^i\, u v_{x_i} + c\, uv \right) dx\,. \tag{9.11}$$

A function $u \in H_0^1$ is a weak solution of (9.2) provided that

$$B[u,v] = (f,v)_{\mathbf{L}^2} \qquad \text{for all } v \in H_0^1\,. \tag{9.12}$$

Here and in the sequel we use the notation

$$(f,g)_{\mathbf{L}^2} \doteq \int_\Omega f\, g\, dx \tag{9.13}$$

for the inner product in $\mathbf{L}^2(\Omega)$, to distinguish it from the inner product in $H^1(\Omega)$

$$(f,g)_{H^1} \doteq \int_\Omega f\, g\, dx + \int_\Omega \sum_{i=1}^n f_{x_i} g_{x_i}\, dx\,. \tag{9.14}$$

Remark 9.4 (Choice of the sign). The minus sign in front of the second-order terms in (9.1) disappears in (9.11), after a formal integration by parts.

As will become apparent later, this sign is chosen so that the corresponding quadratic form $B[u,v]$ can be positive definite.

Remark 9.5 (General boundary conditions). Given a function $g \in H^1(\Omega)$, one can consider the nonhomogeneous boundary value problem

$$\begin{cases} Lu = f, & x \in \Omega, \\ u = g, & x \in \partial\Omega. \end{cases} \tag{9.15}$$

This can be rewritten as a homogeneous problem for the function $\tilde{u} \doteq u - g$, namely

$$\begin{cases} L\tilde{u} = f - Lg, & x \in \Omega, \\ \tilde{u} = 0, & x \in \partial\Omega. \end{cases} \tag{9.16}$$

Assuming that $Lg \in \mathbf{L}^2(\Omega)$, problem (9.16) is exactly of the same type as (9.2).

Remark 9.6 (Operators not in divergence form). A differential operator of the form

$$Lu \doteq -\sum_{i,j=1}^{n} a^{ij}(x)u_{x_ix_j} + \sum_{i=1}^{n} b^i(x)u_{x_i} + c(x)u$$

can be rewritten as

$$Lu = -\sum_{i,j=1}^{n} \left(a^{ij}(x)u_{x_i}\right)_{x_j} + \sum_{i=1}^{n} \Big(b^i(x) + \sum_{j=1}^{n} a^{ij}_{x_j}(x)\Big) u_{x_i} + c(x)u\,.$$

Assuming that $a^{ij}, a^{ij}_{x_j}, b^i, c \in \mathbf{L}^\infty(\Omega)$, a weak solution of the corresponding problem (9.2) can again be obtained by solving (9.12), where the bilinear form B is now defined by

$$B[u,v] \doteq \int_\Omega \left(\sum_{i,j=1}^{n} a^{ij}\, u_{x_i}v_{x_j} + \sum_{i=1}^{n} \Big(b^i + \sum_{j=1}^{n} a^{ij}_{x_j}\Big) u_{x_i}v + c\,uv \right) dx\,.$$

As a first example, consider the boundary value problem

$$\begin{cases} -\Delta u + u = f, & x \in \Omega, \\ u = 0, & x \in \partial\Omega. \end{cases} \tag{9.17}$$

Clearly, the operator $-\Delta u \doteq -\sum_i u_{x_ix_i}$ is uniformly elliptic, because in this case the $n \times n$ matrix $A(x) = (a^{ij}(x))$ is the identity matrix, for every $x \in \Omega$. The existence of a weak solution to (9.17) can be proved by a remarkably concise argument, based on the Riesz representation theorem.

Lemma 9.7. *Let $\Omega \subset \mathbb{R}^n$ be a bounded open set. Then for every $f \in \mathbf{L}^2(\Omega)$ the boundary value problem (9.17) has a unique weak solution $u \in H_0^1(\Omega)$. The corresponding map $f \mapsto u$ is a compact linear operator from $\mathbf{L}^2(\Omega)$ into $H_0^1(\Omega)$.*

Proof. By the Rellich-Kondrachov theorem, the canonical embedding $\iota : H_0^1(\Omega) \mapsto \mathbf{L}^2(\Omega)$ is compact. Hence its dual operator ι^* is also compact. Since H_0^1 and $\mathbf{L}^2$ are Hilbert spaces, they can be identified with their duals. We thus obtain the following diagram:

$$\begin{array}{ccccccc} & & H_0^1(\Omega) & \xrightarrow{\iota} & \mathbf{L}^2(\Omega), & & \\ H_0^1(\Omega) & = & [H_0^1(\Omega)]^* & \xleftarrow{\iota^*} & [\mathbf{L}^2(\Omega)]^* & = & \mathbf{L}^2(\Omega)\,. \end{array} \tag{9.18}$$

For each $f \in \mathbf{L}^2(\Omega)$, the definition of dual operator yields

$$(\iota^* f,\, v)_{H^1} \;=\; (f, \iota v)_{\mathbf{L}^2} \;=\; (f, v)_{\mathbf{L}^2} \qquad \text{for all } f \in \mathbf{L}^2(\Omega),\;\; v \in H_0^1(\Omega)\,.$$

By (9.14) this means that $\iota^* f$ is precisely the weak solution to (9.17). □

9.1.3. Homogeneous second-order elliptic operators. We begin by studying solutions to the elliptic boundary value problem

$$\begin{cases} Lu \;=\; f, & x \in \Omega\,, \\ \;\;u \;=\; 0, & x \in \partial\Omega\,, \end{cases} \tag{9.19}$$

assuming that the differential operator L contains only second-order terms:

$$Lu \;\doteq\; -\sum_{i,j=1}^{n} \big(a^{ij}(x) u_{x_i}\big)_{x_j}\,. \tag{9.20}$$

We recall that a weak solution of (9.19) is a function $u \in H_0^1(\Omega)$ such that

$$B[u,v] \;=\; (f, v)_{\mathbf{L}^2} \qquad \text{for all } v \in H_0^1(\Omega)\,, \tag{9.21}$$

where $B : H_0^1 \times H_0^1 \mapsto \mathbb{R}$ is the continuous bilinear form

$$B[u,v] \;\doteq\; \int_\Omega \sum_{i,j=1}^{n} a^{ij} u_{x_i} v_{x_j}\, dx\,. \tag{9.22}$$

Theorem 9.8 (Unique solution of the elliptic boundary value problem). *Let $\Omega \subset \mathbb{R}^n$ be a bounded open set. Let the operator L in (9.20) be uniformly elliptic, with coefficients $a^{ij} \in \mathbf{L}^\infty(\Omega)$. Then, for every $f \in \mathbf{L}^2(\Omega)$, the boundary value problem (9.19) has a unique weak solution $u \in H_0^1(\Omega)$. The corresponding solution operator, which we denote as $L^{-1} : f \mapsto u$, is a compact linear operator from $\mathbf{L}^2(\Omega)$ into $H_0^1(\Omega)$.*

Proof. The existence and uniqueness of a weak solution to the elliptic boundary value problem (9.19) will be achieved by checking that the bilinear form B in (9.22) satisfies all the assumptions of the Lax-Milgram theorem.

1. The continuity of B is clear. Indeed,

$$\Big|B[u,v]\Big| \;\leq\; \sum_{i,j=1}^n \int_\Omega |a^{ij}u_{x_i}v_{x_j}|\,dx \;\leq\; \sum_{i,j=1}^n \|a^{ij}\|_{\mathbf{L}^\infty}\|u_{x_i}\|_{\mathbf{L}^2}\|v_{x_j}\|_{\mathbf{L}^2}$$

$$\leq\; C\,\|u\|_{H^1}\|v\|_{H^1}\,.$$

2. We claim that B is strictly positive definite, i.e., there exists $\beta > 0$ such that

$$B[u,u] \;\geq\; \beta\,\|u\|^2_{H^1(\Omega)} \qquad \text{for all } u \in H^1_0(\Omega)\,. \tag{9.23}$$

Indeed, since Ω is bounded, Poincaré's inequality yields the existence of a constant κ such that

$$\|u\|^2_{\mathbf{L}^2(\Omega)} \;\leq\; \kappa \int_\Omega |\nabla u|^2\,dx \qquad \text{for all } u \in H^1_0(\Omega)\,.$$

On the other hand, the uniform ellipticity condition implies

$$B[u,u] \;=\; \int_\Omega \sum_{i,j=1}^n a^{ij}u_{x_i}u_{x_j}\,dx \;\geq\; \int_\Omega \theta \sum_{i=1}^n u^2_{x_i}\,dx \;=\; \theta \int_\Omega |\nabla u|^2\,dx\,.$$

Together, the two above inequalities yield

$$\|u\|^2_{H^1} \;=\; \|u\|^2_{\mathbf{L}^2} + \|\nabla u\|^2_{\mathbf{L}^2} \;\leq\; (\kappa+1)\,\|\nabla u\|^2_{\mathbf{L}^2} \;\leq\; \frac{\kappa+1}{\theta}\,B[u,u]\,.$$

This proves (9.23) with $\beta = \theta/(\kappa+1)$.

3. By the Lax-Milgram theorem, for every $\tilde f \in H^1_0(\Omega)$ there exists a unique element $u \in H^1_0$ such that

$$B[u,v] \;=\; (\tilde f, v)_{H^1} \qquad \text{for all } v \in H^1_0(\Omega)\,. \tag{9.24}$$

Moreover, the map $\Lambda : \tilde f \mapsto u$ is continuous, namely

$$\|u\|_{H^1} \;\leq\; \beta^{-1}\,\|\tilde f\|_{H^1}\,.$$

Choosing $\tilde f \doteq \iota^* f \in H^1_0(\Omega)$, defined in (9.18), we thus achieve

$$B[u,v] \;=\; (\iota^* f, v)_{H^1} \;=\; (f,v)_{\mathbf{L}^2} \qquad \text{for all } v \in H^1_0(\Omega)\,. \tag{9.25}$$

By definition, u is a weak solution of (9.19).

4. To prove that the solution operator $L^{-1} : f \mapsto u$ is compact, consider the the diagram

$$\mathbf{L}^2(\Omega) \xrightarrow{\iota^*} H_0^1(\Omega) \xrightarrow{\Lambda} H_0^1(\Omega).$$

By Lemma 9.7, the linear operator ι^* is compact. Moreover, Λ is continuous. Therefore the composition $L^{-1} = \Lambda \circ \iota^*$ is compact. □

9.1.4. Representation of solutions in terms of eigenfunctions. Since $H_0^1(\Omega) \subset \mathbf{L}^2(\Omega)$, the solution operator L^{-1} described in Theorem 9.8 can also be regarded as a compact operator from $\mathbf{L}^2(\Omega)$ into itself. In the symmetric case where $a^{ij} = a^{ji}$, the operator L^{-1} is selfadjoint. Hence, by the Hilbert-Schmidt theorem, it admits a representation in terms of a countable basis of eigenfunctions.

Theorem 9.9 (Representation of solutions as a series of eigenfunctions). *Assume $a^{ij} = a^{ji} \in \mathbf{L}^\infty(\Omega)$. Then, in the setting of Theorem 9.8, the linear operator $L^{-1} : \mathbf{L}^2(\Omega) \mapsto \mathbf{L}^2(\Omega)$ is compact, one-to-one, and selfadjoint.*

The space $\mathbf{L}^2(\Omega)$ admits a countable orthonormal basis $\{\phi_k\,;\ k \geq 1\}$ consisting of eigenfunctions of L^{-1}, and one has the representation

$$L^{-1}f = \sum_{k=1}^{\infty} \lambda_k (f, \phi_k)_{\mathbf{L}^2}\, \phi_k\,. \tag{9.26}$$

The corresponding eigenvalues λ_k satisfy

$$\lim_{k\to\infty} \lambda_k = 0\,, \qquad \lambda_k > 0 \text{ for all } k \geq 1. \tag{9.27}$$

Proof. 1. By Theorem 9.8, L^{-1} is a compact linear operator from $\mathbf{L}^2(\Omega)$ into itself.

To show that L^{-1} is one-to-one, assume $u = L^{-1}f = 0$. Then

$$0 = B[u, v] = (f, v)_{\mathbf{L}^2}$$

for every $v \in H_0^1(\Omega)$. In particular, for every test function $\phi \in \mathcal{C}_c^\infty(\Omega)$ we have

$$\int_\Omega f\,\phi\,dx = 0\,.$$

This implies $f(x) = 0$ for a.e. $x \in \Omega$. Hence $\mathrm{Ker}(L^{-1}) = \{0\}$ and the operator L^{-1} is one-to-one.

2. To prove that L^{-1} is selfadjoint, assume

$$f, g \in \mathbf{L}^2(\Omega)\,, \qquad u = L^{-1}f\,, \qquad v = L^{-1}g\,.$$

By the definition of weak solution in (9.9), this implies $u, v \in H_0^1(\Omega)$ and

$$(L^{-1}f, g)_{\mathbf{L}^2} = \int_\Omega u\, g\, dx = \int_\Omega \sum_{i,j} a^{ij} u_{x_i} v_{x_j}\, dx = \int_\Omega f\, v\, dx = (f, L^{-1}g)_{\mathbf{L}^2}\,.$$

Note that the second equality follows from the fact that $v = L^{-1}g$, using $u \in H_0^1(\Omega)$ as a test function. The third equality follows from the fact that $u = L^{-1}f$, using $v \in H_0^1(\Omega)$ as a test function.

3. Since L^{-1} is a compact and selfadjoint operator on the separable space $\mathbf{L}^2(\Omega)$, by the theorem of Hilbert-Schmidt in Chapter 6, there exists of a countable orthonormal basis consisting of eigenfunctions of L^{-1}. This yields (9.26).

By the compactness of the operator L^{-1}, the eigenvalues satisfy $\lambda_k \to 0$. Finally, choosing $v = \phi_k$ as the test function in (9.21), one obtains

$$B[\lambda_k\phi_k,\, \phi_k] = (\phi_k,\, \phi_k)_{\mathbf{L}^2} = 1.$$

Since the quadratic form B is strictly positive definite, we conclude that

$$\lambda_k = \frac{1}{B[\phi_k,\, \phi_k]} > 0\,. \qquad \square$$

Example 9.10. Let $\Omega \doteq]0, \pi[\subset \mathbb{R}$ and $Lu = -u_{xx}$. Given $f \in \mathbf{L}^2(]0, \pi[)$, consider the elliptic boundary value problem

$$(9.28) \qquad \begin{cases} -u_{xx} = f, & 0 < x < \pi\,, \\ u(0) = u(\pi) = 0\,. \end{cases}$$

As a first step, we compute the eigenfunctions of the operator $Lu = -u_{xx}$. Solving the boundary value problem

$$-u_{xx} = \mu u\,, \qquad u(0) = u(\pi) = 0\,,$$

we find the eigenvalues and the normalized eigenfunctions

$$\mu_k = k^2\,, \qquad \phi_k(x) = \sqrt{\frac{2}{\pi}}\, \sin kx\,.$$

Of course, the inverse operator L^{-1} has the same eigenfunctions ϕ_k, with eigenvalues $\lambda_k = 1/k^2$.

In this special case, formula (9.26) yields the well-known representation of solutions of (9.28) in terms of a Fourier sine series:

$$\begin{aligned} u(x) &= L^{-1}f = \sum_{k=1}^{\infty} \lambda_k\,(f,\phi_k)_{\mathbf{L}^2}\,\phi_k \\ &= \sum_{k=1}^{\infty} \frac{1}{k^2}\left(\int_0^{\pi} f(y)\sqrt{\frac{2}{\pi}}\sin ky\,dy\right)\sqrt{\frac{2}{\pi}}\,\sin\,kx \\ &= \sum_{k=1}^{\infty} \frac{2}{\pi k^2}\left(\int_0^{\pi} f(y)\,\sin ky\,dy\right)\sin\,kx\,. \end{aligned}$$

9.1.5. More general linear elliptic operators. The existence and uniqueness result stated in Theorem 9.8 relied on the fact that the bilinear form B in (9.21) is strictly positive definite on the space $H^1_0(\Omega)$. This property no longer holds for the more general bilinear form B in (9.11), where additional lower-order terms are present. For example, if the function $c(x)$ is large and negative, one may find some $u \in H^1_0(\Omega)$ such that $B[u,u] < 0$.

Example 9.11. Consider the open interval $\Omega \doteq]0,\pi[$. Observe that the operator

$$Lu \doteq -u_{xx} - 4u$$

is uniformly elliptic on Ω. However, the corresponding bilinear form

$$B[u,v] = \int_0^{\pi} u_x\,v_x - 4uv\,dx$$

is not positive definite on $H^1_0(\Omega)$. For example, taking $u(x) = \sin x$, we find

$$B[u,u] = \int_0^{\pi} \cos^2 x - 4\sin^2 x\,dx = -\frac{3\pi}{2}\,.$$

If $f(x) = \sin 2x$, then the boundary value problem

$$\left\{\begin{array}{rcll} -u_{xx} - 4u &=& \sin 2x, & x \in]0,\pi[\,, \\ u(0) &=& u(\pi) = 0 & \end{array}\right. \tag{9.29}$$

has no solutions. Indeed, choosing $v(x) = \sin 2x$, for every $u \in H^1_0(\Omega)$ an integration by parts yields

$$\begin{aligned} B[u,v] &= \int_0^{\pi} u_x v_x - 4uv\,dx = \int_0^{\pi}\Big(2u_x\cos 2x - 4u\,\sin 2x\Big)\,dx \\ &= \int_0^{\pi}\Big(2u\,\sin 2x\Big)_x\,dx = 0 \neq \int_0^{\pi}\sin^2 2x\,dx = (f,v)_{\mathbf{L}^2}\,. \end{aligned}$$

Therefore there is no function $u \in H_0^1(\Omega)$ that satisfies (9.12).

In this connection one should also notice that the corresponding homogeneous problem

$$\begin{cases} -u_{xx} - 4u &= 0, \qquad x \in]0, \pi[\,, \\ u(0) &= u(\pi) \;=\; 0 \end{cases} \tag{9.30}$$

admits infinitely many solutions: $u(x) \;=\; \kappa \sin 2x$, for any constant κ.

We now study the existence and uniqueness of weak solutions to the more general boundary value problem (9.1)–(9.2). Our approach is based on two steps.

STEP 1: By choosing a constant $\gamma > 0$ sufficiently large, the operator

$$L_\gamma u \;\doteq\; Lu + \gamma u \tag{9.31}$$

is strictly positive definite. More precisely, the corresponding bilinear form

$$B_\gamma[u,v] \;\doteq\; \int_\Omega \left(\sum_{i,j=1}^n a^{ij}(x)\, u_{x_i} v_{x_j} - \sum_{i=1}^n b^i(x)\, u v_{x_i} + c(x)\, uv + \gamma\, uv \right) dx \tag{9.32}$$

satisfies

$$B_\gamma[u,u] \;\geq\; \beta\, \|u\|_{H^1}^2 \qquad \text{for all } u \in H_0^1(\Omega)\,, \tag{9.33}$$

for some constant $\beta > 0$. Using the Lax-Milgram theorem, we conclude that for every $f \in \mathbf{L}^2(\Omega)$ the equation

$$L_\gamma u \;=\; f$$

has a unique weak solution $u \in H_0^1(\Omega)$. Moreover, the map $f \mapsto u = L_\gamma^{-1} f$ is a linear compact operator from $\mathbf{L}^2(\Omega)$ into $H_0^1(\Omega) \subset \mathbf{L}^2(\Omega)$. We regard L_γ^{-1} as a compact operator from $\mathbf{L}^2(\Omega)$ into itself.

STEP 2: The original problem (9.2) can now be written as

$$Lu \;=\; L_\gamma u - \gamma u \;=\; f\,.$$

Applying the operator L_γ^{-1} to both sides, one obtains

$$u - L_\gamma^{-1} \gamma u \;=\; L_\gamma^{-1} f\,. \tag{9.34}$$

Introducing the notation

$$K \;\doteq\; \gamma L_\gamma^{-1}\,, \qquad h \;\doteq\; L_\gamma^{-1} f\,, \tag{9.35}$$

we are led to the equation

$$(I - K)u \;=\; h\,. \tag{9.36}$$

Since K is a compact operator from $\mathbf{L}^2(\Omega)$ into itself, Fredholm's theory can be applied. In particular, one has

Fredholm's alternative: either

(i) *for every* $h \in \mathbf{L}^2(\Omega)$ *the equation* $u - Ku = h$ *has a unique solution* $u \in \mathbf{L}^2(\Omega)$,

or else

(ii) *the equation* $u - Ku = 0$ *has a nontrivial solution* $u \in \mathbf{L}^2(\Omega)$.

Calling $K^* : \mathbf{L}^2 \mapsto \mathbf{L}^2$ the adjoint operator, case (ii) occurs if and only if the adjoint equation $v - K^*v = 0$ has a nontrivial solution $v \in \mathbf{L}^2(\Omega)$. Information about the existence and uniqueness of weak solutions to (9.2) can thus also be obtained by studying the adjoint operator

$$L^*v \doteq -\sum_{i,j=1}^{n} (a^{ij}(x)\, v_{x_j})_{x_i} - \sum_{i=1}^{n} b^i(x) v_{x_i} + c(x)\, v\,. \tag{9.37}$$

The remainder of this section will provide detailed proofs of the above claims.

Lemma 9.12 (Estimates on elliptic operators). *Let the operator L in (9.1) be uniformly elliptic, with coefficients $a^{ij}, b^i, c \in \mathbf{L}^\infty(\Omega)$. Then there exist constants $\alpha, \beta, \gamma > 0$ such that*

$$|B[u,v]| \leq \alpha \|u\|_{H^1} \|v\|_{H^1}\,, \tag{9.38}$$

$$\beta \|u\|_{H^1}^2 \leq B[u,u] + \gamma \|u\|_{\mathbf{L}^2}^2\,, \tag{9.39}$$

for all $u, v \in H_0^1(\Omega)$.

Proof. 1. The boundedness of the bilinear form $B : H_0^1 \times H_0^1 \mapsto \mathbb{R}$ follows from

$$\begin{aligned}
\Big|B[u,v]\Big| &= \left| \int_\Omega \left(\sum_{i,j=1}^{n} a^{ij} u_{x_i} v_{x_j} - \sum_{i=1}^{n} b^i u v_{x_i} + c\, uv \right) dx \right| \\
&\leq \sum_{i,j=1}^{n} \|a^{ij}\|_{\mathbf{L}^\infty} \|u_{x_i}\|_{\mathbf{L}^2} \|v_{x_j}\|_{\mathbf{L}^2} + \sum_{i=1}^{n} \|b^i\|_{\mathbf{L}^\infty} \|u\|_{\mathbf{L}^2} \|v_{x_i}\|_{\mathbf{L}^2} \\
&\quad + \|c\|_{\mathbf{L}^\infty} \|u\|_{\mathbf{L}^2} \|v\|_{\mathbf{L}^2} \\
&\leq \alpha \|u\|_{H^1} \|v\|_{H^1}\,.
\end{aligned}$$

2. Concerning the second estimate, using the ellipticity condition (9.4) and the elementary inequality $ab \le \frac{1}{2\theta}a^2 + \frac{\theta}{2}b^2$, we obtain

$$\begin{aligned}
\theta \sum_{i=1}^{n} \|u_{x_i}\|_{\mathbf{L}^2}^2 &= \theta \int_\Omega \sum_{i=1}^{n} u_{x_i}^2\, dx \le \int_\Omega \sum_{i,j=1}^{n} a^{ij} u_{x_i} u_{x_j}\, dx \\
&= B[u,u] + \int_\Omega \left(\sum_{i=1}^{n} b^i u u_{x_i} - cu^2 \right) dx \\
&\le B[u,u] + \sum_{i=1}^{n} \|b^i\|_{\mathbf{L}^\infty} \|u\|_{\mathbf{L}^2} \|u_{x_i}\|_{\mathbf{L}^2} + \|c\|_{\mathbf{L}^\infty} \|u\|_{\mathbf{L}^2}^2 \\
&\le B[u,u] + \left(\frac{1}{2\theta} \sum_{i=1}^{n} \|b^i\|_{\mathbf{L}^\infty}^2 \|u\|_{\mathbf{L}^2}^2 + \frac{\theta}{2} \sum_{i=1}^{n} \|u_{x_i}\|_{\mathbf{L}^2}^2 \right) + \|c\|_{\mathbf{L}^\infty} \|u\|_{\mathbf{L}^2}^2 .
\end{aligned}$$

Therefore

$$B[u,u] \ge \frac{\theta}{2} \sum_{i=1}^{n} \|u_{x_i}\|_{\mathbf{L}^2}^2 - C\|u\|_{\mathbf{L}^2}^2 \qquad \text{for all } u \in H_0^1(\Omega),$$

for a suitable constant C. Taking $\beta = \theta/2$ and $\gamma = C + \theta/2$, the inequality (9.39) is satisfied. □

Remark 9.13. By the above lemma, if the constant $\gamma > 0$ is large enough, then the bilinear form

$$B_\gamma[u,v] \doteq B[u,v] + \gamma\,(u,v)_{\mathbf{L}^2}$$

in (9.32) is strictly positive definite. Notice that, for $\gamma > 0$ large, it would be very easy to show that the bilinear form

$$\widetilde{B}_\gamma \doteq B[u,v] + \gamma\,(u,v)_{H^1}$$

is strictly positive definite on $H_0^1(\Omega)$. However, Lemma 9.12 shows that we can achieve strict positivity by adding the much weaker term $\gamma\,(u,v)_{\mathbf{L}^2}$.

Let γ be as in (9.39) and define the bilinear form B_γ according to (9.32). Since B_γ is strictly positive definite, we can apply the Lax-Milgram theorem and conclude that, for every $f \in \mathbf{L}^2(\Omega)$, there exists a unique $u \in H_0^1(\Omega)$ such that

$$(9.40) \qquad B_\gamma[u,v] = (f,v)_{\mathbf{L}^2} = (\iota^* f, v)_{H^1} \qquad \text{for all } v \in H_0^1(\Omega).$$

Since the map ι^* is compact, the solution operator $f \mapsto u = L_\gamma^{-1} f$ is a linear compact operator from $\mathbf{L}^2(\Omega)$ into $H_0^1(\Omega)$. Therefore, it is also a compact operator from $\mathbf{L}^2(\Omega)$ into itself.

An entirely similar result holds for the adjoint problem

$$\begin{cases} L_\gamma^* v = g, & x \in \Omega, \\ v = 0, & x \in \partial\Omega, \end{cases} \tag{9.41}$$

where L^* is the adjoint operator introduced in (9.37) and $L_\gamma^* u \doteq L^* u + \gamma u$. Given $g \in \mathbf{L}^2(\Omega)$, a weak solution of (9.41) is defined to be a function $v \in H_0^1(\Omega)$ such that

$$B_\gamma^*[v, u] \doteq B_\gamma[u, v] = (u, g)_{\mathbf{L}^2} \qquad \text{for all } u \in H_0^1(\Omega). \tag{9.42}$$

Since B_γ^* is strictly positive definite, for every $g \in \mathbf{L}^2$ the Lax-Milgram theorem yields a unique weak solution v of (9.41). The map $g \mapsto v = (L_\gamma^*)^{-1} g$ is a linear, compact operator from $\mathbf{L}^2(\Omega)$ into itself.

Lemma 9.14 (Adjoint operator). *In the above setting, the operator $(L_\gamma^*)^{-1}$ is the adjoint of the operator L_γ^{-1}.*

Proof. By definition, for every $f, g \in \mathbf{L}^2(\Omega)$ and $u, v \in H_0^1(\Omega)$ one has

$$(f, v)_{\mathbf{L}^2} = B_\gamma\left[L_\gamma^{-1} f,\, v\right], \qquad (u, g)_{\mathbf{L}^2} = B_\gamma\left[u,\, (L_\gamma^{-1})^* g\right].$$

In particular, choosing $v = (L_\gamma^*)^{-1} g$ and $u = L_\gamma^{-1} f$, we obtain

$$\left(f,\, (L_\gamma^*)^{-1} g\right)_{\mathbf{L}^2} = B_\gamma\left[L_\gamma^{-1} f,\, (L_\gamma^{-1})^* g\right] = \left(L_\gamma^{-1} f,\, g\right)_{\mathbf{L}^2}. \qquad \square$$

Lemma 9.15 (Representation of weak solutions). *Given any $f \in \mathbf{L}^2(\Omega)$, a function $u \in \mathbf{L}^2(\Omega)$ is a weak solution of (9.2) if and only if*

$$(I - K)u = h, \qquad \textit{with} \quad K = \gamma L_\gamma^{-1}, \quad h = L_\gamma^{-1} f. \tag{9.43}$$

Proof. 1. Let u be a weak solution of (9.2). By definition, this means that $u \in H_0^1(\Omega)$ and

$$B_\gamma[u, v] = B[u, v] + \gamma(u, v)_{\mathbf{L}^2} = (f + \gamma u, v)_{\mathbf{L}^2} \qquad \text{for all } v \in H_0^1(\Omega).$$

Therefore

$$u = L_\gamma^{-1}(f + \gamma u) = h + Ku.$$

2. To prove the converse, let (9.43) hold. Then

$$u = \gamma L_\gamma^{-1} u + L_\gamma^{-1} f \in H_0^1(\Omega).$$

Moreover, for every $v \in H_0^1(\Omega)$ we have

$$B[u, v] = B_\gamma[u, v] - \gamma(u, v)_{\mathbf{L}^2} = (f + \gamma u, v)_{\mathbf{L}^2} - \gamma(u, v)_{\mathbf{L}^2} = (f, v)_{\mathbf{L}^2}.$$

$\square$

In order to apply Fredholm's theory (Theorem 6.1), together with (9.2) we consider the homogeneous problem

$$(9.44) \qquad \begin{cases} Lu = 0, & x \in \Omega, \\ u = 0, & x \in \partial\Omega, \end{cases}$$

and the adjoint problem

$$(9.45) \qquad \begin{cases} L^*v = 0, & x \in \Omega, \\ v = 0, & x \in \partial\Omega, \end{cases}$$

where L^* is the adjoint linear operator defined at (9.37).

Theorem 9.16 (Unique solutions to the elliptic problem). *Under the basic hypotheses* **(H)**, *the following statements are equivalent:*

(i) *For every $f \in \mathbf{L}^2(\Omega)$, the elliptic boundary value problem* (9.2) *has a unique weak solution.*

(ii) *The homogeneous boundary value problem* (9.44) *has the only solution $u(x) \equiv 0$.*

(iii) *The adjoint homogeneous problem* (9.45) *has the only solution $v(x) \equiv 0$.*

Proof. 1. Since $K = \gamma L_\gamma^{-1}$ is a compact operator from $\mathbf{L}^2(\Omega)$ into itself, Fredholm's theorem can be applied. As a consequence, the linear operator $I - K$ is surjective if and only if it is one-to-one, i.e., if and only if $\mathrm{Ker}(I - K) = \{0\}$.

2. By Lemma 9.15, $u - Ku = 0$ if and only if u is a weak solution of (9.44). An entirely similar argument shows that $v - K^*v = 0$ if and only if v is a weak solution of (9.45).

By Fredholm's theorem, $\mathrm{Ker}(I - K)$ and $\mathrm{Ker}(I - K^*)$ have the same dimension. We thus obtain a chain of equivalent statements:

$$\begin{aligned} &I - K \text{ is surjective,} \\ &\mathrm{Ker}(I - K) = \{0\}, \\ &\mathrm{Ker}(I - K^*) = \{0\}, \\ &u \equiv 0 \text{ is the unique solution of (9.44),} \\ &v \equiv 0 \text{ is the unique solution of (9.45).} \end{aligned}$$

□

Theorem 9.16 covers the situation where $I - K$ is one-to-one and Fredholm's first alternative holds. In the case where $I - K$ is not necessarily one-to-one, the existence of solutions to

$$u - Ku = L_\gamma^{-1} f$$

can be determined using the identity

$$\text{Range}(I-K) \;=\; [\text{Ker}(I-K^*)]^\perp . \tag{9.46}$$

Theorem 9.17 (Existence of solutions to the elliptic problem). *Under the hypotheses* **(H)**, *problem* (9.2) *has at least one weak solution if and only if*

$$(f,v)_{\mathbf{L}^2} \;=\; 0 \tag{9.47}$$

for every weak solution $v \in H_0^1(\Omega)$ of the adjoint problem (9.45).

Proof. The boundary value problem (9.2) has a weak solution provided that $L_\gamma^{-1} f \in \text{Range}(I-K)$. By (9.46), this holds if and only if $L_\gamma^{-1} f$ is orthogonal to every $v \in \text{Ker}(I-K^*)$, i.e., to every solution v of the adjoint problem (9.45).

We claim that this holds if and only if f itself is orthogonal to every solution v of (9.45). Indeed, if $v - K^*v = 0$, one has

$$(f,v)_{\mathbf{L}^2} \;=\; (f,K^*v)_{\mathbf{L}^2} \;=\; (Kf,v)_{\mathbf{L}^2} \;=\; \gamma\,(L_\gamma^{-1}f,v)_{\mathbf{L}^2} .$$

Since $\gamma > 0$, this proves our claim. □

9.2. Parabolic equations

Let $\Omega \subset \mathbb{R}^n$ be a bounded open set and let L be the operator in (9.1). In addition to the standard hypotheses **(H)** stated at the beginning of the chapter, we now assume that the coefficients a^{ij} satisfy the stronger regularity condition

$$a^{ij} \in W^{1,\infty}(\Omega) . \tag{9.48}$$

In this section we study the parabolic initial-boundary value problem

$$\begin{cases} u_t + Lu \;=\; 0, & t>0\,,\ x\in\Omega\,,\\ u(t,x) \;=\; 0, & t>0\,,\ x\in\partial\Omega\,,\\ u(0,x) \;=\; g(x), & x\in\Omega\,. \end{cases} \tag{9.49}$$

It is convenient to reformulate (9.49) as a Cauchy problem in the Hilbert space $X = \mathbf{L}^2(\Omega)$, namely

$$\frac{d}{dt}u \;=\; Au, \qquad u(0) \;=\; g\,, \tag{9.50}$$

for a suitable (unbounded) linear operator $A : \mathbf{L}^2(\Omega) \mapsto \mathbf{L}^2(\Omega)$. More precisely

$$A \;\doteq\; -L\,, \qquad \text{Dom}(A) = \Big\{u \in H_0^1(\Omega)\,;\ Lu \in \mathbf{L}^2(\Omega)\Big\}. \tag{9.51}$$

In other words, $u \in \text{Dom}(A)$ if u is the solution to the elliptic boundary value problem (9.2), for some $f \in \mathbf{L}^2(\Omega)$. In this case, $Au = -f$.

Our eventual goal is to construct solutions of (9.50) using semigroup theory. We first consider the case where the operator L is strictly positive definite on $H_0^1(\Omega)$. More precisely, we assume that there exists $\beta > 0$ such that the bilinear form $B : H_0^1(\Omega) \times H_0^1(\Omega) \mapsto \mathbb{R}$ defined at (9.11) is strictly positive definite: there exists $\beta > 0$ such that

$$B[u, u] \geq \beta \|u\|_{H^1}^2 \qquad \text{for all } u \in H_0^1(\Omega)\,. \tag{9.52}$$

Notice that this is certainly true if $b^i \equiv 0$ and $c \geq 0$.

Theorem 9.18 (Semigroup of solutions of a parabolic equation. I). *Let the standard hypotheses* **(H)** *hold, together with* (9.48). *Moreover, assume that the corresponding bilinear form B in* (9.11) *is strictly positive definite, so that* (9.52) *holds.*

Then the operator $A = -L$ generates a contractive semigroup $\{S_t\,;\ t \geq 0\}$ of linear operators on $\mathbf{L}^2(\Omega)$.

Proof. To prove that A generates a contraction semigroup on $X = \mathbf{L}^2(\Omega)$, we need to check the following:

(i) $\text{Dom}(A)$ is dense in $\mathbf{L}^2(\Omega)$.

(ii) The graph of A is closed.

(iii) Every real number $\lambda > 0$ is in the resolvent set of A, and $\|(\lambda I - A)^{-1}\| \leq \lambda^{-1}$.

1. To prove (i), we observe that, if $\varphi \in \mathcal{C}_c^2(\Omega)$, then the regularity assumptions (9.48) imply $f \doteq L\varphi \in \mathbf{L}^2(\Omega)$. This proves that $\text{Dom}(A)$ contains the subspace $\mathcal{C}_c^2(\Omega)$ and therefore it is dense in $\mathbf{L}^2(\Omega)$.

2. If (9.52) holds, then, by the Lax-Milgram theorem, for every $f \in \mathbf{L}^2(\Omega)$ there exists a unique $u \in H_0^1(\Omega)$ such that

$$B[u, v] = (f, v)_{\mathbf{L}^2} \qquad \text{for all } v \in H_0^1(\Omega)\,.$$

The map $f \mapsto u \doteq L^{-1}f$ is a bounded linear operator from $\mathbf{L}^2(\Omega)$ into $\mathbf{L}^2(\Omega)$.

We observe that the pair of functions (u, f) lies in the graph of A if and only if $(-f, u)$ is in the graph L^{-1}. Since L^{-1} is a continuous operator, its graph is closed. Hence the graph of A is closed as well.

3. According to the definition of A, to prove (iii) we need to show that, for every $\lambda > 0$, the operator $\lambda I - A$ has a bounded inverse with operator norm $\|(\lambda I - A)^{-1}\| \leq \lambda^{-1}$. Equivalently, for every $f \in \mathbf{L}^2(\Omega)$, we need to show that the problem

$$\begin{cases} \lambda u + Lu = f, & x \in \Omega, \\ u = 0, & x \in \partial\Omega, \end{cases} \tag{9.53}$$

has a weak solution satisfying

$$\|u\|_{\mathbf{L}^2} \leq \frac{1}{\lambda}\|f\|_{\mathbf{L}^2}. \tag{9.54}$$

By the Lax-Milgram theorem, there exists a unique $u \in H_0^1(\Omega)$ such that

$$(\lambda u, v)_{\mathbf{L}^2} + B[u,v] = (f,v)_{\mathbf{L}^2} \qquad \text{for all } v \in H_0^1(\Omega). \tag{9.55}$$

Taking $v = u$ in (9.55), we obtain

$$\lambda\|u\|^2_{\mathbf{L}^2} + B[u,u] = (f,u)_{\mathbf{L}^2} \leq \|f\|_{\mathbf{L}^2} \cdot \|u\|_{\mathbf{L}^2}.$$

Since we are assuming $B[u,u] \geq 0$, this yields

$$\lambda\,\|u\|_{\mathbf{L}^2} \leq \|f\|_{\mathbf{L}^2},$$

proving (9.54).

4. We can now use Theorem 7.13 and conclude that the linear operator A generates a contractive semigroup. □

9.2.1. Representation of solutions in terms of eigenfunctions. Consider the special case where $a^{ij} = a^{ji}$ and L is the operator in (9.20), containing only second-order terms. In this case, by Poincaré's inequality, the bilinear form B in (9.22) is strictly positive definite and Theorem 9.18 can be applied.

Using Theorem 9.9, one obtains a representation of the semigroup trajectories in terms of an orthonormal basis $\{\phi_k\,;\ k \geq 1\}$ of eigenfunctions of the compact selfadjoint operator L^{-1}. By construction, for every $k \geq 1$ one has

$$L^{-1}\phi_k = \lambda_k \phi_k,$$

where $\lambda_k > 0$ is the corresponding eigenvalue. Therefore

$$\phi_k \in \mathrm{Dom}(L), \qquad L\phi_k = \mu_k\,\phi_k, \qquad \mu_k = \frac{1}{\lambda_k}. \tag{9.56}$$

Notice that $\lambda_k \to 0$ and $\mu_k \to \infty$, as $k \to \infty$. For every $k \geq 1$, the function

$$u(t) \doteq e^{-\mu_k t}\phi_k$$

provides a $\mathcal{C}^1$ solution to the Cauchy problem

$$\frac{d}{dt}u(t) = -Lu(t), \qquad u(0) = \phi_k.$$

Hence, by the uniqueness of semigroup trajectories, one must have

$$S_t\phi_k \;=\; e^{-\mu_k t}\phi_k\,.$$

By linearity, for any coefficients $c_1, \ldots, c_N \in \mathbb{R}$ one has

$$S_t\left(\sum_{k=1}^{N} c_k\phi_k\right) \;=\; \sum_{k=1}^{N} c_k e^{-\mu_k t}\phi_k\,.$$

Since S_t is a bounded linear operator, decomposing an arbitrary function $g \in \mathbf{L}^2(\Omega)$ along the orthonormal basis $\{\phi_k\,;\ k \geq 1\}$, we thus obtain

$$S_t g \;=\; \sum_{k=1}^{\infty} e^{-\mu_k t}(g, \phi_k)_{\mathbf{L}^2}\,\phi_k\,. \tag{9.57}$$

The above representation of semigroup trajectories is valid for every $g \in \mathbf{L}^2(\Omega)$ and every $t \geq 0$.

Lemma 9.19. *Let L be the operator in* (9.20)*. Then for every $g \in \mathbf{L}^2(\Omega)$ the formula* (9.57) *defines a map $t \mapsto u_t = S_t g$ from $[0, \infty[$ into $\mathbf{L}^2(\Omega)$. This map is continuous for $t \in [0, \infty[$ and continuously differentiable for $t > 0$. Moreover, $u(t) \in \mathrm{Dom}(L) \subseteq H_0^1(\Omega)$ for every $t > 0$ and*

$$\frac{d}{dt}u(t) \;=\; Lu(t) \qquad \textit{for all } t > 0\,. \tag{9.58}$$

Proof. 1. Let $g \in \mathbf{L}^2(\Omega)$. Since $\mu_k > 0$ for every k, it is clear that

$$\left|e^{-\mu_k t}(g, \phi_k)_{\mathbf{L}^2}\right|^2 \;\leq\; (g, \phi_k)^2_{\mathbf{L}^2}.$$

Therefore,

$$\sum_{k\geq 1}\left|e^{-\mu_k t}(g, \phi_k)_{\mathbf{L}^2}\right|^2 \;\leq\; \sum_{k\geq 1}(g, \phi_k)^2_{\mathbf{L}^2} \;=\; \|g\|^2_{\mathbf{L}^2} \;<\; \infty\,.$$

Hence the series in (9.57) is convergent, uniformly for $t \geq 0$. In particular, since the partial sums are continuous functions of time, the map $t \mapsto S_t g$ is continuous as well.

2. We claim that, even if $g \notin H_0^1(\Omega)$, one always has

$$S_t g \;\in\; \mathrm{Dom}(L) \;\subseteq\; H_0^1(\Omega) \qquad \text{for all } t > 0\,. \tag{9.59}$$

Indeed, a function $u = \sum_k c_k\phi_k$ lies in $\mathrm{Dom}(L)$ if and only if the coefficients c_k satisfy

$$\sum_k (c_k\mu_k)^2 \;<\; \infty\,.$$

In the case where $c_k(t) \doteq e^{-\mu_k t}(g, \phi_k)_{\mathbf{L}^2}$, we have the estimate

$$\sum_k (\mu_k\, c_k(t))^2 \;\leq\; \sup_k \left(\mu_k\, e^{-\mu_k t}\right)^2 \cdot \sum_k (g, \phi_k)^2_{\mathbf{L}^2}\,. \tag{9.60}$$

An elementary calculation now shows that, for $\xi \geq 0$ and t fixed, the function $\xi \mapsto \xi\, e^{-t\xi}$ attains its global maximum at $\xi = 1/t$. Therefore,

$$\mu_k\, e^{-\mu_k t} \;\leq\; \max_{\xi\geq 0}\, \xi e^{-t\xi} \;=\; \frac{1}{te}\,.$$

Using this bound in (9.60), we obtain

$$\sum_{k=1}^{\infty} \Big(\mu_k c_k(t)\Big)^2 \;\leq\; \frac{1}{t^2 e^2}\, \|g\|^2_{\mathbf{L}^2}\,.$$

Hence the series defining $Lu(t)$ is convergent. This implies $u(t) \in \mathrm{Dom}(L)$, for each $t > 0$.

3. Differentiating the series (9.57) term by term and observing that the series of derivatives is also convergent, we achieve (9.58). □

Example 9.20. As in Example 9.10, let $\Omega \doteq]0, \pi[\subset \mathbb{R}$ and $Lu = -u_{xx}$. Given $g \in \mathbf{L}^2(]0, \pi[)$, consider the parabolic initial-boundary value problem

$$\begin{cases} u_t &= u_{xx}, & t > 0,\ 0 < x < \pi\,, \\ u(0, x) &= g(x), & 0 < x < \pi\,, \\ u(t, 0) &= u(t, \pi) \;=\; 0\,. \end{cases}$$

In this special case, the formula (9.57) yields the solution as the sum of a Fourier sine series:

$$u(t, x) \;=\; \sum_{k=1}^{\infty} \frac{2}{\pi}\, e^{-k^2 t} \left(\int_0^{\pi} g(y)\, \sin ky\, dy\right) \sin kx\,.$$

9.2.2. More general operators. To motivate the following construction, we begin with a finite-dimensional example. Let A be an $n \times n$ matrix and consider the linear ODE on $\mathbb{R}^n$

$$\frac{d}{dt} x(t) \;=\; -Ax(t)\,. \tag{9.61}$$

If A is positive definite, i.e., if $\langle Ax,\, x\rangle \geq 0$ for all $x \in \mathbb{R}^n$, then $-A$ generates a contractive semigroup. Indeed

$$\frac{d}{dt}\, |x(t)|^2 \;=\; 2 \left\langle \frac{d}{dt}\, x(t)\,,\ x(t)\right\rangle \;=\; 2\,\langle -Ax(t),\, x(t)\rangle \;\leq\; 0\,,$$

showing that the Euclidean norm of a solution does not increase in time.

Next, let A be an arbitrary matrix. We can then find a number $\gamma \geq 0$ large enough so that $A + \gamma I$ is positive definite, and hence the matrix $-(A+\gamma I)$ generates a contractive semigroup. In this case, if $x(t) = e^{-tA}x(0)$ is a solution to (9.61), writing $-A = \gamma I - (A + \gamma I)$, one obtains

$$
\begin{aligned}
|x(t)| &= |e^{-At}x(0)| \\
&= |e^{\{\gamma I-(A+\gamma I)\}t}x(0)| \\
&= e^{\gamma t}|e^{-(A+\gamma I)t}x(0)| \\
&\leq e^{\gamma t}|x(0)|.
\end{aligned} \tag{9.62}
$$

According to (9.62), the operator $-A$ generates a semigroup of type γ.

We shall work out a similar construction in the case where L is a general elliptic operator, as in (9.1), and the corresponding bilinear form $B[u, v]$ in (9.11) is not necessarily positive definite. According to Lemma 9.12, there exists a constant $\gamma > 0$ large enough so that the bilinear form

$$
B_\gamma[u, v] \;=\; B[u, v] + \gamma(u, v)_{\mathbf{L}^2} \tag{9.63}
$$

is strictly positive definite on $H_0^1(\Omega)$. We can thus define

$$
L_\gamma u \;\doteq\; Lu + \gamma u\,, \qquad B_\gamma[u, v] \;\doteq\; B[u, v] + \gamma(u, v)_{\mathbf{L}^2}\,.
$$

The parabolic equation in (9.49) can now be written as

$$
u_t \;=\; -L_\gamma u + \gamma u\,.
$$

By the previous analysis, the operator $A_\gamma \doteq -(L + \gamma I)$ generates a contractive semigroup of linear operators, say $\{S_t^{(\gamma)}\,;\ t \geq 0\}$. Therefore, the operator $A \doteq -L = \gamma I - L_\gamma$ defined as in (9.51) generates a semigroup of type γ. Namely $\{S_t\,;\ t \geq 0\}$, with

$$
S_t = e^{\gamma t}\, S_t^{(\gamma)}, \qquad t \geq 0.
$$

Summarizing the above analysis, we have

Theorem 9.21 (Semigroup of solutions of a parabolic equation. II). *Let $\Omega \subset \mathbb{R}^n$ be a bounded open set. Assume that the operator L in* (9.1) *satisfies the regularity conditions* (9.48) *and the uniform ellipticity condition* (9.4).

Then the operator $A = -L$ defined at (9.51) *generates a semigroup $\{S_t\,;\ t \geq 0\}$ of linear operators on $\mathbf{L}^2(\Omega)$.*

Having constructed a semigroup $\{S_t\,;\ t \geq 0\}$ generated by the operator A, one needs to understand in which sense a trajectory of the semigroup $t \mapsto u(t) = S_t f$ provides a solution to the parabolic equation (9.49). In the case where L is the symmetric operator defined in (9.20), the representation

(9.57) yields all the needed information. Indeed, according to Lemma 9.19, for every initial data $g \in \mathbf{L}^2(\Omega)$ the solution $t \mapsto u(t) = S_t g$ is a $\mathcal{C}^1$ map, which takes values in $\mathrm{Dom}(L)$ and satisfies (9.58) for every $t > 0$.

A similar result can be proved for general elliptic operators of the form (9.1). However, this analysis is beyond the scope of the present notes. Here we shall only make a few remarks:

(1) Initial condition. The map $t \mapsto u(t) \doteq S_t g$ is continuous from $[0, \infty[$ into $\mathbf{L}^2(\Omega)$ and satisfies $u(0) = g$. The initial condition in (9.49) is thus satisfied as an identity between functions in $\mathbf{L}^2(\Omega)$.

(2) Regular solutions. If $g \in \mathrm{Dom}(A)$, then $u(t) = S_t g \in \mathrm{Dom}(A)$ for all $t \geq 0$. Moreover, the map $t \mapsto u(t)$ is continuously differentiable and satisfies the ODE (9.50) at every time $t > 0$. Since $\mathrm{Dom}(A) \subset H_0^1(\Omega)$, this also implies that $u(t)$ satisfies the correct boundary conditions, for all $t \geq 0$.

(3) Distributional solutions. Given a general initial condition $f \in \mathbf{L}^2(\Omega)$, one can construct a sequence of initial data $f_m \in \mathrm{Dom}(A)$ such that $\|f_m - f\|_{\mathbf{L}^2} \to 0$ as $m \to \infty$. In this case, if the semigroup is of type γ, we have

$$\|S_t f_m - S_t f\|_{\mathbf{L}^2} \leq e^{\gamma t} \|f_m - f\|_{\mathbf{L}^2} .$$

Therefore the trajectory $t \mapsto u(t) = S_t f$ is the limit of a sequence of $\mathcal{C}^1$ solutions $t \mapsto u_m(t) \doteq S_t f_m$. The convergence is uniform for t in bounded sets.

Relying on these approximations, we now show that the function $u = u(t, x)$ provides a solution to the parabolic equation

$$u_t = \sum_{i,j=1}^{n} (a^{ij}(x) u_{x_i})_{x_j} - \sum_{i=1}^{n} b^i(x) u_{x_i} - c(x) u \tag{9.64}$$

in the distributional sense. Namely, for every test function $\varphi \in \mathcal{C}_c^\infty(\Omega \times]0, \infty[)$, one has

$$\iint \left\{ u\varphi_t + \sum_{i,j=1}^{n} u(a^{ij}\varphi_{x_j})_{x_i} + \sum_{i=1}^{n} u(b^i\varphi)_{x_i} - c\,u\varphi \right\} dx\, dt = 0 . \tag{9.65}$$

To prove (9.65), consider a sequence of initial data $f_m \in \mathrm{Dom}(A)$ such that $\|f_m - f\|_{\mathbf{L}^2} \to 0$. Then, for any fixed time interval $[0, T]$, the trajectories $t \mapsto u_m(t) \doteq S_t f_m$ converge to the continuous trajectory $t \mapsto u(t) = S_t f$ in $\mathcal{C}^0([0, T]\,;\, \mathbf{L}^2(\Omega))$. Since each u_m is clearly a solution in the distributional sense, writing

$$\iint \left\{ u_m \varphi_t + \sum_{i,j=1}^{n} u_m (a^{ij} \varphi_{x_j})_{x_i} + \sum_{i=1}^{n} u_m (b^i \varphi)_{x_i} - c\, u_m \varphi \right\} dx\, dt \;=\; 0$$

and letting $m \to \infty$, we obtain (9.65).

9.3. Hyperbolic equations

In this last section we consider the linear hyperbolic initial-boundary value problem

$$(9.66) \qquad \begin{cases} u_{tt} + Lu &= 0, & t \in \mathbb{R},\ x \in \Omega, \\ u(t,x) &= 0, & t \in \mathbb{R},\ x \in \partial\Omega, \\ u(0,x) &= f(x),\ u_t(0,x) = g(x), & x \in \Omega. \end{cases}$$

Compared with (9.49), notice that here we are taking two derivatives with respect to time. The system (9.66) can thus be regarded as a second-order evolution equation in the space $\mathbf{L}^2(\Omega)$. For simplicity, we shall only treat the case where L is the homogeneous second-order elliptic operator

$$(9.67) \qquad Lu \;=\; -\sum_{i,j=1}^{n} (a^{ij}(x) u_{x_i})_{x_i},$$

assuming that the coefficients a^{ij} satisfy
(9.68)

$$a^{ij} = a^{ji} \in W^{1,\infty}(\Omega), \qquad \sum_{i,j=1}^{n} a^{ij}(x)\xi_i\xi_j \;\geq\; \theta|\xi|^2 \qquad \text{for all } x \in \Omega,\ \xi \in \mathbb{R}^n.$$

According to Theorem 9.9, the space $\mathbf{L}^2(\Omega)$ admits an orthonormal basis $\{\phi_k\,;\ k \geq 1\}$ consisting of eigenfunctions of L, so that

$$(9.69) \qquad \phi_k \in \mathrm{Dom}(L), \qquad L\phi_k \;=\; \mu_k\, \phi_k,$$

for a sequence of strictly positive eigenvalues $\mu_k \to +\infty$ as $k \to \infty$. It is thus natural to construct a solution of (9.66) in the form

$$(9.70) \qquad u(t,x) \;=\; \sum_{k=1}^{\infty} c_k(t)\, \phi_k(x).$$

Taking the inner product of both sides of (9.70) with ϕ_k, we see that each coefficient $c_k(\cdot)$ should satisfy the linear second-order ODE

$$(9.71) \qquad c_k'' + \mu_k c_k \;=\; 0, \qquad \begin{cases} c_k(0) &= (\phi_k, g)_{\mathbf{L}^2}, \\ c_k'(0) &= (\phi_k, h)_{\mathbf{L}^2}. \end{cases}$$

The following analysis will show that the formal expansion (9.70)–(9.71) is indeed valid, provided that the initial data satisfy

$$(9.72) \qquad g \,\in\, H_0^1(\Omega), \qquad h \,\in\, \mathbf{L}^2(\Omega).$$

As a first step, let us rewrite (9.66) as a first-order system, setting $v \doteq u_t$. On the product space

$$X \doteq H_0^1(\Omega) \times \mathbf{L}^2(\Omega) \tag{9.73}$$

we thus consider the evolution problem

$$\frac{d}{dt}\begin{pmatrix} u \\ v \end{pmatrix} = \begin{pmatrix} 0 & I \\ -L & 0 \end{pmatrix}\begin{pmatrix} u \\ v \end{pmatrix}, \qquad \begin{pmatrix} u \\ v \end{pmatrix}(0) = \begin{pmatrix} f \\ g \end{pmatrix}. \tag{9.74}$$

In the special case where

$$f = a_k \phi_k, \qquad g = b_k \phi_k, \tag{9.75}$$

for a given $k \geq 1$ and $a_k, b_k \in \mathbb{R}$, an explicit solution of (9.74) is found in the form

$$u(t) = c_k(t)\phi_k, \qquad v(t) = u_t(t) = c_k'(t)\phi_k,$$

where the coefficient $c_k(t)$ satisfies

$$c_k''(t) + \mu_k c_k(t) = 0, \qquad c_k(0) = a_k, \qquad c_k'(0) = b_k.$$

An elementary computation yields

$$c_k(t) = a_k \cos(\sqrt{\mu_k}\, t) + \frac{b_k}{\sqrt{\mu_k}} \sin(\sqrt{\mu_k}\, t).$$

Hence

$$\begin{pmatrix} u(t) \\ v(t) \end{pmatrix} = \begin{pmatrix} \cos(\sqrt{\mu_k}\, t) & \frac{1}{\sqrt{\mu_k}} \sin(\sqrt{\mu_k}\, t) \\ -\sqrt{\mu_k} \sin(\sqrt{\mu_k}\, t) & \cos(\sqrt{\mu_k}\, t) \end{pmatrix}\begin{pmatrix} a_k \phi_k \\ b_k \phi_k \end{pmatrix}. \tag{9.76}$$

Observe that $t \mapsto \big(u(t), v(t)\big)$ is a continuously differentiable map from $\mathbb{R}$ into $H_0^1(\Omega) \times \mathbf{L}^2(\Omega)$ which satisfies the initial conditions and the differential equation in (9.74).

By taking linear combinations of solutions of the form (9.76), we now obtain a group of linear operators:

$$S_t \begin{pmatrix} f \\ g \end{pmatrix} \doteq \sum_{k=1}^{\infty} \begin{pmatrix} \cos(\sqrt{\mu_k}\, t) & \frac{1}{\sqrt{\mu_k}} \sin(\sqrt{\mu_k}\, t) \\ -\sqrt{\mu_k} \sin(\sqrt{\mu_k}\, t) & \cos(\sqrt{\mu_k}\, t) \end{pmatrix}\begin{pmatrix} (f, \phi_k)_{\mathbf{L}^2}\, \phi_k \\ (g, \phi_k)_{\mathbf{L}^2}\, \phi_k \end{pmatrix}. \tag{9.77}$$

The next theorem shows that $\{S_t\,;\ t \in \mathbb{R}\}$ is actually a group of linear isometries on the product space $X = H_0^1(\Omega) \times \mathbf{L}^2(\Omega)$, with the equivalent norm

$$\|(u, v)\|_X \doteq \Big(B[u, u] + \|v\|_{\mathbf{L}^2}^2\Big)^{1/2}. \tag{9.78}$$

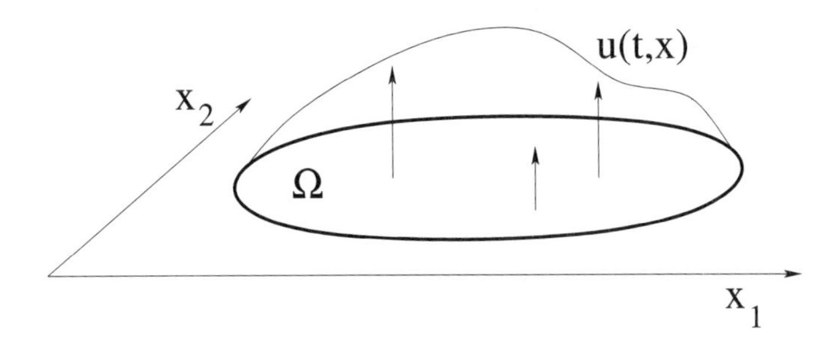

Figure 9.3.1. An elastic membrane, clamped along the boundary of the domain Ω, whose points vibrate in the vertical direction.

Remark 9.22. Consider an elastic membrane which occupies a region Ω in the plane, is clamped along the boundary $\partial\Omega$, and is subject to small vertical vibrations (Figure 9.3.1). Let $u(t,x)$ denote the vertical displacement of a point x on this membrane, at time t. Then the quantity $\|(u,u_t)\|_X^2$ can be regarded as the *total energy* of the vibrating membrane. Indeed, the term $B[u,u]$ describes an *elastic potential energy*, while $\|u_t\|_{\mathbf{L}^2}^2$ yields the *kinetic energy.*

Theorem 9.23 (Solutions of a linear hyperbolic problem). *In the above setting, the formula* (9.77) *defines a strongly continuous group of bounded linear operators* $\{S_t\,;\ t\in\mathbb{R}\}$ *on the space* $X \doteq H_0^1(\Omega)\times\mathbf{L}^2(\Omega)$. *Each operator* $S_t : X \mapsto X$ *is an isometry with respect to the equivalent norm* (9.78).

Proof. 1. The equivalence between the norm (9.78) and the standard product norm

$$\|(u,v)\|_{H^1\times\mathbf{L}^2} \doteq \Big(\|u\|_{H^1}^2 + \|v\|_{\mathbf{L}^2}^2\Big)^{1/2}$$

is an immediate consequence of (9.23).

2. Let the functions f,g be given by

$$f = \sum_{k=1}^{\infty} a_k\phi_k\,, \qquad g = \sum_{k=1}^{\infty} b_k\phi_k \qquad \text{with} \quad a_k = (f,\phi_k)_{\mathbf{L}^2}\,,\ b_k = (g,\phi_k)_{\mathbf{L}^2}\,.$$

Then

$$B[f,f] = \sum_{k=1}^{\infty}\Big(\mu_k\,a_k\phi_k\,,\ a_k\phi_k\Big)_{\mathbf{L}^2} = \sum_{k=1}^{\infty}\mu_k\,a_k^2\,, \qquad \|g\|_{\mathbf{L}^2}^2 = \sum_{k=1}^{\infty} b_k^2\,.$$

Therefore $\begin{pmatrix} f \\ g \end{pmatrix} \in X$ if and only if

$$\sum_{k=1}^{\infty} \mu_k\, a_k^2 = \sum_{k=1}^{\infty} \mu_k\, (f, \phi_k)^2_{\mathbf{L}^2} < \infty\,, \qquad \sum_{k=1}^{\infty} b_k^2 = \sum_{k=1}^{\infty} (g, \phi_k)^2_{\mathbf{L}^2} < \infty\,. \tag{9.79}$$

If $\begin{pmatrix} f \\ g \end{pmatrix} \in X$, then for every $t \in \mathbb{R}$ the series defining $S_t \begin{pmatrix} f \\ g \end{pmatrix}$ in (9.77) is convergent. Introducing the coefficients

$$\begin{cases} a_k(t) &= \cos(\sqrt{\mu_k}\, t) a_k + \dfrac{1}{\sqrt{\mu_k}} \sin(\sqrt{\mu_k}\, t)\, b_k\,, \\ \\ b_k(t) &= a_k'(t) \;=\; -\sqrt{\mu_k}\, \sin(\sqrt{\mu_k}\, t)\, a_k + \cos(\sqrt{\mu_k}\, t)\, b_k\,, \end{cases} \tag{9.80}$$

at any time t we have

$$\left\| S_t \begin{pmatrix} f \\ g \end{pmatrix} \right\|_X^2 = \sum_{k=1}^{\infty} \mu_k\, |a_k(t)|^2 + \sum_{k=1}^{\infty} |b_k(t)|^2 = \left\| \begin{pmatrix} f \\ g \end{pmatrix} \right\|_X^2 . \tag{9.81}$$

This shows that each linear operator S_t is an isometry with respect to the equivalent norm $\|\cdot\|_X$.

3. We claim that the family of linear operators $\{S_t\ ;\ t \in \mathbb{R}\}$ satisfies the group properties

$$S_0 \begin{pmatrix} f \\ g \end{pmatrix} = \begin{pmatrix} f \\ g \end{pmatrix}, \qquad S_t S_s \begin{pmatrix} f \\ g \end{pmatrix} = S_{t+s} \begin{pmatrix} f \\ g \end{pmatrix}, \qquad t, s \in \mathbb{R}\,. \tag{9.82}$$

Indeed, (9.82) is clearly satisfied for initial data f, g of the special form (9.75). By linearity and continuity, it must hold for all initial data.

To complete the proof, we need to show that, for f, g fixed, the map $t \mapsto S_t \begin{pmatrix} f \\ g \end{pmatrix}$ is continuous from $\mathbb{R}$ into X. But this is clear, because the above map is the uniform limit of the continuous maps

$$t \mapsto S_t \begin{pmatrix} f_m \\ g_m \end{pmatrix}, \qquad f_m \doteq \sum_{k=1}^{m} (f, \phi_k)_{\mathbf{L}^2}\, \phi_k\,, \qquad g_m \doteq \sum_{k=1}^{m} (g, \phi_k)_{\mathbf{L}^2}\, \phi_k\,, \tag{9.83}$$

as $m \to \infty$. □

Having constructed the group of linear operators $\{S_t\,;\ t \in \mathbb{R}\}$, we still need to explain in which sense the trajectories

$$t \mapsto \begin{pmatrix} u(t) \\ v(t) \end{pmatrix} = S_t \begin{pmatrix} f \\ g \end{pmatrix} \tag{9.84}$$

provide a solution to the hyperbolic initial-boundary value problem (9.66).

(1) The initial and boundary conditions are satisfied. Consider an arbitrary initial data $\begin{pmatrix} f \\ g \end{pmatrix} \in X = H_0^1(\Omega) \times \mathbf{L}^2(\Omega)$. By the continuity of the map in (9.84), it follows that

$$\|u(t) - f\|_{H^1} \to 0, \qquad \|v(t) - g\|_{\mathbf{L}^2} \to 0 \qquad \text{as } t \to 0.$$

Hence the initial conditions in (9.66) are satisfied.

Moreover, by the definition of the space X, we have $u(t) \in H_0^1(\Omega)$ for all $t \geq 0$. Hence the boundary condition $u = 0$ on $\partial\Omega$ is also satisfied.

(2) The hyperbolic equation is satisfied in the distributional sense. If both functions f and g are finite linear combinations of the eigenfunctions ϕ_k, then the corresponding trajectory (9.84) is a continuously differentiable map from $\mathbb{R}$ into X, and it satisfies (9.74) at all times $t > 0$.

More generally, given initial data $f \in H_0^1(\Omega)$ and $g \in \mathbf{L}^2(\Omega)$, one can construct a sequence of approximations f_m, g_m as in (9.83), so that $\|f_m - f\|_{H^1} \to 0$, $\|g_m - g\|_{\mathbf{L}^2} \to 0$ as $m \to \infty$. The corresponding semigroup trajectories $t \mapsto \begin{pmatrix} u_m(t) \\ v_m(t) \end{pmatrix} = S_t \begin{pmatrix} f_m \\ g_m \end{pmatrix}$ converge to the trajectory (9.84), uniformly for $t \in \mathbb{R}$. Relying on these approximations, we now show that the function $u = u(t,x)$ provides a solution to the hyperbolic equation

$$u_{tt} = \sum_{i,j=1}^{n} (a^{ij}(x) u_{x_i})_{x_j}$$

in the distributional sense. Indeed, consider any test function $\varphi \in \mathcal{C}_c^\infty(]0,\infty[\times\Omega)$. Since each $u_m = u_m(t,x)$ is a distributional solution of (9.66),

$$\iint \left\{ u_m \varphi_{tt} + \sum_{i,j=1}^{n} a^{ij}(x)\,(u_m)_{x_i} \varphi_{x_j} \right\} dx\, dt = 0.$$

Letting $m \to \infty$ and using the convergence $\|u_m(t) - u(t)\|_{H^1} \to 0$, uniformly for $t \in \mathbb{R}$, we obtain

$$\iint \left\{ u\, \varphi_{tt} + \sum_{i,j=1}^{n} a^{ij}(x)\, u_{x_i} \varphi_{x_j} \right\} dx = 0.$$

Example 9.24. Let $\Omega \doteq]0,\pi[\subset \mathbb{R}$ and $Lu = -u_{xx}$. Given $f \in H_0^1(\Omega)$ and $g \in \mathbf{L}^2(]0,\pi[)$, consider the hyperbolic initial-boundary value problem

$$\begin{cases} u_{tt} = u_{xx}, & t > 0,\ 0 < x < \pi, \\ u(0,x) = f(x),\ u_t(0,x) = g(x), & 0 < x < \pi, \\ u(t,0) = u(t,\pi) = 0. & \end{cases}$$

In this special case, the formula (9.77) yields the solution in terms of a Fourier sine series:

$$\begin{aligned} u(t,x) &= \sum_{k=1}^{\infty} \frac{2}{\pi}\Bigg[\cos kt\left(\int_0^{\pi} f(y)\,\sin ky\,dy\right) \\ &\qquad + \frac{\sin kt}{k}\left(\int_0^{\pi} g(y)\,\sin ky\,dy\right)\Bigg]\sin kx\,. \end{aligned}$$

9.4. Problems

1. Let $\Omega \doteq \{(x,y)\,;\ x^2+y^2<1\}$ be the open unit disc in $\mathbb{R}^2$. Prove that, for every bounded measurable function $f=f(x,y)$, the problem

$$\begin{cases} u_{xx} + x\,u_{xy} + u_{yy} = f & \text{on } \Omega\,, \\ u = 0 & \text{on } \partial\Omega \end{cases}$$

has a unique weak solution.

2. Let $\Omega \doteq \{(x,y)\,;\ x^2+y^2<1\}$ be the unit disc in $\mathbb{R}^2$. On the space $X = H_0^1(\Omega)$, consider the inner product

$$\langle u,v\rangle_\diamond = \int_\Omega \Big[u_x v_x + 2u_y v_y + y(u_x v_y + u_y v_x)\Big]\,dxdy\,.$$

(i) Prove that $\langle\cdot,\cdot\rangle_\diamond$ is indeed an inner product on X, which makes X a Hilbert space.

(ii) Given $f \in \mathbf{L}^2(\Omega)$, show that there exists a unique $u \in X$ such that

$$\langle u,v\rangle_\diamond = \int_\Omega f\,v\,dx \qquad \text{for all } v \in X = H_0^1(\Omega)\,.$$

What elliptic equation does u solve ?

3. Consider the differential operator on $\mathbb{R}^2$

$$Lu = -(xu_x)_x - (yu_y)_y + 2u_{xy} + 3(u_x+u_y) - 6u.$$

Determine for which bounded open sets $\Omega \subset \mathbb{R}^2$ it is true that the operator L is uniformly elliptic on Ω.

4. Let $\Omega \subset \mathbb{R}^n$ be a bounded open set. Let $\beta > 0$ be the best constant in Poincaré's inequality, namely

$$\beta \doteq \sup\Big\{\|u\|_{\mathbf{L}^2(\Omega)}\,;\ u \in H_0^1(\Omega),\ \|\nabla u\|_{\mathbf{L}^2(\Omega)} \le 1\Big\}.$$

(i) Establish a lower bound on the eigenvalues of the operator $Lu = -\Delta u$. More precisely, if $\phi \in H_0^1(\Omega)$ provides a nontrivial weak solution to

$$\begin{cases} -\Delta\phi &= \mu\,\phi, & x \in \Omega\,, \\ \phi &= 0, & x \in \partial\Omega\,, \end{cases}$$

prove that $\mu \geq 1/\beta^2$.

(ii) Prove that the solution of the parabolic initial-boundary value problem

$$\begin{cases} u_t &= \Delta u, & t > 0\,,\ x \in \Omega\,, \\ u(t,x) &= 0, & t > 0\,,\ x \in \partial\Omega\,, \\ u(0,x) &= g(x), & x \in \Omega\,, \end{cases}$$

decays to zero as $t \to \infty$. Indeed, $\|u(t)\|_{\mathbf{L}^2} \leq e^{-t/\beta^2}\|g\|_{\mathbf{L}^2}$ for every $t \geq 0$.

5. Let $\Omega \subseteq \widetilde{\Omega} \subset \mathbb{R}^n$ be bounded open sets. Let $0 < \mu_1 \leq \mu_2 \leq \cdots$ be the eigenvalues of the operator $-\Delta$ on $H_0^1\Omega$, and let $0 < \tilde{\mu}_1 \leq \tilde{\mu}_2 \leq \cdots$ be the eigenvalues of the operator $-\Delta$ on $H_0^1(\widetilde{\Omega})$. Prove that $\tilde{\mu}_1 \leq \mu_1$.

6. On the interval $[0,T]$, consider the Sturm-Liouville eigenvalue problem

$$\begin{cases} (p(t)u')' + q(t)u &= \mu\,u, \qquad 0 < t < T\,, \\ u(0) \;=\; u(T) &= 0\,. \end{cases} \tag{9.85}$$

Assume that

$$p \in \mathcal{C}^1(]0,T[)\,, \qquad q \in \mathcal{C}^0([0,T]), \qquad p(t) \geq \theta > 0 \qquad \text{for all } t\,.$$

Prove that the space $\mathbf{L}^2([0,T])$ admits an orthonormal basis $\{\phi_k\,;\ k \geq 1\}$ where each ϕ_k satisfies (9.85), for a suitable eigenvalue μ_k. Moreover, $\mu_k \to -\infty$ as $k \to \infty$.

7. In Theorem 9.9, take $Lu = -\Delta u$. Let $\{\phi_k\,;\ k \geq 1\}$ be an orthonormal basis of $\mathbf{L}^2(\Omega)$ consisting of eigenfunctions of L^{-1}. Show that in this special case the eigenfunctions ϕ_k are also mutually orthogonal with respect to the inner product in H^1, namely $(\phi_j, \phi_k)_{H^1} = 0$ whenever $j \neq k$.

8. Consider the open rectangle $Q \doteq \{(x,y)\,;\ 0 < x < a\,,\ 0 < y < b\}$. Define the functions

$$\phi_{m,n}(x,y) \;\doteq\; \sqrt{\frac{2}{ab}}\,\sin\frac{m\pi\,x}{a}\,\sin\frac{n\pi\,y}{b}, \qquad m,n \geq 1\,.$$

(i) Check that $\phi_{m,n} \in H_0^1(Q)$. Moreover, prove that the countable set of functions $\{\phi_{m,n}\,;\ m,n \geq 1\}$ is an orthonormal basis of $\mathbf{L}^2(Q)$ consisting of eigenfunctions of the elliptic operator $Lu = -\Delta u$. Compute the corresponding eigenvalues $\mu_{m,n}$.

(ii) If $\Omega \subset \mathbb{R}^2$ is an open domain contained inside a rectangle Q with sides a, b, prove that

$$\|u\|_{\mathbf{L}^2(\Omega)} \;\leq\; \frac{ab}{\pi\sqrt{a^2+b^2}}\,\|\nabla u\|_{\mathbf{L}^2(\Omega)} \qquad \text{for all } u \in H_0^1(\Omega)\,.$$

9. In the setting of Theorem 9.16, assume that, for every $f \in \mathbf{L}^2(\Omega)$, the elliptic boundary value problem (9.2) has a unique solution. Prove that the solution map $f \mapsto u$ is a compact operator from $\mathbf{L}^2(\Omega)$ into itself.

10. (Galerkin approximations) Let $\Omega \subset \mathbb{R}^n$ be a bounded open set. Let $\{\varphi_k\,;\ k \geq 1\}, \varphi_k \in H_0^1(\Omega)$, be an orthonormal basis of $\mathbf{L}^2(\Omega)$ consisting of eigenfunctions of the Laplace operator Δ.

Let the operator L in (9.1) be uniformly elliptic and assume that the corresponding bilinear form $B[\cdot,\cdot]$ in (9.11) is strictly positive definite.

Given $f \in \mathbf{L}^2(\Omega)$, construct a sequence of approximate solutions u_m to the boundary value problem (9.2) as follows.

(i) For a fixed $m \geq 1$, define

$$u_m(x) = \sum_{k=1}^{m} c_k\, \varphi_k(x) \tag{9.86}$$

choosing the coefficients $c_1, \ldots, c_m$ so that

$$B[u_m, \varphi_j] = (f, \varphi_j)_{\mathbf{L}^2}, \qquad j = 1, \ldots, m. \tag{9.87}$$

Show that (9.87) yields an algebraic system of m linear equations for the m variables $c_1, \ldots, c_m$. Prove that this system has a unique solution.

(ii) Letting $m \to \infty$, prove that the sequence u_m is uniformly bounded in $H_0^1(\Omega)$, hence it admits a weakly convergent subsequence, say $u_{m_j} \rightharpoonup u$. Prove that u is a weak solution to (9.2). By uniqueness, show that the entire sequence converges: $u_m \rightharpoonup u$ as $m \to \infty$.

11. On the open interval $\Omega =]0,3[$, consider the boundary value problem

$$\begin{cases} -u_{xx} = 1, & 0 < x < 3, \\ u(0) = u(3) = 0. \end{cases} \tag{9.88}$$

Consider the two linearly independent functions $\varphi_1, \varphi_2 \in H_0^1(\,]0,3[\,)$, defined by

$$\varphi_1(x) = \begin{cases} x & \text{if } x \in [0,1], \\ 2-x & \text{if } x \in [1,2], \\ 0 & \text{if } x \in [2,3], \end{cases} \qquad \varphi_2(x) = \begin{cases} 0 & \text{if } x \in [0,1], \\ x-1 & \text{if } x \in [1,2], \\ 3-x & \text{if } x \in [2,3]. \end{cases}$$

Explicitly compute the Galerkin approximation $U(x) = c_1\varphi_1(x) + c_2\varphi_2(x)$ such that

$$B[U, \varphi_i] = \int_0^3 U_x \cdot \varphi_{i,x}\, dx = \int_0^3 1 \cdot \varphi_i\, dx = (1, \varphi_i)_{\mathbf{L}^2}, \qquad i = 1, 2.$$

Compare U with the exact solution of (9.88).

12. Let $\Omega = \{(x,y)\,;\ x^2 + y^2 < 1\}$ be the open unit disc in $\mathbb{R}^2$, and let u be a smooth solution to the equation

$$\begin{aligned} u_{tt} &= 2u_{xx} + y\, u_{xy} + 3u_{yy} + \frac{1}{2}u_x, \qquad \text{on } \Omega \times [0,T], \\ u &= 0, \qquad \text{on } \partial\Omega \times [0,T]. \end{aligned} \tag{9.89}$$

(i) Write the equation (9.89) in the form $u_{tt} + Lu = 0$, proving that the operator L is uniformly elliptic on the domain Ω.

(ii) Define a suitable energy $e(t)$ = [kinetic energy] + [elastic potential energy], and check that it is constant in time.

13. Consider the homogenous linear elliptic operator L in (9.20) assuming that (9.4) holds, together with $a^{ij} = a^{ji} \in \mathbf{L}^\infty(\Omega)$. Extending the argument used in Lemma 9.7, work out the following alternative proof of Theorem 9.8.

(i) Show that the bilinear functional B in (9.22) is an inner product on $H_0^1(\Omega)$. The corresponding norm

$$\|u\|_\diamond \doteq \left(\sum_{i,j=1}^{n} a^{ij}(x) u_{x_i} u_{x_j}\right)^{1/2}$$

is equivalent to the H^1 norm. Namely,

$$\frac{1}{C} \cdot \|u\|_{H^1} \leq \|u\|_\diamond \leq C\|u\|_{H^1} \qquad \text{for all } u \in H_0^1(\Omega)\,.$$

(ii) Call $H_\diamond$ the Hilbert space H_0^1 endowed with this equivalent norm. Following the proof of Lemma 9.7, construct the solution of (9.19) as $u = \iota^* f$, using the following diagram:

$$H_\diamond(\Omega) \quad \stackrel{\iota}{\longrightarrow} \quad \mathbf{L}^2(\Omega),$$

$$H_\diamond(\Omega) = [H_\diamond(\Omega)]^* \quad \stackrel{\iota^*}{\longleftarrow} \quad [\mathbf{L}^2(\Omega)]^* = \mathbf{L}^2(\Omega)\,.$$

Here ι is the canonical immersion of $H_\diamond$ into $\mathbf{L}^2$, while ι^* is its adjoint operator.

14. Let $\Omega \subset \mathbb{R}^N$ be a bounded open set. In the same setting as in Lemma 9.19, prove that the map $t \mapsto u(t)$ is $\mathcal{C}^\infty$ from $]0, T]$ into $\mathbf{L}^2(\Omega)$.

15. (Neumann's problem) Let Ω be a bounded connected open set with smooth boundary $\partial\Omega$. By definition, a function $u \in H^1(\Omega)$ is a weak solution of Neumann's problem[1]

$$\begin{cases} -\Delta u = f, & x \in \Omega, \\ \dfrac{\partial u}{\partial \nu} = 0, & x \in \partial\Omega, \end{cases} \tag{9.90}$$

if

$$\int_\Omega \nabla u \cdot \nabla v \, dx = \int_\Omega f\, v\, dx \qquad \text{for all } v \in H^1(\Omega). \tag{9.91}$$

[1] Here and in the sequel, $\frac{\partial u}{\partial \nu}$ denotes the derivative of u in the direction of the outer normal to the boundary of Ω.

Given $f \in \mathbf{L}^2(\Omega)$, prove that Neumann's problem (9.90) has a weak solution if and only if

$$\int_\Omega f\,dx \;=\; 0.$$

Hint: As a first step, show that for $\gamma > 0$ the bilinear form

$$B_\gamma[u,v] \;\doteq\; \int_\Omega (\nabla u \cdot \nabla v + \gamma uv)\,dx$$

is strictly positive definite on $H^1(\Omega)$. Express the weak solution of (9.90) in terms of the inverse operator L_γ^{-1}, where $L_\gamma u = -\Delta u + \gamma u$.

16. (Biharmonic equation) Let $\Omega \subset \mathbb{R}^n$ be a bounded open set with smooth boundary. A function $u \in H_0^2(\Omega)$ is a weak solution of the biharmonic equation

$$\text{(9.92)} \qquad \left\{ \begin{array}{rll} \Delta^2 u & = f, & x \in \Omega\,, \\[2mm] u = \dfrac{\partial u}{\partial \nu} & = 0, & x \in \partial\Omega\,, \end{array} \right.$$

if

$$\text{(9.93)} \qquad \int_\Omega \Delta u\,\Delta v\,dx \;=\; \int_\Omega f\,v\,dx \qquad \text{for all } v \in H_0^2(\Omega).$$

Given $f \in \mathbf{L}^2(\Omega)$, prove that the boundary value problem (9.92) has a unique weak solution. Hint: Show that the bilinear form $B[u,v]$ on $H_0^2(\Omega)$ defined by the left-hand side of (9.93) is strictly positive definite on $H_0^2(\Omega)$.

Appendix

Background Material

A.1. Partially ordered sets

A set S is **partially ordered** by a binary relation $\preceq$ if, for every $a, b, c \in S$, one has

(i) $a \preceq a$,

(ii) $a \preceq b$ and $b \preceq a$ implies $a = b$,

(iii) $a \preceq b$ and $b \preceq c$ implies $a \preceq c$.

A subset $S' \subseteq S$ of a partially ordered set S is said to be **totally ordered** if, for every $a, b \in S'$, one has either $a \preceq b$ or $b \preceq a$. We say that the subset S' is **maximal** (with respect to the property of being totally ordered) if S' is not strictly contained in any other totally ordered set.

Using Zorn's lemma, or the axiom of choice, one can prove

Theorem A.1 (Hausdorff Maximality Principle). *If S is any partially ordered set, every totally ordered subset $S' \subseteq S$ is contained in a maximal totally ordered subset.*

A.2. Metric and topological spaces

A **distance** on a set X is a map $d : X \times X \mapsto \mathbb{R}_+$ satisfying the following three properties:

(1) positivity: $d(x, y) \geq 0$, $d(x, y) = 0$ if and only if $x = y$,

(2) symmetry: $d(x, y) = d(y, x)$,

(3) triangle inequality: $d(x,z) \leq d(x,y) + d(y,z)$.

A set X endowed with a distance function $d(\cdot,\cdot)$ is called a **metric space**.

The **open ball** centered at a point x with radius $r > 0$ is the set

$$B(x,r) \doteq \{y \in X\,;\ d(y,x) < r\}.$$

In turn, a metric $d(\cdot,\cdot)$ determines a topology on X.

A subset $A \subseteq X$ is **open** if, for every $x \in A$, there exists a radius $r > 0$ such that $B(x,r) \subseteq A$.

A subset $C \subseteq X$ is **closed** if its complement $X \setminus C = \{x \in X\,;\ x \notin C\}$ is open.

If $B(x,r) \subseteq A$ for some $r > 0$, we say that A is a **neighborhood** of the point x, or equivalently that x is an **interior point** of A.

• The union of any family of open sets is open. The intersection of finitely many open sets is open.

• The intersection of any family of closed sets is closed. The union of finitely many closed sets is closed.

The entire space X and the empty set $\emptyset$ are always both open and closed. If there exists no other subset $S \subset X$ which is at the same time open and closed, we say that the space X is **connected**.

A sequence $(x_n)_{n\geq 1}$ **converges** to a point $x \in X$ if

$$\lim_{n\to\infty} d(x_n, x) = 0\,.$$

In this case, we write $\lim_{n\to\infty} x_n = x$ or simply $x_n \to x$.

The **closure** of a set A, denoted by $\overline{A}$, is the smallest closed set containing A. This is obtained as the intersection of all closed sets containing A. A point x lies in the closure of the set A if and only if there exists a sequence of points $x_n \in A$ that converges to x.

A subset $S \subseteq X$ is **dense** in X if $\overline{S} = X$. This is the case if and only if S intersects every nonempty open subset of X. The space X is **separable** if it contains a countable dense subset.

A sequence $(x_n)_{n\geq 1}$ is a **Cauchy sequence** if for every $\varepsilon > 0$ one can find an integer N large enough so that

$$d(x_m, x_n) \leq \varepsilon \qquad \text{whenever } m, n \geq N.$$

The metric space X is **complete** if every Cauchy sequence converges to some limit point $x \in X$.

If X, Y are two metric spaces, a map $f : X \mapsto Y$ is **continuous** if, for every open set $A \subseteq Y$, the pre-image $f^{-1}(A) \doteq \{x \in X\,;\ f(x) \in A\}$ is an open subset of X.

• A map $f : X \mapsto Y$ is continuous if and only if, for every $x_0 \in X$ and $\varepsilon > 0$, there exists $\delta > 0$ such that

$$d(x, x_0) < \delta \qquad \text{implies} \qquad d(f(x), f(x_0)) < \varepsilon.$$

We say that a function $f : X \mapsto Y$ is **Lipschitz continuous** if there exists a constant $C \geq 0$ such that

$$d\big(f(x),\, f(x')\big) \leq C \cdot d(x, x') \qquad \text{for all } x, x' \in X\,.$$

More generally, we say that $f : X \mapsto Y$ is **Hölder continuous** of exponent $0 < \alpha \leq 1$ if there exists a constant C such that

$$d\big(f(x),\, f(x')\big) \leq C \cdot \big[d(x, x')\big]^{\alpha} \qquad \text{for all } x, x' \in X\,.$$

A collection of open sets $\{A_i\,;\ i \in \mathcal{I}\}$ such that $K \subseteq \bigcup_{i \in \mathcal{I}} A_i$ is called an **open covering** of the set K. Here the set of indices $\mathcal{I}$ may well be infinite. A set $K \subseteq X$ is **compact** if, from every open covering of K, one can extract a finite subcovering.

A set S is **relatively compact** if its closure $\overline{S}$ is compact.

A set S is **precompact** if, for every $\varepsilon > 0$, it can be covered by finitely many balls with radius ε.

Theorem A.2 (Compact subsets of $\mathbb{R}^n$). *A subset $S \subset \mathbb{R}^n$ is compact if and only if it is closed and bounded.*

Theorem A.3 (Equivalent characterizations of compactness). *Let S be a metric space. The following are equivalent:*

(i) *S is compact.*

(ii) *S is precompact and complete.*

(iii) *From every sequence $(x_k)_{k \geq 1}$ of points in S one can extract a subsequence converging to some limit point $x \in S$.*

A.2.1. Fixed points of contractive maps. Let $\phi : X \mapsto X$ be a map from a complete metric space X into itself. A point x^* such that $\phi(x^*) = x^*$ is called a **fixed point** of ϕ. For a strictly contractive map, the fixed point is unique and can be obtained by a simple iterative procedure.

Theorem A.4 (Contraction Mapping Theorem). *Let X be a complete metric space, and let $\phi : X \mapsto X$ be a continuous mapping such that, for some $\kappa < 1$,*

$$d(\phi(x),\, \phi(y)) \leq \kappa d(x, y) \qquad \textit{for all } x, y \in X. \tag{A.1}$$

Then there exists a unique point $x^* \in X$ *such that*

$$x^* = \phi(x^*). \tag{A.2}$$

Moreover, for any $y \in X$ *one has*

$$d(y, x^*) \leq \frac{1}{1-\kappa} d(y, \phi(y)). \tag{A.3}$$

Proof. Fix any point $y \in X$ and consider the sequence

$$y_0 = y, \quad y_1 = \phi(y_0), \quad \ldots, \quad y_{n+1} = \phi(y_n), \ \ldots.$$

By induction, we have

$$\begin{aligned} d(y_2, y_1) &\leq \kappa\, d(y_1, y_0), \\ d(y_3, y_2) &\leq \kappa d(y_2, y_1) \leq \kappa^2 d(y_1, y_0), \\ &\cdots \\ d(y_{n+1}, y_n) &\leq \kappa\, d(y_n, y_{n-1}) \leq \kappa^n d(y_1, y_0) = \kappa^n d(\phi(y), y). \end{aligned}$$

For $m < n$ we have
(A.4)

$$d(y_n, y_m) \leq \sum_{j=m}^{n-1} d(y_{j+1}, y_j) \leq \sum_{j=m}^{n-1} \kappa^j d(y, \phi(y)) \leq \frac{\kappa^m}{1-\kappa} d(y, \phi(y)).$$

Since $\kappa < 1$, the right-hand side of (A.4) approaches zero as $m \to \infty$. Hence the sequence $(y_n)_{n \geq 1}$ is Cauchy. Since X is complete, this sequence converges to some limit point x^*. By the continuity of ϕ one has

$$x^* = \lim_{n\to\infty} y_n = \lim_{n\to\infty} \phi(y_{n-1}) = \phi\Big(\lim_{n\to\infty} y_{n-1}\Big) = \phi(x^*);$$

hence (A.2) holds. The uniqueness of the fixed point x^* is proved observing that, if

$$x_1 = \phi(x_1), \quad x_2 = \phi(x_2),$$

then by (A.1) it follows that

$$d(x_1, x_2) = d\big(\phi(x_1), \phi(x_2)\big) \leq \kappa d(x_1, x_2).$$

The assumption $\kappa < 1$ thus implies $d(x_1, x_2) = 0$ and hence $x_1 = x_2$.

Finally, using (A.4) with $m = 0$, for every $n \geq 1$ we obtain

$$d(y_n, y) \leq \frac{1}{1-\kappa} d(y, \phi(y)).$$

Letting $n \to \infty$, we obtain (A.3). □

A.2.2. The Baire category theorem. Let X be a metric space. Among all subsets of X we would like to define a family of "large sets" and a family of "small sets" with the following natural properties:

(i) A set $S \subseteq X$ is large if and only if its complement $X \setminus S$ is small.

(ii) The intersection of countably many large sets is large.

(iii) A large set is nonempty.

If a probability measure μ on X is given, one can call "large sets" the sets having probability one, and "small sets" those with probability zero. With such definition, all properties (i)–(iii) are clearly satisfied.

If the metric space X is complete, relying on Baire's category theory, one can still introduce a concept of "large sets" and "small sets", based exclusively on the topological structure. Namely, we say that a set $S \subseteq X$ is of **second category** (i.e., "topologically large") if S is the intersection of countably many open dense sets. On the other hand, a set $S \subseteq X$ is said to be of **first category**, or equivalently **meager** (i.e., "topologically small"), if S is the union of countably many closed sets with empty interior. From the definition, it is clear that these topologically large or small sets satisfy the above properties (i) and (ii). The fact that (iii) also holds is an important consequence of the following theorem.

Theorem A.5 (Baire). *Let $(V_k)_{k\geq 1}$ be a sequence of open, dense subsets of a complete metric space X. Then the intersection $V \doteq \bigcap_{k=1}^{\infty} V_k$ is a nonempty, dense subset of X.*

Proof. Let $\Omega \subset X$ be any open set. We need to show that $\left(\bigcap_{k=1}^{\infty} V_k\right) \cap \Omega$ is nonempty.

Choose a point x_0 and a radius $r_0 < 1$ such that $B(x_0, 3r_0) \subset \Omega$.

By induction, for every $k \geq 1$ we choose a point x_k and a radius r_k such that $B(x_k, 3r_k) \subset V_k \cap B(x_{k-1}, r_{k-1})$. This is possible because V_k is open and dense. For each $k \geq 1$, the above choice implies

$$r_{k+1} \leq \frac{r_k}{3}, \qquad d(x_{k+1}, x_k) \leq r_k \leq \frac{1}{3^k}.$$

Therefore, the sequence $(x_k)_{k\geq 1}$ is Cauchy. Since X is complete, this sequence has a limit: $x_k \to x^*$ for some $x^* \in X$. We now observe that

$$d(x^*, x_k) \leq \sum_{j=k}^{\infty} d(x_{j+1}, x_j) \leq \sum_{j=k}^{\infty} r_j \leq \sum_{j=k}^{\infty} 3^{k-j} r_k = \frac{3}{2} r_k.$$

Therefore, for every $k \geq 1$,

$$x^* \in B(x_k, 3r_k) \subset V_k.$$

When $k = 0$, this same argument yields $x^* \in B(x_0, 3r_0) \subset \Omega$. Hence $x^* \in \left(\bigcap_{k=1}^{\infty} V_k\right) \cap \Omega$. □

A.3. Review of Lebesgue measure theory

A.3.1. Measurable sets. A family $\mathcal{F}$ of subsets of $\mathbb{R}^n$ is called a **σ-algebra** if

(i) $\emptyset \in \mathcal{F}$ and $\mathbb{R}^n \in \mathcal{F}$,

(ii) if $A \in \mathcal{F}$, then $\mathbb{R}^n \setminus A \in \mathcal{F}$,

(iii) if $A_k \in \mathcal{F}$ for every $k \geq 1$, then $\bigcup_{k=1}^{\infty} A_k \in \mathcal{F}$ and $\bigcap_{k=1}^{\infty} A_k \in \mathcal{F}$.

Theorem A.6 (**Existence of Lebesgue measure on $\mathbb{R}^n$**). *There exists a σ-algebra $\mathcal{F}$ of subsets of $\mathbb{R}^n$ and a mapping $A \mapsto m_n(A)$, from $\mathcal{F}$ into $[0, +\infty]$, with the following properties.*

(i) *$\mathcal{F}$ contains every open subset of $\mathbb{R}^n$ and hence also every closed subset of $\mathbb{R}^n$.*

(ii) *If B is a ball in $\mathbb{R}^n$, then $m_n(B)$ equals the n-dimensional volume of B.*

(iii) *If $A_k \in \mathcal{F}$ for every $k \geq 1$ and if the sets A_k are pairwise disjoint, then*

$$m_n\left(\bigcup_{k=1}^{\infty}\right) = \sum_{k=1}^{\infty} m_n(A_k) \qquad \textit{(countable additivity).}$$

(iv) *If $A \subseteq B$ with $B \in \mathcal{F}$ and $m_n(B) = 0$, then also $A \in \mathcal{F}$ and $m_n(A) = 0$.*

The sets contained in the σ-algebra $\mathcal{F}$ are called **Lebesgue measurable sets**, while $m_n(A)$ is the **n-dimensional Lebesgue measure** of the set $A \in \mathcal{F}$.

If a property $P(x)$ is true for all points $x \in \mathbb{R}^n$, except for those in a measurable set $\mathcal{N}$ with $m_n(\mathcal{N}) = 0$, we say that the property P holds **almost everywhere** (a.e.).

A function $f : \mathbb{R}^n \mapsto \mathbb{R}$ is **measurable** if

$$f^{-1}(U) \doteq \{x \in \mathbb{R}^n ;\ f(x) \in U\} \in \mathcal{F}$$

for every open set $U \subseteq \mathbb{R}$.

Every continuous function is measurable. If f, g are measurable, then $f + g$ and $f \cdot g$ are measurable. Given a uniformly bounded sequence of measurable functions $(f_k)_{k\geq 1}$, the functions defined as

$$f^*(x) \doteq \limsup_{k\to\infty} f_k(x), \qquad f_*(x) \doteq \liminf_{k\to\infty} f_k(x)$$

are both measurable. The **essential supremum** of a measurable function f is defined as

$$\text{ess sup} f \doteq \inf \Big\{ \alpha \in \mathbb{R}\,;\ f(x) < \alpha \text{ for a.e. } x \in \mathbb{R}^n \Big\}.$$

Theorem A.7 (Egoroff). *Let $(f_k)_{k\geq 1}$ be a sequence of measurable functions, and assume the pointwise convergence*

$$f_k(x) \to f(x) \qquad \text{for a.e. } x \in A,$$

for some measurable function f and a measurable set $A \subset \mathbb{R}^n$ with $m_n(A) < \infty$. Then for each $\varepsilon > 0$ there exists a subset $E \subseteq A$ such that

(i) $m_n(A \setminus E) \leq \varepsilon$,

(ii) *$f_k \to f$ uniformly on E.*

A.3.2. Lebesgue integration. In order to define the Lebesgue integral, one begins with a special class of functions. The **characteristic function** of a set A is

$$\chi_A(x) = \begin{cases} 1 & \text{if } x \in A, \\ 0 & \text{if } x \notin A. \end{cases}$$

A function taking finitely many values, i.e., having the form

$$g(x) = \sum_{i=1}^{N} c_i \chi_{A_i}(x), \tag{A.5}$$

for some disjoint measurable sets $A_1, \ldots, A_N \subset \mathbb{R}^n$ and constants $c_1, \ldots, c_N \in \mathbb{R}$, is called a **simple function**. If the function g in (A.5) is nonnegative, its Lebesgue integral is defined by

$$\int_{\mathbb{R}^n} g\, dx \doteq \sum_{i=1}^{N} c_i\, m_n(A_i).$$

As before, $m_n(A_i)$ denotes the n-dimensional Lebesgue measure of the set A_i. More generally, if $f : \mathbb{R}^n \mapsto \mathbb{R}$ is a nonnegative measurable function, its Lebesgue integral is defined as

$$\int_{\mathbb{R}^n} f\, dx \doteq \sup\left\{ \int_{\mathbb{R}^n} g\, dx\,;\ g \text{ is simple, } g \leq f \right\}.$$

For any measurable function $f : \mathbb{R}^n \mapsto \mathbb{R}$, its positive and negative parts are denoted as

$$f_+ \doteq \max\{f, 0\}, \qquad f_- \doteq \max\{-f, 0\}.$$

We then define the **Lebesgue integral** of f as

$$\int_{\mathbb{R}^n} f\,dx \doteq \int_{\mathbb{R}^n} f_+\,dx - \int_{\mathbb{R}^n} f_-\,dx \tag{A.6}$$

provided that at least one of the terms on the right-hand side is finite. In this case we say that f is **integrable**. Notice that the integral in (A.6) may well be $+\infty$ or $-\infty$.

A measurable function $f : \mathbb{R}^n \mapsto \mathbb{R}$ is **summable** if

$$\int_{\mathbb{R}^n} |f|\,dx < \infty.$$

We say that f is **locally summable** if the product $f \cdot \chi_K$ is summable for every compact set $K \subset \mathbb{R}^n$.

The Lebesgue integral has useful convergence properties.

Theorem A.8 (Fatou's lemma). *Let $(f_k)_{k\geq 1}$ be a sequence of functions which are nonnegative and summable. Then*

$$\int_{\mathbb{R}^n} \Big(\liminf_{k\to\infty} f_k\Big)\,dx \leq \liminf_{k\to\infty} \int_{\mathbb{R}^n} f_k\,dx\,.$$

Theorem A.9 (Monotone convergence). *Let $(f_k)_{k\geq 1}$ be a sequence of measurable functions such that f_1 is summable and $f_1 \leq f_2 \leq \cdots \leq f_k \leq f_{k+1} \leq \cdots$. Then*

$$\int_{\mathbb{R}^n} \Big(\lim_{k\to\infty} f_k\Big)\,dx = \lim_{k\to\infty} \int_{\mathbb{R}^n} f_k\,dx\,.$$

Theorem A.10 (Lebesgue dominated convergence). *Let $(f_k)_{k\geq 1}$ be a sequence of measurable functions such that*

$$f_k(x) \to f(x) \qquad \text{for a.e. } x \in \mathbb{R}^n.$$

Moreover, assume that there exists a summable function g such that

$$|f_k(x)| \leq g(x) \qquad \text{for every } k \geq 1 \text{ and a.e. } x \in \mathbb{R}^n.$$

Then

$$\lim_{k\to\infty} \int_{\mathbb{R}^n} f_k\,dx = \int_{\mathbb{R}^n} f\,dx\,.$$

Given a Lebesgue measurable set $U \subset \mathbb{R}^n$, the integral of a measurable function $f : U \mapsto \mathbb{R}$ with respect to Lebesgue measure can be defined as

$$\int_U f\,dx \doteq \int_{\mathbb{R}^n} \tilde{f}\,dx\,, \qquad \text{where} \quad \tilde{f}(x) \doteq \begin{cases} f(x) & \text{if } x \in U\,, \\ 0 & \text{if } x \notin U\,. \end{cases}$$

For $1 \leq p < \infty$, $\mathbf{L}^p(U)$ denotes the space of all Lebesgue measurable functions $f : U \mapsto \mathbb{R}$ such that

$$\|f\|_{\mathbf{L}^p(U)} \doteq \left(\int_U |f|^p \, dx \right)^{1/p} < \infty. \tag{A.7}$$

Moreover, $\mathbf{L}^\infty(U)$ denotes the space of all measurable functions $f : U \mapsto \mathbb{R}$ which are essentially bounded, i.e., such that

$$\|f\|_{\mathbf{L}^\infty(U)} \doteq \text{ess-}\sup_{x \in U} |f(x)| < \infty. \tag{A.8}$$

Two functions whose values coincide outside a set of measure zero are regarded to be the same element of $\mathbf{L}^p$, or $\mathbf{L}^\infty$.

Given an open set $\Omega \subseteq \mathbb{R}^n$ and $1 \leq p \leq \infty$, by $\mathbf{L}^p_{\text{loc}}(\Omega)$ we denote the space of all measurable functions $f : \Omega \mapsto \mathbb{R}$ such that $f \in \mathbf{L}^p(U)$ for every bounded open set U whose closure is contained in Ω. If $0 < m_n(\Omega) < \infty$, the average value of f on the set Ω is defined as

$$⨍_\Omega f \, dx \doteq \frac{1}{m_n(\Omega)} \int_\Omega f \, dx.$$

Observe that a function f is continuous at the point x_0 if and only if

$$\lim_{r \to 0+} \sup_{x \in B(x_0, r)} \big| f(x) - f(x_0) \big| = 0. \tag{A.9}$$

Replacing the supremum with an average, we say that f is **quasi-continuous** at the point x_0 if

$$\lim_{r \to 0+} ⨍_{B(x_0, r)} \big| f(x) - f(x_0) \big| \, dx = 0. \tag{A.10}$$

Theorem A.11 (Lebesgue). *Let $f : \mathbb{R}^n \mapsto \mathbb{R}$ be locally summable. Then f is quasi-continuous at a.e. point $x_0 \in \mathbb{R}^n$.*

A point x_0 where (A.10) holds is called a **Lebesgue point** of f.

Given an interval $[a, b] \subset \mathbb{R}$, a function $F : [a, b] \mapsto \mathbb{R}$ is **absolutely continuous** if, for every $\varepsilon > 0$, there exists $\delta > 0$ such that, for any finite family of disjoint intervals $[a_1, b_1], \ldots, [a_n, b_n]$ contained in $[a, b]$ one has

$$\sum_{i=1}^n (b_i - a_i) < \delta \qquad \text{implies} \qquad \sum_{i=1}^n \big| F(b_i) - F(a_i) \big| < \varepsilon.$$

Theorem A.12 (Absolutely continuous functions). *The following are equivalent.*

(i) $F : [a, b] \mapsto \mathbb{R}$ *is absolutely continuous.*

(ii) *There exists a function* $f \in \mathbf{L}^1([a, b])$ *such that*

$$F(x) = F(a) + \int_{[a,x]} f(x)\, dx \qquad \text{for all } x \in [a, b]\,.$$

If (i) and (ii) hold, then F *is a.e. differentiable, with derivative* $F'(x) = f(x)$ *for a.e.* $x \in [a, b]$.

Next, consider two measurable subsets $X \subseteq \mathbb{R}^m$ and $Y \subseteq \mathbb{R}^n$, so that the Cartesian product

$$X \times Y \doteq \Big\{(x, y)\,;\ x \in X,\ y \in Y\Big\}$$

is a measurable subset of $\mathbb{R}^{m+n}$. Given a function of two variables $f : X \times Y \mapsto \mathbb{R}$, for each fixed x we consider the function $y \mapsto f^x(y) \doteq f(x, y)$ of the variable y alone. Similarly, for each fixed y we consider the function $x \mapsto f^y(x) \doteq f(x, y)$ of the variable x alone.

Theorem A.13 (Fubini). *Let* $X \subseteq \mathbb{R}^m$, $Y \subseteq \mathbb{R}^n$ *be measurable sets, and assume* $f \in \mathbf{L}^1(X \times Y)$. *Then*

- $f^y \in \mathbf{L}^1(X)$ *for a.e.* $y \in Y$,
- $f^x \in \mathbf{L}^1(Y)$ *for a.e.* $x \in X$,
- *the integral function* $F(y) \doteq \int_X f^y(x)dx$ *is in* $\mathbf{L}^1(Y)$,
- *the integral function* $G(x) \doteq \int_Y f^x(y)dy$ *is in* $\mathbf{L}^1(X)$.

Moreover, one has the identity

$$\iint_{X\times Y} f(x, y)\, dxdy = \int_Y \left[\int_X f^y(x)dx\right] dy = \int_X \left[\int_Y f^x(y)\, dy\right] dx\,.$$

For detailed proofs of all the above theorems we refer to [**F**].

A.4. Integrals of functions taking values in a Banach space

Let X be a Banach space with norm $\|\cdot\|$. Let X^* be its dual space. Every $x^* \in X^*$ thus determines a linear continuous mapping $x \mapsto \langle x^*, x\rangle$ from X into $\mathbb{R}$. In this section we show how to construct the integral of a function $f : [0, T] \mapsto X$.

A function $g : [0, T] \mapsto X$ is **simple** if it has the form

$$g(t) = \sum_{i=1}^{N} u_i \chi_{A_i}(t), \qquad t \in [0, T], \tag{A.11}$$

where, for each $i = 1, \ldots, N$, $u_i \in X$ and $A_i \subseteq [0, T]$ is a measurable set.

A function $f : [0, T] \mapsto X$ is **strongly measurable** if there exists a sequence of simple functions $g_k : [0, T] \mapsto X$ such that

$$g_k(t) \to f(t) \qquad \text{for a.e. } t \in [0, T].$$

Moreover, we say that f is **summable** if there exists a sequence of simple functions g_k such that

$$\int_0^T \big\| g_k(t) - f(t) \big\| \, dt \ \to \ 0. \tag{A.12}$$

A function $f : [0, T] \mapsto X$ is **weakly measurable** if, for each $x^* \in X^*$, the scalar function $t \mapsto \langle x^*, f(t) \rangle$ is measurable.

A function $f : [0, T] \mapsto X$ is **almost separably valued** if there exists a subset $\mathcal{N} \subset [0, T]$ with zero measure such that the set of images $\{f(t)\,;\ t \in [0, T] \setminus \mathcal{N}\}$ is separable (i.e., it admits a countable dense subset).

Theorem A.14 (Pettis). *A map $f : [0, T] \mapsto X$ is strongly measurable if and only if f is weakly measurable and almost separably valued.*

The integral of a function f with values in a Banach space is defined in two steps.

If g is the simple function in (A.11), we define

$$\int_0^T g \, dt \ \doteq \ \sum_{i=1}^{N} u_i \, m_1(A_i).$$

Here m_1 is the one-dimensional Lebesgue measure on the interval $[0, T]$.

If f is summable, we define

$$\int_0^T f \, dt \ \doteq \ \lim_{k \to \infty} \int_0^T g_k(t) \, dt$$

where $(g_k)_{k \geq 1}$ is any sequence of simple functions for which (A.12) holds.

Theorem A.15 (Bochner). *A strongly measurable function $f : [0, T] \mapsto X$ is summable if and only if the scalar function $t \mapsto \|f(t)\|$ is summable. In* *this case one has*

$$\left\| \int_0^T f(t)\,dt \right\| \leq \int_0^T \|f(t)\|\,dt\,.$$

Moreover, for every $x^* \in X^*$,

$$\left\langle x^*, \int_0^T f(t)\,dt \right\rangle = \int_0^T \langle x^*, f(t)\rangle\,dt\,.$$

Proofs of the above theorems can be found in [**Y**].

A.5. Mollifications

In the analysis of Sobolev spaces, an important technique is the approximation of a general function with smooth functions. This can be done by means of mollifications.

The **standard mollifier** on $\mathbb{R}^n$ is defined as

$$J(x) \doteq \begin{cases} C_n \exp\left\{\frac{1}{|x|^2-1}\right\} & \text{if } |x| < 1\,, \\ 0 & \text{if } |x| \geq 1\,, \end{cases} \tag{A.13}$$

where the constant C_n is chosen so that $\int_{\mathbb{R}^n} J(x)\,dx = 1$. For each $\varepsilon > 0$ we also define the rescaled function

$$J_\varepsilon(x) \doteq \frac{1}{\varepsilon^n}\, J\left(\frac{x}{\varepsilon}\right). \tag{A.14}$$

Notice that $J_\varepsilon \in \mathcal{C}_c^\infty(\mathbb{R}^n)$ and that

$$\int_{\mathbb{R}^n} J_\varepsilon = 1\,, \qquad \text{Supp}(J_\varepsilon) = \{|x| \leq \varepsilon\}.$$

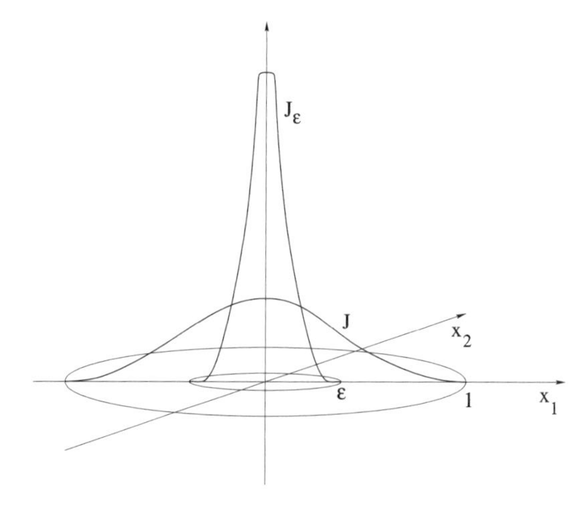

Figure A.5.1. A standard mollifier J and its rescaling J_ε.

Now let $\Omega \subseteq \mathbb{R}^n$ be an open set and let $f \in \mathbf{L}^1_{\text{loc}}(\Omega)$ be a locally integrable function. We then define the **mollification** of f in terms of a convolution:

$$f_\varepsilon \doteq J_\varepsilon * f.$$

Since f is defined on Ω and J_ε has support in the ball centered at the origin with radius ε, this convolution is well defined at all points in the subset

$$\Omega_\varepsilon \doteq \left\{x \in \Omega\,;\ \overline{B(x,\varepsilon)} \subseteq \Omega\right\}.$$

Indeed

$$f_\varepsilon(x) \doteq \int_{B(x,\varepsilon)} J_\varepsilon(x-y)f(y)\,dy = \int_{B(0,\varepsilon)} J_\varepsilon(z)f(x-z)\,dz\,.$$

Theorem A.16 (Properties of mollifiers). *Let $\Omega \subseteq \mathbb{R}^n$ be an open set and let $f \in \mathbf{L}^1_{\text{loc}}(\Omega)$. Then:*

(i) *For every $\varepsilon > 0$ one has $f_\varepsilon \in \mathcal{C}^\infty(\Omega_\varepsilon)$.*

(ii) *As $\varepsilon \to 0$, one has the pointwise convergence $f_\varepsilon(x) \to f(x)$ for a.e. $x \in \Omega$.*

(iii) *If f is continuous, then $f_\varepsilon \to f$ uniformly on compact subsets of Ω.*

(iv) *If $1 \leq p < \infty$ and $f \in \mathbf{L}^p_{\text{loc}}(\Omega)$, then $f_\varepsilon \to f$ in $\mathbf{L}^p_{\text{loc}}(\Omega)$.*

Proof. 1. Notice that each Ω_ε is open. Indeed, if the closed ball $\overline{B(x_0,\varepsilon)}$ is entirely contained in the open set Ω, the same holds for the ball $\overline{B(x,\varepsilon)}$, whenever $|x - x_0|$ is sufficiently small.

Let $\{e_1, \ldots, e_n\}$ be the standard orthonormal basis of $\mathbb{R}^n$. Fix a point $x \in \Omega_\varepsilon$ and $i \in \{1, \ldots, n\}$. If h is so small that $x + he_i \in \Omega_\varepsilon$, then the difference quotient is computed as

$$\begin{aligned}&\frac{f_\varepsilon(x+he_i) - f_\varepsilon(x)}{h} \\ &\quad = \frac{1}{\varepsilon^n}\int_{B(x,\varepsilon)} \frac{1}{h}\left[J\left(\frac{x+he_i-y}{\varepsilon}\right) - J\left(\frac{x-y}{\varepsilon}\right)\right] f(y)\,dy\,.\end{aligned} \tag{A.15}$$

Since the closed ball $\overline{B(x,\varepsilon)}$ is entirely contained in Ω, we have $f \in \mathbf{L}^1(B(x,\varepsilon))$. Moreover, since $J \in \mathcal{C}^\infty$, we have the uniform convergence

$$\lim_{h\to 0}\frac{1}{h}\left[J\left(\frac{x+he_i-y}{\varepsilon}\right) - J\left(\frac{x-y}{\varepsilon}\right)\right] = \frac{1}{\varepsilon}\frac{\partial}{\partial x_i}J\left(\frac{x-y}{\varepsilon}\right).$$

Letting $h \to 0$ in (A.15), we obtain the existence of the partial derivative

$$\frac{\partial}{\partial x_i}f_\varepsilon(x) = \int_{B(x,\varepsilon)} \frac{\partial}{\partial x_i}J_\varepsilon(x-y)\,f(y)\,dy\,.$$

A similar argument shows that, for every multi-index α, the derivative $D^\alpha f_\varepsilon$ exists and

$$D^\alpha f_\varepsilon(x) = \int_{B(x,\varepsilon)} D^\alpha J_\varepsilon(x-y)\, f(y)\, dy\,.$$

This proves (i).

2. By the Lebesgue differentiation theorem, for a.e. $x \in \Omega$ we have

$$\lim_{r\to 0} ⨍_{B(x,r)} |f(y) - f(x)|\, dy = 0\,. \tag{A.16}$$

If x is a point for which (A.16) holds, then

$$\begin{aligned} |f_\varepsilon(x) - f(x)| &= \left| \int_{B(x,\varepsilon)} J_\varepsilon(x-y)\big[f(y) - f(x)\big]\, dy \right| \\ &\leq \frac{1}{\varepsilon^n} \int_{B(x,\varepsilon)} J\left(\frac{x-y}{\varepsilon}\right) \big|f(y) - f(x)\big|\, dy \\ &\leq C ⨍_{B(x,\varepsilon)} \big|f(y) - f(x)\big|\, dy\,. \end{aligned}$$

As $\varepsilon \to 0$, the right-hand side goes to zero because of (A.16). This proves (ii).

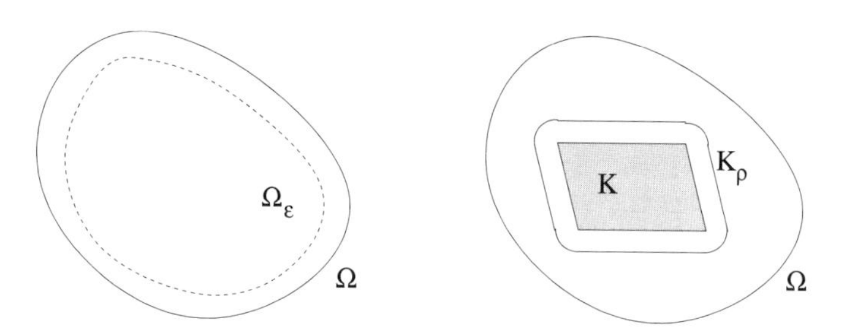

Figure A.5.2. The sets $\Omega_\varepsilon \subset \Omega$ and the compact sets $K \subset\subset K_\rho \subset\subset \Omega$, used in the analysis of mollifiers.

3. Assume that f is continuous, and let $K \subset \Omega$ be a compact subset. Then we can choose $\delta > 0$ small enough so that the compact neighborhood

$$K_\rho \doteq \{x \in \mathbb{R}^n\,;\ d(x, K) \leq \rho\} \tag{A.17}$$

is still contained inside Ω. Since f is uniformly continuous on the compact set K_ρ, the previous calculations show that $f_\varepsilon(x) \to f(x)$ uniformly for $x \in K$. This proves (iii).

4. To prove (iv), assume $1 \le p < \infty$ and $f \in \mathbf{L}^p_{\mathrm{loc}}(\Omega)$. Let $K \subset \Omega$ be a compact subset and choose $\rho > 0$ so that the compact neighborhood K_ρ in (A.17) is still contained in Ω. We claim that, for every $0 < \varepsilon < \rho$,

$$\|f_\varepsilon\|_{\mathbf{L}^p(K)} \le \|f\|_{\mathbf{L}^p(K_\rho)} . \tag{A.18}$$

Indeed, for $x \in K$ an application of Hölder's inequality with $q = \frac{p}{p-1}$ yields

$$\begin{aligned}
|f_\varepsilon(x)| &= \left| \int_{B(x,\varepsilon)} J_\varepsilon(x-y)\, f(y)\, dy \right| \\
&\le \int_{B(x,\varepsilon)} \Big(J_\varepsilon(x-y)\Big)^{\frac{p-1}{p}} \Big(J_\varepsilon(x-y)\Big)^{\frac{1}{p}} |f(y)|\, dy \\
&\le \left(\int_{B(x,\varepsilon)} J_\varepsilon(x-y)\, dy \right)^{\frac{p-1}{p}} \left(\int_{B(x,\varepsilon)} J_\varepsilon(x-y) |f(y)|^p\, dy \right)^{\frac{1}{p}} .
\end{aligned}$$

Recalling that $\int_{B(x,\varepsilon)} J_\varepsilon(x-y)\, dy = 1$, for $0 < \varepsilon < \rho$ the above inequality yields

$$\begin{aligned}
\int_K |f_\varepsilon(x)|^p\, dx &\le \int_K \left(\int_{B(x,\varepsilon)} J_\varepsilon(x-y) |f(y)|^p\, dy \right) dx \\
&\le \int_{K_\rho} |f(y)|^p \left(\int_{B(y,\varepsilon)} J_\varepsilon(x-y)\, dx \right) dy = \int_{K_\rho} |f(y)|^p\, dy .
\end{aligned}$$

This proves (A.18).

Next, for any $\delta > 0$, we choose $g \in \mathcal{C}(K_\rho)$ such that

$$\|f - g\|_{\mathbf{L}^p(K_\rho)} < \delta .$$

Together with (A.18), this yields

$$\begin{aligned}
\|f_\varepsilon - f\|_{\mathbf{L}^p(K)} &\le \|f_\varepsilon - g_\varepsilon\|_{\mathbf{L}^p(K)} + \|g_\varepsilon - g\|_{\mathbf{L}^p(K)} + \|g - f\|_{\mathbf{L}^p(K)} \\
&\le \|f - g\|_{\mathbf{L}^p(K_\rho)} + \|g_\varepsilon - g\|_{\mathbf{L}^p(K)} + \|g - f\|_{\mathbf{L}^p(K)} \\
&\le \delta + \|g_\varepsilon - g\|_{\mathbf{L}^p(K)} + \delta .
\end{aligned}$$

Since g is continuous, by (ii) it follows that $|g_\varepsilon - g| \to 0$ uniformly on the compact set K_ρ. Hence $\limsup_{\varepsilon \to 0} \|f_\varepsilon - f\|_{\mathbf{L}^p(K)} \le 2\delta$. Since $\delta > 0$ was arbitrary, this proves (iv). $\square$

Corollary A.17. *Let $f \in \mathbf{L}^1_{\mathrm{loc}}(\Omega)$ and assume*

$$\int_\Omega f\,\phi\,dx = 0 \qquad \text{for every } \phi \in \mathcal{C}_c^\infty(\Omega).$$

Then $f(x) = 0$ for a.e. $x \in \Omega$.

Indeed, let $x \in \Omega$ be a Lebesgue point of f and let J_ε be the standard mollifier. Taking $\phi(y) = J_\varepsilon(x-y)$ and letting $\varepsilon \to 0$, one obtains

$$0 \;=\; \int_\Omega J_\varepsilon(x-y) f(y)\,dy \;=\; \int_{B(x,\varepsilon)} J(x-y) f(y)\,dy \;\to\; f(x).$$

Hence $f(x) = 0$ for a.e. $x \in \Omega$.

A.5.1. Partitions of unity. Let $S \subseteq \mathbb{R}^n$ and let $V_1, V_2, \ldots$ be open sets that cover S, so that

$$S \subseteq \bigcup_{k\geq 1} V_k\,.$$

We say that a family of functions $\{\phi_k\,;\; k \geq 1\}$ is a **smooth partition of unity** subordinate to the sets V_k if the following hold.

(i) For every $k \geq 1$, $\phi_k : \mathbb{R}^n \mapsto [0,1]$ is a $\mathcal{C}^\infty$ function with support contained inside the open set V_k.

(ii) Each point $x \in S$ has a neighborhood which intersects the support of finitely many functions ϕ_k. Moreover

$$\sum_k \phi_k(x) \;=\; 1 \qquad \text{for all } x \in S\,. \tag{A.19}$$

Note that, by assumption **(ii)**, at each point $x \in S$ the summation in (A.19) contains only finitely many nonzero terms.

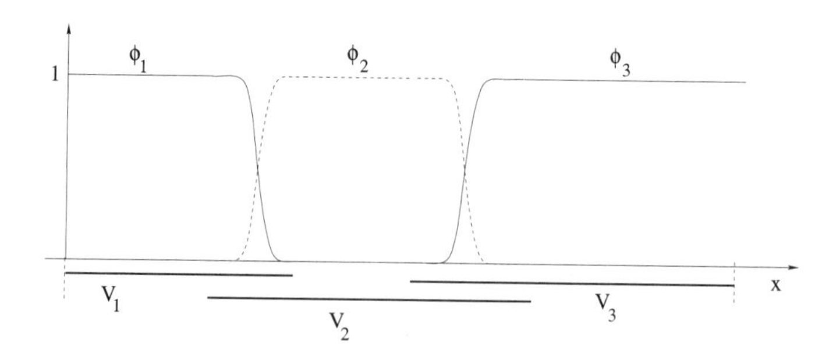

Figure A.5.3. A partition of unity subordinate to the sets V_1, V_2, V_3.

Theorem A.18 (Existence of a smooth partition of unity). *Let $S \subseteq \mathbb{R}^n$ and let $\{V_k\,;\ k \geq 1\}$ be a family of open sets covering S. Then there exists a smooth partition of unity subordinate to the sets V_k.*

A.6. Inequalities

A.6.1. Convex sets and convex functions.

A set $\Omega \subseteq \mathbb{R}^n$ is **convex** if

$$x, y \in \Omega\,,\ \theta \in [0,1] \qquad \text{implies} \qquad \theta x + (1-\theta)y \ \in\ \Omega.$$

In other words, if Ω contains two points x, y, then it also contains the entire segment joining x with y.

Let $\Omega \subset \mathbb{R}^n$ be a convex set. We say that a function $f : \Omega \mapsto \mathbb{R}$ is **convex** if

$$f(\theta x + (1-\theta)y) \ \leq\ \theta f(x) + (1-\theta)f(y) \tag{A.20}$$

whenever $x, y \in \Omega$ and $\theta \in [0,1]$. Notice that (A.20) holds if and only if the **epigraph** of f, i.e., the set

$$\Big\{(x,z) \in \Omega \times \mathbb{R};\ z \geq f(x)\Big\},$$

is a convex subset of $\mathbb{R}^n \times \mathbb{R}$.

A twice differentiable function $f : \mathbb{R}^n \mapsto \mathbb{R}$ is **uniformly convex** if its Hessian matrix of second derivatives is uniformly positive definite. This means that, for some constant $\kappa > 0$,

$$\sum_{i,j=1}^{n} f_{x_i x_j}(x)\ \xi_i \xi_j \ \geq\ \kappa \sum_{i=1}^{n} \xi_i^2 \qquad \text{for all } x, \xi \in \mathbb{R}^n.$$

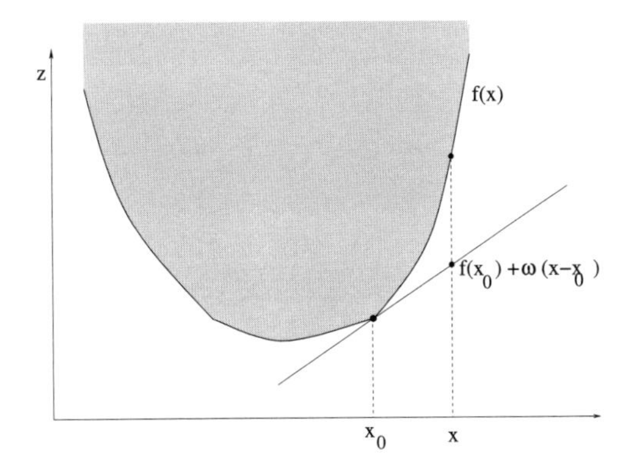

Figure A.6.1. A convex function f and its epigraph. A support hyperplane at the point x_0.

Theorem A.19 (Supporting hyperplanes). *Let $f : \mathbb{R}^n \mapsto \mathbb{R}^n$ be a convex function. Then for every $x_0 \in \mathbb{R}^n$ there exists a (subgradient) vector ω such that*

$$f(x) \ \geq \ f(x_0) + \langle \omega, x - x_0 \rangle \qquad \text{for all } x \in \mathbb{R}^n.$$

The hyperplane $\{(x, z)\,;\ z = \langle \omega, x - x_0 \rangle\} \ \subset \ \mathbb{R}^{n+1}$ touches the graph of f at the point x_0 and remains below this graph at all points $x \in \mathbb{R}^n$. It is thus called a **supporting hyperplane** to f at x_0.

Theorem A.20 (Jensen's inequality). *Let $f : \mathbb{R} \mapsto \mathbb{R}$ be a convex function, and let $\Omega \subset \mathbb{R}^n$ be a bounded open set. If $u \in \mathbf{L}^1(\Omega)$ is any integrable function, then*

$$f\left(\fint_\Omega u\,dx\right) \ \leq \ \fint_\Omega f(u)\,dx\,. \tag{A.21}$$

Here $\fint_\Omega u\,dx \doteq \frac{1}{\text{meas}(\Omega)} \int_\Omega u dx$ is the average value of u on the set Ω, while $\fint_\Omega f(u)\,dx \doteq \frac{1}{\text{meas}(\Omega)} \int_\Omega f(u) dx$ is the average value of $f(u)$. Notice that we do not require that the set Ω be convex.

Proof. Set $u_0 \doteq \fint_\Omega u(y)\,dy$. Then there exists a support hyperplane to the graph of f at the point u_0, say $f(u) \ \geq \ f(u_0) + \langle \omega, u - u_0 \rangle$ for some constant $\omega \in \mathbb{R}$ and all $u \in \mathbb{R}$. Hence

$$f(u(x)) \ \geq \ f\left(\fint_\Omega u(y)\,dy\right) dx + \left\langle \omega, u(x) - \fint_\Omega u(y)\,dy \right\rangle.$$

Taking the average value of both sides on the set Ω, we obtain (A.21). □

A.6.2. Basic inequalities.

1. Cauchy's inequality.

$$ab \ \leq \ \frac{a^2 + b^2}{2} \qquad \text{for all } a, b \in \mathbb{R}\,. \tag{A.22}$$

Indeed, $0 \ \leq \ (a - b)^2 \ = \ a^2 + b^2 - 2ab$.

For any $\varepsilon > 0$, replacing a with $\sqrt{2\varepsilon}\,a$ and b with $b/\sqrt{2\varepsilon}$, from (A.22) we obtain the slightly more general inequality

$$ab \ \leq \ \varepsilon a^2 + \frac{b^2}{4\varepsilon} \qquad (a, b \in \mathbb{R}\,,\ \varepsilon > 0)\,. \tag{A.23}$$

2. Young's inequality.

$$ab \ \leq \ \frac{a^p}{p} + \frac{b^q}{q} \qquad \left(a, b > 0\,,\ 1 < p < q < \infty\,,\ \frac{1}{p} + \frac{1}{q} \ = \ 1\right). \tag{A.24}$$

Indeed, since the map $f(u) = e^u$ is convex, one has $\exp\{\frac{u}{p} + \frac{v}{q}\} \leq \frac{e^u}{p} + \frac{e^v}{q}$. Therefore

$$ab = e^{\ln a + \ln b} = \exp\left\{\frac{1}{p}\ln a^p + \frac{1}{q}\ln b^q\right\} \leq \frac{1}{p}e^{\ln a^p} + \frac{1}{q}e^{\ln b^q} = \frac{a^p}{p} + \frac{b^q}{q}.$$

3. Hölder's inequality. If $f \in \mathbf{L}^p(\Omega)$, $g \in \mathbf{L}^q(\Omega)$ with $1 \leq p, q \leq \infty$ and $\frac{1}{p} + \frac{1}{q} = 1$, then

$$\int_\Omega |fg|\, dx \leq \|f\|_{\mathbf{L}^p(\Omega)} \|g\|_{\mathbf{L}^q(\Omega)}. \tag{A.25}$$

Indeed, in the special case where $\|f\|_{\mathbf{L}^p(\Omega)} = \|g\|_{\mathbf{L}^q(\Omega)} = 1$, Young's inequality yields

$$\int_\Omega |fg|\, dx \leq \frac{1}{p}\int_\Omega |f|^p\, dx + \frac{1}{q}\int_\Omega |g|^q\, dx = \frac{1}{p} + \frac{1}{q} = 1 = \|f\|_{\mathbf{L}^p} \|g\|_{\mathbf{L}^q}.$$

To cover the general case, we simply replace the functions f, g with $\tilde{f} = f/\|f\|_{\mathbf{L}^p}$ and $\tilde{g} = g/\|g\|_{\mathbf{L}^q}$, respectively.

By induction, one can establish the following more general version of Hölder's inequality. Let $1 \leq p_1, \ldots, p_m \leq \infty$, with $\frac{1}{p_1} + \cdots + \frac{1}{p_m} = 1$. Assume $f_k \in \mathbf{L}^{p_k}(\Omega)$ for $k = 1, \ldots, m$. Then

$$\int_\Omega |f_1 f_2 \cdots f_m|\, dx \leq \prod_{k=1}^m \|f_k\|_{\mathbf{L}^{p_k}(\Omega)}. \tag{A.26}$$

4. Minkowski's inequality. For any $1 \leq p \leq \infty$ and $f, g \in \mathbf{L}^p(\Omega)$, one has

$$\|f + g\|_{\mathbf{L}^p(\Omega)} \leq \|f\|_{\mathbf{L}^p(\Omega)} + \|g\|_{\mathbf{L}^p(\Omega)}. \tag{A.27}$$

Indeed, in the case $p = 1$ the result is trivial. If $p > 1$, applying Hölder's inequality with exponents p and $q = \frac{p}{p-1}$, one obtains

$$\|f + g\|^p_{\mathbf{L}^p(\Omega)} = \int_\Omega |f + g|^p\, dx = \int_\Omega |f + g|^{p-1}(|f| + |g|)\, dx$$

$$\leq \left(\int_\Omega |f + g|^{(p-1)\cdot\frac{p}{p-1}} dx\right)^{\frac{p-1}{p}} \left[\left(\int_\Omega |f|^p dx\right)^{\frac{1}{p}} + \left(\int_\Omega |g|^p dx\right)^{\frac{1}{p}}\right]$$

$$= \|f + g\|^{p-1}_{\mathbf{L}^p(\Omega)} \left(\|f\|_{\mathbf{L}^p(\Omega)} + \|g\|_{\mathbf{L}^p(\Omega)}\right).$$

Dividing both sides by $\|f + g\|^{p-1}_{\mathbf{L}^p(\Omega)}$, we obtain (A.27).

5. Interpolation inequality. Let $1 \leq p \leq r \leq q \leq \infty$, with

$$\frac{1}{r} = \frac{\theta}{p} + \frac{1-\theta}{q} \qquad \text{for some } \theta \in [0,1].$$

If $f \in \mathbf{L}^p(\Omega) \cap \mathbf{L}^q(\Omega)$, then we also have $f \in \mathbf{L}^r(\Omega)$ and

$$\|f\|_{\mathbf{L}^r(\Omega)} \leq \|f\|^{\theta}_{\mathbf{L}^p(\Omega)} \|f\|^{1-\theta}_{\mathbf{L}^q(\Omega)}. \tag{A.28}$$

Indeed, observing that $\frac{\theta r}{p} + \frac{(1-\theta)r}{q} = 1$, by Hölder's inequality one obtains

$$\begin{aligned} \int_\Omega |f|^r dx &= \int_\Omega |f|^{\theta r} |f|^{(1-\theta)r}\, dx \\ &\leq \left(\int_\Omega |f|^{\theta r \cdot \frac{p}{\theta r}}\, dx\right)^{\frac{\theta r}{p}} \left(\int_\Omega |f|^{(1-\theta)r \cdot \frac{q}{(1-\theta)r}}\, dx\right)^{\frac{(1-\theta)r}{q}}. \end{aligned}$$

6. Discrete Hölder and Minkowski inequalities. The inequalities (A.25) and (A.27) hold, more generally, when Ω is any measure space. In particular, one can take $\Omega = \{1, \ldots, n\}$ with the counting measure. For any collection of numbers $a_1, \ldots, a_n$ and $b_1, \ldots, b_n$ and for $1 \leq p, q < \infty$ with $\frac{1}{p} + \frac{1}{q} = 1$, one has the **discrete Hölder inequality**

$$\sum_{k=1}^n |a_k b_k| \leq \left(\sum_{k=1}^n |a_k|^p\right)^{\frac{1}{p}} \left(\sum_{k=1}^n |b_k|^q\right)^{\frac{1}{q}} \tag{A.29}$$

and the **discrete Minkowski inequality**

$$\left(\sum_{k=1}^n |a_k + b_k|^p\right)^{\frac{1}{p}} \leq \left(\sum_{k=1}^n |a_k|^p\right)^{\frac{1}{p}} + \left(\sum_{k=1}^n |b_k|^p\right)^{\frac{1}{p}}. \tag{A.30}$$

Given two vectors $x = (x_1, \ldots, x_n)$ and $y = (y_1, \ldots, y_n)$, using the discrete Hölder inequality (A.29) with $p = q = 2$, one obtains the **Cauchy-Schwarz inequality** for the inner product on $\mathbb{R}^n$:

$$|\langle x, y\rangle| \leq |x|\,|y|. \tag{A.31}$$

Indeed,

$$|\langle x, y\rangle| = \left|\sum_{k=1}^n x_k\, y_k\right| \leq \sum_{k=1}^n |x_k\, y_k| \leq \left(\sum_{k=1}^n |x_k|^2\right)^{\frac{1}{2}} \left(\sum_{k=1}^n |y_k|^2\right)^{\frac{1}{2}} = |x|\,|y|.$$

A.6.3. A differential inequality.

Theorem A.21 (Gronwall's inequality). *Let $t \mapsto z(t)$ be a nonnegative, absolutely continuous function defined for $t \in [0,T]$. Assume that the time derivative $z' = \frac{d}{dt}z$ satisfies*

$$z'(t) \leq \phi(t)\,z(t) + \psi(t) \qquad \textit{for a.e. } t \in [0,T],$$

where $\phi, \psi \in \mathbf{L}^1([0,T])$ are nonnegative functions. Then

$$z(t) \leq e^{\int_0^t \phi(s)\,ds} z(0) + \int_0^t e^{\int_s^t \phi(\sigma)\,d\sigma} \psi(s)\,ds \qquad \textit{for all } t \in [0,T]. \tag{A.32}$$

Notice that the right-hand side of (A.32) is precisely the solution to the Cauchy problem

$$Z'(t) = \phi(t)\,Z(t) + \psi(t), \qquad Z(0) = z(0).$$

To prove (A.32), we write

$$\frac{d}{ds}\left(e^{-\int_0^s \phi(\sigma)\,d\sigma} z(s)\right) = e^{-\int_0^s \phi(\sigma)\,d\sigma}\Big(z'(s) - \phi(s)z(s)\Big) \leq e^{-\int_0^s \phi(\sigma)\,d\sigma}\psi(s).$$

Integrating over the interval $[0,t]$, one finds

$$e^{-\int_0^t \phi(\sigma)\,d\sigma} z(t) - z(0) \leq \int_0^t e^{-\int_0^s \phi(\sigma)\,d\sigma}\psi(s)\,ds\,.$$

Multiplying by $e^{\int_0^t \phi(\sigma)\,d\sigma}$, we thus obtain (A.32).

A.7. Problems

1. Let $\{A_i\,;\ i \in \mathcal{I}\}$ be an open covering of a compact metric space K. Prove that there exists $\rho > 0$ such that, for every $x \in K$, the ball $B(x,\rho)$ is entirely contained in one of the sets A_i.

2. Let $(x_n)_{n\geq 1}$ be a sequence of points in a metric space E. Prove the following statements.

(i) The sequence converges to a point $\bar{x}$ if and only if from every subsequence $(x_{n_j})_{j\geq 1}$ one can extract a further subsequence converging to $\bar{x}$.

(ii) If $d(x_m, x_n) \geq \delta > 0$ for all $m \neq n$, then no convergent subsequence can exist.

(iii) Let E be complete and assume that, for every $\varepsilon > 0$, from any sequence one can extract a further subsequence $(x_{n_j})_{j\geq 1}$ such that

$$\limsup_{j,k\to\infty} d(x_{n_j}, x_{n_k}) < \varepsilon.$$

Then the sequence admits a convergent subsequence.

3. Consider the function

$$f(x) \;=\; \begin{cases} \dfrac{1}{x(\ln x)^2} & \text{if } 0 < x < \dfrac{1}{2}, \\[2ex] \qquad 0 & \text{otherwise.} \end{cases}$$

Let $f_\varepsilon \doteq J_\varepsilon * f$ be the corresponding mollifications, and let

$$F(x) \;\doteq\; \sup_{0<\varepsilon<1} f_\varepsilon(x).$$

Prove that $f \in \mathbf{L}^1(\mathbb{R})$ but $F \notin \mathbf{L}^1(\mathbb{R})$. As a consequence, although $f_\varepsilon \to f$ pointwise, one cannot use the Lebesgue dominated convergence theorem to prove that $\|f_\varepsilon - f\|_{\mathbf{L}^1(\mathbb{R})} \to 0$.

4. Let $f_n : \mathbb{R} \mapsto \mathbb{R}$, $n \geq 1$, be a sequence of absolutely continuous functions such that

(i) at the point $x = 0$, the sequence $f_n(0)$ is bounded,

(ii) there exists a function $g \in \mathbf{L}^1(\mathbb{R})$ such that the derivatives f_n' satisfy $|f_n'(x)| \;\leq\; g(x)$ for every $n \geq 1$ and a.e. $x \in \mathbb{R}$.

Prove that there exists a subsequence $(f_{n_j})_{j\geq 1}$ which converges uniformly on the entire real line.

5. Consider a sequence of functions $f_n \in \mathbf{L}^1(\mathbb{R})$ with $\|f_n\|_{\mathbf{L}^1} \leq C$ for every $n \geq 1$. Define

$$f(x) \;\doteq\; \begin{cases} \lim_{n\to\infty} f_n(x) & \text{if the limit exists,} \\ \qquad 0 & \text{otherwise.} \end{cases}$$

Prove that f is Lebesgue measurable and $\|f\|_{\mathbf{L}^1} \leq C$.

6. Let $f : \mathbb{R} \to \mathbb{R}$ be an absolutely continuous function. Prove that f maps sets of Lebesgue measure zero into sets of Lebesgue measure zero.

7. (i) If $(f_n)_{n\geq 1}$ is a sequence of functions in $\mathbf{L}^1([0,1])$ such that $\|f_n\|_{\mathbf{L}^1} \to 0$, prove that there exists a subsequence that converges pointwise for a.e. $x \in [0,1]$.

(ii) Construct a sequence of measurable functions $f_n : [0,1] \mapsto [0,1]$ such that $\|f_n\|_{\mathbf{L}^1} \to 0$ but, for each $x \in [0,1]$, the sequence $f_n(x)$ has no limit.

8. For every (nonempty) open set $\Omega \subseteq \mathbb{R}^n$ and for $1 \leq p \leq \infty$, prove that the space $\mathbf{L}^p(\Omega)$ is infinite-dimensional. Construct a sequence of functions $(f_j)_{j\geq 1}$ such that

$$\|f_j\|_{\mathbf{L}^p} \;=\; 1, \qquad \|f_i - f_j\|_{\mathbf{L}^p} \geq 1 \qquad \text{for all } i,j \geq 1,\ i \neq j\,.$$

9. Consider the set $\mathbb{R}^2$ with the partial ordering

$$x \preceq y \qquad \text{if and only if} \qquad x_1 \leq y_1 \text{ and } x_2 \leq y_2 .$$

Let $f : \mathbb{R} \mapsto \mathbb{R}$ be a continuous, nondecreasing function. Show that the set

$$S \doteq \text{Graph}(f) = \Big\{(t,\, f(t))\,;\ t \in \mathbb{R}\Big\}$$

is a maximal totally ordered subset of $\mathbb{R}^2$. Is every maximal totally ordered subset obtained in this way?

10. Give a proof of the generalized Hölder inequality (A.26).

Summary of Notation

$\mathbb{R}$, the field of real numbers.

$\mathbb{C}$, the field of complex numbers.

$\mathbb{K}$, a field of numbers, either $\mathbb{R}$ or $\mathbb{C}$.

Re z and Im z, the real and imaginary parts of a complex number z.

$\bar{z} = a - ib$, the complex conjugate of the number $z = a + ib \in \mathbb{C}$.

$[a, b]$, a closed interval; $]a, b[$, an open interval; $]a, b]$, $[a, b[$ half-open intervals.

$\mathbb{R}^n$, the n-dimensional Euclidean space.

$\langle \cdot, \cdot \rangle$, scalar product on the Euclidean space $\mathbb{R}^n$.

$|v| \doteq \sqrt{\langle v, v \rangle}$, the Euclidean length of a vector $v \in \mathbb{R}^n$.

$A \setminus B \doteq \{x \in A\,,\ x \notin B\}$, a set-theoretic difference.

$\overline{A}$, the closure of a set A.

∂A, the boundary of a set A.

$\Omega' \subset\subset \Omega$, the closure of Ω' is a compact subset of Ω.

χ_A, the indicator function of a set A. $\chi_A(x) = \begin{cases} 1 & \text{if } x \in A, \\ 0 & \text{if } x \notin A. \end{cases}$

$f : A \mapsto B$, a mapping from a set A into a set B.

$a \mapsto b = f(a)$, the function f maps the element $a \in A$ to the element $b \in B$.

$\doteq$, equal by definition.

$\Longleftrightarrow$, if and only if.

$\mathcal{C}(E) = \mathcal{C}(E, \mathbb{R})$, the vector space of all continuous, real-valued functions on the metric space E.

$\mathcal{C}(E, \mathbb{C})$, the vector space of all continuous, complex-valued functions on the metric space E.

$\mathcal{BC}(E)$, the space of all bounded, continuous, real-valued functions $f : E \mapsto \mathbb{R}$, with norm $\|f\| = \sup_{x \in E} |f(x)|$.

ℓ^1, ℓ^p, ℓ^∞, spaces of sequences of real (or complex) numbers.

$\mathbf{L}^1(\Omega)$, $\mathbf{L}^p(\Omega)$, $\mathbf{L}^\infty(\Omega)$, Lebesgue spaces.

$W^{k,p}(\Omega)$, the Sobolev space of functions whose weak partial derivatives up to order k lie in $\mathbf{L}^p(\Omega)$, for some open set $\Omega \subseteq \mathbb{R}^n$.

$H^k(\Omega) = W^{k,2}(\Omega)$, Hilbert-Sobolev space.

$\mathcal{C}^{k,\gamma}(\Omega)$, the Hölder space of functions $u : \Omega \mapsto \mathbb{R}$ whose derivatives up to order k are Hölder continuous with exponent $\gamma \in]0,1]$.

$\|\cdot\| = \|\cdot\|_X$, the norm on a vector space X.

$(\cdot,\cdot) = (\cdot,\cdot)_H$, the inner product on a Hilbert space H.

X^*, the dual space of X, i.e., the space of all continuous linear functionals $x^* : X \mapsto \mathbb{K}$.

$\langle x^*, x\rangle = x^*(x)$, the duality product of $x^* \in X^*$ and $x \in X$.

$x_n \to x$, strong convergence in norm; this means $\|x_n - x\| \to 0$.

$x_n \rightharpoonup x$, weak convergence.

$\varphi_n \overset{*}{\rightharpoonup} \varphi$, weak-star convergence.

$f * g$, the convolution of two functions $f, g : \mathbb{R}^n \mapsto \mathbb{R}$.

$\nabla u = (u_{x_1}, u_{x_2}, \ldots, u_{x_n})$, the gradient of a function $u : \mathbb{R}^n \mapsto \mathbb{R}$.

$D^\alpha = \left(\frac{\partial}{\partial x_1}\right)^{\alpha_1}\left(\frac{\partial}{\partial x_2}\right)^{\alpha_2}\cdots\left(\frac{\partial}{\partial x_n}\right)^{\alpha_n} = \partial_{x_1}^{\alpha_1}\partial_{x_2}^{\alpha_2}\cdots\partial_{x_n}^{\alpha_n}$, a partial differential operator of order $|\alpha| \doteq \alpha_1 + \alpha_2 + \cdots + \alpha_n$.

$\mathrm{meas}(\Omega)$, the Lebesgue measure of a set $\Omega \subset \mathbb{R}^n$.

$\fint_\Omega f\,dx = \frac{1}{\mathrm{meas}(\Omega)}\int_\Omega f\,dx$, the average value of f over the set Ω.

Bibliography

[B] H. Brezis, *Functional Analysis, Sobolev Spaces and Partial Differential Equations*, Universitext, Springer-Verlag, New York, 2011.

[Ba] V. Barbu, *Nonlinear Semigroups and Differential Equations in Banach Spaces*, Nordhoff, 1976.

[C] J. B. Conway, *A Course in Functional Analysis*, second edition, Springer-Verlag, 1990.

[D] K. Deimling, *Nonlinear Functional Analysis*, Dover, 2010.

[E] L. C. Evans, *Partial Differential Equations*, American Mathematical Society, Providence, RI, 1998.

[EG] L. C. Evans and R. F. Gariepy, *Measure Theory and Fine Properties of Functions*, CRC Press, 1992.

[F] G. B. Folland, *Real Analysis. Modern Techniques and Their Applications*, second edition, Wiley, New York, 1999.

[GT] D. Gilbarg and S. N. Trudinger, *Elliptic Partial Differential Equations of Second Order*, reprint of the 1998 edition, Springer-Verlag, Berlin, 2001.

[H] D. Henry, *Geometric Theory of Semilinear Parabolic Equations*, Lecture Notes in Mathematics **840**, Springer-Verlag, 1981.

[HPC] V. Hutson, J. S. Pym, and M. J. Cloud, *Applications of Functional Analysis and Operator Theory*, second edition, Elsevier, Amsterdam, 2005.

[K] S. Kesavan, *Topics in Functional Analysis and Applications*, Wiley, New York, 1989.

[L] P. Lax, *Functional Analysis*, Wiley-Interscience, New York, 2002.

[Lu] A. Lunardi, *Analytic Semigroups and Optimal Regularity in Parabolic Problems*, Birkhäuser, Basel, 1995.

[Ma] R. H. Martin, *Nonlinear Operators and Differential Equations in Banach Spaces*, Wiley, New York, 1976.

[McO] R. McOwen, *Partial Differential Equations: Methods and Applications*, Prentice Hall, 2001.

[Mi] M. Miklavcic, *Applied Functional Analysis and Partial Differential Equations*, World Scientific, River Edge, NJ, 1998.

[MU] D. Mitrovic and D. Ubrini, *Fundamentals of Applied Functional Analysis*, Pitman, Longman, Harlow, 1998.

[P] A. Pazy, *Semigroups of Linear Operators and Applications to Partial Differential Equations*, Springer-Verlag, New York, 1983.

[PW] M. Protter and H. Weinberger, *Maximum Principles in Differential Equations*, Prentice-Hall, 1967.

[RN] F. Riesz and B. Sz.-Nagy, *Functional Analysis*, F. Unger, New York, 1955.

[RR] M. Renardy and R. C. Rogers, *An Introduction to Partial Differential Equations*, second edition, Springer, 2004,

[R] W. Rudin, *Functional Analysis*, McGraw-Hill, 1973.

[S] J. Smoller, *Shock Waves and Reaction-Diffusion Equations*, second edition, Springer-Verlag, 1994.

[T] M. E. Taylor, *Partial Differential Equations I. Basic Theory*, second edition, Springer-Verlag, New York, 2011.

[Y] K. Yosida, *Functional Analysis*, reprint of the sixth (1980) edition, Springer-Verlag, Berlin, 1995.

Index

adjoint operator, 66, 69, 198
advection, 186
algebra of functions, 47
approximation, 48
  backward Euler, 117, 125, 128, 135
  by polynomials, 51
  by trigonometric polynomials, 52, 58
  forward Euler, 117
  of Sobolev functions, 157
  with smooth functions, 157
Ascoli, 54

Baire, 62, 221
ball
  closed, 13
  open, 13
Banach space, 13
Banach-Alaoglu, 34
Banach-Steinhaus, 61
basis
  orthonormal, 85, 88
Bessel, 87
biharmonic equation, 216
Bochner, 227
boundary condition, 189

Cantor, 141
Cartesian product, 36
Cauchy, 78, 234
Cauchy problem, 115
closed graph theorem, 64
closure, 218
contraction mapping theorem, 115, 219
convergence
  of weak derivatives, 146
  pointwise, 46
  strong, 33
  uniform, 46
  weak, 33, 92
  weak star, 33
convex hull, 37

derivative
  distributional, 140
  of a distribution, 144
  pointwise, 140
  strong, 159
  weak, 140, 144, 156, 159
diffusion, 186
Dini, 46
Dirac, 141
Dirichlet's boundary condition, 186
distance, 11, 217
  induced by seminorms, 24
distribution, 143
  order of, 143
domain of an operator, 16

Egoroff, 223
eigenfunction, 192
eigenvalue, 106
eigenvector, 106
elliptic equation, 185
embedding, 163
  compact, 175, 178
  Gagliardo-Nirenberg, 172

Morrey, 168
Sobolev, 172
energy, 209
epigraph, 233
equicontinuity, 53
essential spectrum, 106
essential supremum, 223
Euler, 117
exponential
of a linear operator, 118
of a matrix, 118
extension
of a linear functional, 27
of a Sobolev function, 161
extreme point, 97

Fatou, 224
fixed point, 219
Fourier, 88
Fréchet, 25
Fredholm, 101
alternative, 105, 196
Fubini, 226
function
absolutely continuous, 148, 225
almost separably valued, 227
Banach space valued, 227
Cantor, 141
constant, 147
continuous, 219
Heaviside, 140
Hölder continuous, 56, 219
integrable, 224
Lipschitz continuous, 115, 181, 219
locally summable, 139, 142, 224
measurable, 222
simple, 223, 227
strongly measurable, 227
summable, 224
test, 142
uniformly convex, 233
weakly differentiable, 149
weakly measurable, 227

Gagliardo-Nirenberg, 163
generator of a semigroup, 122
Gram determinant, 98
Gram-Schmidt orthogonalization, 85
graph, 64
Gronwall, 237

Hahn-Banach, 27
Hausdorff Maximality Principle, 28, 217
Heaviside, 140
Hilbert, 79, 109
Hölder, 56, 235
hyperbolic equation, 207
hyperplane
supporting, 234

inequality
Bessel, 87
Cauchy, 234
Cauchy-Schwarz, 78
discrete Hölder, 236
discrete Minkowski, 236
Gagliardo-Nirenberg, 169
Gronwall, 237
Hölder, 235
interpolation, 236
Jensen, 234
Minkowski, 78, 235
Morrey, 164
Poincaré, 156, 178
Young, 234
inner product, 78

Jensen, 234

kernel of an operator, 17

Lax-Milgram, 91
Lebesgue, 222, 224
Lebesgue point, 225
limit
pointwise, 62
weak, 33
linear combination, 15
linear semigroup, 118
Lipschitz, 115, 219

map
bilinear, 73
continuous, 13
open, 63
matrix, 18
positive definite, 2
measure, 222
Dirac, 141
metric space, 218
mild solution, 136
Minkowski, 78, 235
mollification, 146, 228
monotone convergence, 224
Morrey, 163

multi-index, 143, 149

net smoothness, 172
Neumann's boundary condition, 216
norm, 11
  equivalent, 20
  Hölder, 56

open covering, 219
operator
  adjoint, 66, 69, 198
  backward Euler, 124, 126
  bounded, 17
  closed, 64, 123
  compact, 68, 69, 93, 102, 106, 109
  continuous, 17
  diagonal, 18, 38
  differential, 19
  elliptic, 196
  elliptic homogeneous, 190
  integral, 20, 70
  linear, 16
  multiplication, 20
  partial differential, 143
  Picard, 116
  positive definite, 89
  resolvent, 126
  shift, 19
  symmetric, 107, 109
  uniformly elliptic, 186
ordinary differential equation, 115
  in a Banach space, 115
  linear, 118
orthogonal projection, 80
orthogonality, 80
orthonormal basis, 85, 88
orthonormal set, 84

parabolic equation, 200
partial differential equation
  elliptic, 185
  hyperbolic, 207
  parabolic, 200
partial differential operator, 143
partition of unity, 232
perpendicular projection, 80, 99
Pettis, 227
Picard, 116
Poincaré, 156, 178
point spectrum, 106
positively invariant set, 136
product space, 64

Rademacher, 181
range of an operator, 16
Rellich-Kondrachov, 175
resolvent identities, 126
resolvent integral formula, 127
resolvent operator, 126
resolvent set, 106, 126
Riesz, 82

Schmidt, 109
Schwarz, 78
semigroup, 201, 205
  contractive, 122
  generation of, 128
  of linear operators, 120
  of type ω, 121
  strongly continuous, 121
semigroup property, 121
semilinear equation, 136
seminorm, 24
separable space, 218
separation of convex sets, 31
sequence, 13
  Cauchy, 13, 218
  convergent, 218
  weakly convergent, 67
series, 13
  Fourier, 88
  orthogonal, 87
set
  closed, 218
  compact, 219
  compactly contained, 152
  connected, 218
  convex, 233
  dense, 218
  measurable, 222
  open, 218
  orthonormal, 84
  partially ordered, 217
  positively invariant, 136
  precompact, 219
  relatively compact, 219
sigma-algebra, 222
Sobolev, 151
Sobolev conjugate exponent, 169
solution
  classical, 187
  weak, 188
space
  complete, 13, 26, 218
  dual, 32, 66

Euclidean, 14
finite-dimensional, 14, 20
Fréchet, 25
Hilbert, 79
Hilbert-Sobolev, 152
Hölder, 56
Lebesgue, 225
locally compact, 22
metric, 26, 218
normed, 12, 13
of bounded continuous functions, 45
of bounded linear operators, 17
of continuous functions, 14, 45
of sequences, 15
orthogonal, 80
reflexive, 33
separable, 218
Sobolev, 151
spectrum, 106
of a compact operator, 106
of a symmetric operator, 108
Stone-Weierstrass, 48
strong convergence, 33
subalgebra, 47
support, 139

theorem
a.e. differentiability, 179
Ascoli, 54, 59
Baire category, 62, 221
Banach-Alaoglu, 34
Banach-Steinhaus, 61
Bochner, 227
closed graph, 64
contraction mapping, 115, 219
Dini, 46
dominated convergence, 224
Egoroff, 223
Fredholm, 101
Fubini, 226
Gagliardo-Nirenberg embedding, 172
Hahn-Banach, 27
Hilbert-Schmidt, 109
Lax-Milgram, 91
Lebesgue, 225
monotone convergence, 224
Morrey embedding, 168
open mapping, 63
Pettis, 227
Rademacher, 181
Rellich-Kondrachov compactness, 175
Riesz representation, 82
Sobolev embedding, 172
Stone-Weierstrass, 48, 111

uniform boundedness principle, 61

weak convergence, 33, 92
weak derivative, 140, 144, 156
weak limit, 33
weak solution, 198
weak star convergence, 33

Young, 234

Selected Published Titles in This Series